STATISTICAL EXPERIMENTS AND DECISIONS

Asymptotic Theory

ADVANCED SERIES ON STATISTICAL SCIENCE & APPLIED PROBABILITY

Editor: Ole E. Barndorff-Nielsen

Published

Vol. 1: Random Walks of Infinitely Many Particles
by P. Revesz

Vol. 3: Essentials of Stochastic Finance: Facts, Models, Theory
by Albert N. Shiryaev

Vol. 4: Principles of Statistical Inference from a Neo-Fisherian Perspective
by L. Pace and A. Salvan

Vol. 5: Local Stereology
by Eva B. Vedel Jensen

Vol. 6: Elementary Stochastic Calculus — With Finance in View
by T. Mikosch

Vol. 7: Stochastic Methods in Hydrology: Rain, Landforms and Floods
eds. O. E. Barndorff-Nielsen et al.

Vol. 8: Statistical Experiments and Decisions: Asymptotic Theory
by A. N. Shiryaev and V. G. Spokoiny

Forthcoming

Vol. 2: Ruin Probability
by S. Asmussen

Advanced Series on
Statistical Science &
Applied Probability
Vol. 8

STATISTICAL EXPERIMENTS AND DECISIONS

Asymptotic Theory

A N Shiryaev
Steklov Institute, Russia

V G Spokoiny
Weierstrass Institute, Germany

World Scientific
Singapore • New Jersey • London • Hong Kong

Published by

World Scientific Publishing Co. Pte. Ltd.
5 Toh Tuck Link, Singapore 596224
USA office: 27 Warren Street, Suite 401-402, Hackensack, NJ 07601
UK office: 57 Shelton Street, Covent Garden, London WC2H 9HE

Library of Congress Cataloging-in-Punlication Data
Shiriaev, Al'bert Nikolaevich.
Statistical experiments and decisions: asymptotic theory/A.N. Shiryaev, V. G. Spokoiny.
p. cm. - (Advanced series on statistical science & applied probability; vol. 8)
Includes bibliographical references and index.
ISBN 9810241011 (alk. paper)
1. Mathematical statistics . 2. Experimental design. 3. Asymptotic expansions.
I. Spokoiny, V. G. II. Title. III. Advanced series on statistical science & applied probability; vol. 8.

QA276.S4827 2000 00-035186
519.5--dc21

British Library Cataloguing-in-Publication Data
A catalogue record for this book is available from the British Library.

ISBN 13-978-981-02-4101-8
10-981-02-4101-1

to my family

A. N. Shiryaev

to my wife Natalia

V. G. Spokoiny

Contents

Preface

The goal of this book is to give an exposition of some fundamental aspects of the *asymptotic theory of statistical experiments.* The most important of them is "how to construct asymptotically optimal decisions if we know the structure of optimal decisions for the limit experiment".

The notion of a *statistical experiment* was introduced by Blackwell (1951) and it is now wide-spread. It is used for the mathematical description of observed data in the framework of the probabilistic and statistical approach. Due to Blackwell (1951) and Le Cam (1964) and in the spirit of Kolmogorov's axiomatic system of the probability theory a statistical experiment is considered as a triple

$$\mathcal{E} = (\Omega, \mathcal{F}; \mathbf{P}_\theta \,, \theta \in \Theta)$$

where Ω is treated as a sample space of the statistical data, $\mathcal{F}$ as the set of observed events on Ω, and $\mathbf{P}_\theta$ is a probability measure on $(\Omega, \mathcal{F})$ depending on the value of an unknown parameter θ from a set Θ of possible "theories".

The main problem of the mathematical statistics is to construct for a given experiment $\mathcal{E}$ optimal or close to optimal statistical decisions based on the observed data. The classical examples of such problems are the *estimation problem* and *hypothesis testing* (with respect to the parameter θ).

In order to give an accurate mathematical formulation of such problems, Wald (1950) and Blackwell (1951, 1953) proposed the mathematical *decision theory* based on ideas of game theory (the game of a statistician with nature). Now this approach is mostly used and, for example, Bayes and minimax solutions have a clear meaning in this framework.

The *statistical decision theory* deals with the problems of construction of optimal decisions for a given statistical experiment, whereas the questions of comparison, approximation, convergence, etc., for different statistical experiments are the subject of the general *theory of statistical experiments.* The main ideas of this theory are formulated by Le Cam (1964, 1986).

The material presented in the first chapter of the book is related both to the statistical decision theory (the notions of non-randomized and randomized decisions,

Bayes and minimax risks, distribution of a decision, etc.) and to the theory of statistical experiments (the notions of randomization, sufficiency, deficiency, Δ-distance between experiments, etc.). In general, this chapter can be considered as an introduction to the main concepts of the theory of statistical decisions and the theory of statistical experiments.

For practical problems of mathematical statistics *non-asymptotic models* are typical, that is, "the best statistical decisions" are searched for a "finite amount" of statistical data. Unfortunately an explicit solution of a finite sample statistical problem is available only for relatively simple models. At the same time, the theory and practice of statistical inference show that for a "large amount" of statistical data the probabilistic and statistical laws (of the type of the law of large numbers or the central limit theorem) begin to work. This allows us to speak of the possibility of an approximation of the considered models by simpler "limit models". The general approach to develop this idea can be explained and described in the following way. Let $\mathbb{E}^n = (\Omega^n, \mathcal{F}^n; \mathbf{P}^n_\theta, \theta \in \Theta)$, $n \geq 1$, be statistical experiments where n is treated as "the size" ("dimension") of statistical data. For the classical framework of independent and identically distributed (i.i.d.) observations we have $\Omega^n = R^n$, $\mathcal{F}^n = \mathcal{B}(R^n)$, $\mathbf{P}^n_\theta = \mathbf{P}_\theta \times \ldots \times \mathbf{P}_\theta$.

The typical situation for the statistical problems is that for different values $\theta' \neq \theta''$ the sequences of measures $(\mathbf{P}^n_{\theta'})$ and $(\mathbf{P}^n_{\theta''})$ are *asymptotically singular*, i.e., the "distance" between the measures $\mathbf{P}^n_{\theta'}$ and $\mathbf{P}^n_{\theta''}$ increases as the size of the data increases. In such a case one cannot speak in any reasonable sense about a convergence of the experiments $\mathbb{E}^n$ to a *non-trivial* experiment $\mathbb{E}$. Some precise assertions can be derived and the corresponding approach, which is based on results from the theory of large derivations, follows mainly Bahadur (1960) and Chernoff (1952)).

There exists, however, another approach initiated by Fisher, Blackwell, Wald, and in a modern form by Le Cam, which allows us to state convergence of experiments (even in the situation of the asymptotic singularity), maybe not for original experiments but for experiments constructed from the original ones in a special manner. This construction is based on the ideas of *localization* and *reparametrization*. To illustrate the essence of this approach it is convenient to introduce first the ideas of *centering* and *normalizing*. These ideas are permanently used in probability theory, especially for the central limit problem.

Let $\xi_1, \xi_2, \ldots$ be an i.i.d. sequence of Bernoulli random variables on some probability space $(\Omega, \mathcal{F}, \mathbf{P})$ with $\mathbf{P}(\xi_k = 1) = p$, $\mathbf{P}(\xi_k = 0) = q$, $p + q = 1$, $S_n = \xi_1 + \ldots + \xi_n$, $n \geq 1$. Consider the problem of finding the probability $F_n(x) = \mathbf{P}(S_n \leq x)$. In the case under consideration the value $F_n(x)$ can be found, of course, by direct calculation

$$F_n(x) = \sum_{k \leq x} C_n^k \, p^k \, q^{n-k},$$

since $\mathbf{P}(S_n = k) = C_n^k \, p^k \, q^{n-k}$. However, we can use Moivre-Laplace Theorem, and $F_n(x)$ can be calculated by the following approximate formulae:

$$F_n(x) = \mathbf{P}(S_n \le x) = \mathbf{P}\left(\frac{S_n - np}{\sqrt{npq}} \le \frac{x - np}{\sqrt{npq}}\right) \approx \Phi\left(\frac{x - np}{\sqrt{npq}}\right),$$

$\Phi(\cdot)$ being the standard normal distribution. We notice here that the approximation of $F_n(x)$ is obtained by *centering* (we consider the deviation of S_n from the average $\mathbf{E}S_n = np$) and *normalizing* (we pass from $S_n - np$ to $(S_n - np)/\sqrt{npq}$).

Coming back to the statistical framework we assume that $\mathbb{E}^n = (R^n, \mathcal{B}(R^n); \mathbf{P}_\theta^n, \theta \in R)$ with $\mathbf{P}_\theta^n = \mathbf{P}_\theta \times \ldots \times \mathbf{P}_\theta$ and

$$\mathbf{P}_\theta(dx) = p(x - \theta)dx, \quad p(x) > 0, \quad x \in R. \tag{0.1}$$

Let $l(x) = -\log p(x)$. If $p(x) = (2\pi)^{-1/2} \exp\{-x^2/2\}$, then $l(x) = \frac{1}{2}\log(2\pi) + x^2/2$.

One possible way to define the notion of a convergence of the experiments $\mathbb{E}^n$ is to consider, instead of the *families of measures* $(\mathbf{P}_\theta^n)$ and $(\mathbf{P}_\theta)$, the *random functions* (likelihood ratios) $Z^n(\theta, \theta_0)$ and $Z(\theta, \theta_0)$ that are, given θ_0, the Radon-Nykodim derivatives of $\mathbf{P}_\theta^n$ with respect to $\mathbf{P}_{\theta_0}^n$ and the derivative $\mathbf{P}_\theta$ with respect to $\mathbf{P}_{\theta_0}$.

Further, the convergence of the experiments $\mathbb{E}^n$ to some "limit" experiment $\mathbb{E} = (\Omega, \mathcal{F}; \mathbf{P}_\theta, \theta \in \Theta)$ can be treated as a convergence (in some sense) of the random function $Z^n(\theta, \theta_0)$ to $Z(\theta, \theta_0)$.

For the case of (0.1) one has

$$\begin{aligned} Z^n(\theta, \theta_0; x_1, \ldots, x_n) &= d\mathbf{P}_\theta^n / d\mathbf{P}_{\theta_0}^n(x_1, \ldots, x_n) \\ &= \exp\left\{-\sum_{i=1}^{n} [l(x_i - \theta) - l(x_i - \theta_0)]\right\} \end{aligned} \tag{0.2}$$

($= \exp\left\{(\theta - \theta_0)\, n \left[\frac{1}{n}S_n - \frac{1}{2}(\theta + \theta_0)\right]\right\}$ for the normal case). It is not difficult to see that $Z^n(\theta, \theta_0) \to 0$ under both the measure $\mathbf{P}_{\theta_0}^n$ and the measure $\mathbf{P}_\theta^n$ as $n \to \infty$. Thus, in the case of (0.1) the "limit" experiment is *trivial* in the sense that for the different values of the parameter θ the corresponding probability measures are singular. By (0.2) we see that a non-trivial limit for the likelihoods $Z^n(\theta, \theta_0)$ can be obtained if we also "draw together" the values θ and θ_0 as n increases. In fact, let θ_0 be fixed and

$$\theta = \theta_0 + \frac{\alpha}{\sqrt{n}}. \tag{0.3}$$

Then (in the normal case)

$$Z^n(\theta_0 + \frac{\alpha}{\sqrt{n}}, \theta_0) = \exp\left\{\alpha \, \frac{S_n - \theta_0 \, n}{\sqrt{n}} - \frac{1}{2}\alpha^2\right\}.$$

Since $(S_n - \theta_0 n)/\sqrt{n}$ has the standard normal distribution $\mathcal{N}(0,1)$ under $\mathbf{P}^n_{\theta_0}$, then

$$Z^n(\theta_0 + \frac{\alpha}{\sqrt{n}}, \theta_0) \stackrel{d}{=} \exp\left\{\alpha\gamma - \frac{1}{2}\alpha^2\right\}, \tag{0.4}$$

the symbol "d" being used to denote the equality by distribution, and $\gamma \sim \mathcal{N}(0,1)$. In the general case, under the assumption that the function $l = l(x)$ is smooth enough, one has

$$\begin{aligned} Z^n(\theta_0 + \frac{\alpha}{\sqrt{n}}, \theta_0) &= \exp\left\{-\sum_{i=1}^n \left[l(x_i - \theta_0 - \frac{\alpha}{\sqrt{n}}) - l(x_i - \theta_0)\right]\right\} \\ &= \exp\left\{\frac{\alpha}{\sqrt{n}}\sum_{i=1}^n l'(x_i - \theta_0) - \frac{\alpha^2}{2n}\sum_{i=1}^n l''(x_i - \theta_0) + R_n\right\} \\ &\to \exp\left\{\alpha I^{1/2}(\theta_0)\gamma - \frac{1}{2}\alpha^2 I(\theta_0)\right\} \end{aligned} \tag{0.5}$$

with $I(\theta_0) = \mathbf{E}_{\theta_0} l''(x_1 - \theta_0)$ being the *Fisher information* at the point θ_0, $\gamma \sim \mathcal{N}(0,1)$, and the convergence is treated as the convergence under the measures $\mathbf{P}^n_{\theta_0}$. Now we notice that the expression $\exp\{\alpha I^{1/2}(\theta_0)\gamma - \frac{1}{2}\alpha^2 I(\theta_0)\}$ from the right-hand sides of (0.4) and (0.5) is just the *likelihood ratio* for the normal (Gaussian) "shift experiment" $\mathcal{E}(\theta_0) = (R, \mathcal{B}(R); P_\alpha, \alpha \in R)$ with $P_\alpha \sim \mathcal{N}(I^{1/2}(\theta_0)\alpha, 1)$. Therefore, if the convergence of experiments is treated as the convergence of the corresponding likelihood ratios, then the following conclusion holds: although there is no convergence for the original experiments to a non-trivial experiment, the procedures of the *localization* (near the point θ_0) and the *reparametrization* (changing the θ-scale to the α-scale due to (0.3)) allows us to state the convergence (in the sense mentioned above) of the new "localized" experiments $\mathcal{E}^n(\theta_0) = (R^n, \mathcal{B}(R^n); P^n_\alpha(\theta_0), \alpha \in R)$ with $P^n_\alpha(\theta_0) = \mathbf{P}^n_{\theta_0+\alpha/\sqrt{n}}$ to the Gaussian shift experiment $\mathcal{E}(\theta_0)$. Notice also that the asymptotic assertions like (0.5) in the case of i.i.d. observations follow directly from "regularity conditions" on the density function $p(x)$ and the central limit theorem.

In 1960 Le Cam introduced the so-called *Local Asymptotic Normality* (LAN) condition which just assumes that after the reparametrization $\theta = \theta_0 + \varphi(n)\alpha$ with $\varphi(n) \downarrow 0$ the likelihoods $Z^n(\theta_0 + \varphi(n)\alpha, \theta_0)$ have the following structure (compare with (0.5)):

$$Z^n(\theta_0 + \varphi(n)\alpha, \theta_0) = \exp\left\{\alpha I^{1/2}(\theta_0)\gamma_n - \frac{1}{2}\alpha^2 I(\theta_0) + R_n\right\} \tag{0.6}$$

with $\mathbf{P}^n_{\theta_0}(|R_n| > \varepsilon) \to 0$, $n \to \infty$, $\varepsilon > 0$, and the $\mathbf{P}^n_{\theta_0}$-distributions of γ_n tending to the normal law $\mathcal{N}(0,1)$. In other words, this condition means that the likelihood ratios of the "localized" experiments are well approximated by the likelihood ratio of the Gaussian shift experiment.

The LAN condition appeared to be very helpful and fruitful in asymptotic statistics and many important results were obtained within this framework. The most famous among these results are the ***Minimax Hajek-Le Cam Theorem***, the ***Hajek Convolution Theorem***, the series of results on the parameter estimation, given for example in the book of of Ibragimov and Khasminskii (1981). The Local Asymptotic Normality condition has been playing the exclusive role in the forming of the general principles of asymptotic statistics.

Meanwhile, it would be useful to mention that for many statistical problems limit experiments are *non Gaussian.* (This book contains many such examples.) Extending the LAN condition Le Cam introduced other definitions of convergence of statistical experiments. The most important among them are the notions of *weak* and *strong* convergence for which a limit experiment may have an arbitrary structure. These notions are found to be rather useful from the conceptual point of view although their practical application meets a number of problems. In fact, firstly, the verification of these convergence conditions (especially for the strong convergence) is a very difficult task (contrary to the LAN condition). Secondly, there is no direct way to construct asymptotically efficient decisions even if a weak or strong convergence of experiments has been stated.

In Chapter 2 we introduce a new definition of convergence of statistical experiments, the so-called *λ-convergence.* This condition can be considered as a natural generalization of LAN-type conditions. The definition is given in a form that is an abstraction of the procedures of localization and reparametrization (that are present in the LAN condition) as well as the concrete probability structure of the limit experiment, and it allows us to embrace a great number of particular problems by a common method of investigation.

Besides the questions related to the problem of convergence of statistical experiments, the following questions are under consideration in Chapter 2. Let experiments $\mathcal{E}^n$, $n \geq 1$, converge in some sense to a limit experiment $\mathcal{E}$ and let, for a given "decision problem" with a "loss function" W , the value $\mathcal{R}(\mathcal{E}, W)$ of the minimax risk (or the value $\mathcal{R}_\pi(\mathcal{E}, W)$ of the π-Bayes risk) be known, as well as the structure of a minimax decision $\rho^*(\mathcal{E})$ (or a Bayes decision $\rho^*_\pi(\mathcal{E})$). What conclusions with respect to the risks $\mathcal{R}(\mathcal{E}^n, W)$ (or $\mathcal{R}_\pi(\mathcal{E}^n, W)$) can be expected for the experiments $\mathcal{E}^n$ if one or another kind of convergence "$\mathcal{E}^n \to \mathcal{E}$" holds? How can one use the knowledge about the decisions $\rho^*(\mathcal{E})$ (or $\rho^*_\pi(\mathcal{E})$) for the limit experiment $\mathcal{E}$ in order to construct optimal (asymptotically) decisions $\rho^*(\mathcal{E}^n)$ (or $\rho^*_\pi(\mathcal{E}^n)$) for the experiments $\mathcal{E}^n$?

Of course, the answers to these questions depend mostly on the kind of convergence "$\mathcal{E}^n \to \mathcal{E}$". Thus, the weak convergence "$\mathcal{E}^n \xrightarrow{w} \mathcal{E}$" provides the inequality of the form

$$\liminf_n \mathcal{R}(\mathcal{E}^n, W) \geq \mathcal{R}(\mathcal{E}, W), \tag{0.7}$$

but the strong convergence "$\mathcal{E}^n \xrightarrow{\Delta} \mathcal{E}$" gives us already the equality in (0.7) (at least for bounded and continuous loss functions W) and, moreover, it allows us to estimate the rate of convergence of $\mathcal{R}(\mathcal{E}^n, W)$ to $\mathcal{R}(\mathcal{E}, W)$:

$$|\mathcal{R}(\mathcal{E}^n, W) - \mathcal{R}(\mathcal{E}, W)| \leq \|W\|\Delta(\mathcal{E}^n, \mathcal{E}) \tag{0.8}$$

with $\Delta(\mathcal{E}^n, \mathcal{E})$ being the so-called Δ-distance between the experiments $\mathcal{E}^n$ and $\mathcal{E}$.

The notion proposed in this book of λ-convergence " $\mathcal{E}^n \xrightarrow{\lambda} \mathcal{E}$ " has a number of advantages compared with the notions of weak and strong convergence. First of all, because it is formulated almost in the same generality as for the weak and strong convergence but it is easier to verify. Moreover, the random elements λ^n that enter in the definition of λ-convergence are ***asymptotically sufficient statistics*** for the original experiments $\mathcal{E}^n$ and they are the elements that give us the possibility to construct asymptotically optimal decisions ρ_n^* for the experiments $\mathcal{E}^n$ from optimal decisions $\rho^* = \rho^*(\lambda)$ for the limit experiment $\mathcal{E}$, by the simple procedure, roughly speaking, of substituting λ^n instead of λ in $\rho^*(\lambda)$.

The essential role in the justification of this procedure is played by a technical result called Lemma about "Reconstruction". In particular, this result permits us to establish (for bounded loss functions) the asymptotical equality $\lim \mathcal{R}(\mathcal{E}^n, W) = \mathcal{R}(\mathcal{E}, W)$ under the assumption of λ-convergence " $\mathcal{E}^n \xrightarrow{\lambda} \mathcal{E}$ ".

Section 2.8 deals with the questions of relations between the different kinds of convergence. We show that the λ-convergence implies the weak convergence but the uniform variant of λ-convergence already guarantees the strong convergence. We also give sufficient conditions for all three kinds of converge to be equivalent.

Clearly the results like $\underline{\lim}\, \mathcal{R}(\mathcal{E}^n, W) \geq \mathcal{R}(\mathcal{E}, W)$ or $\lim \mathcal{R}(\mathcal{E}^n, W) = \mathcal{R}(\mathcal{E}, W)$ are meaningful only if one can find, or at least estimate, the risk $\mathcal{R}(\mathcal{E}, W)$ for the limit experiment $\mathcal{E}$. Under the LAN condition the limit experiment is a Gaussian shift experiment which is a particular case of shift experiments. For such kind of experiments there are a number of results due to Pitman (1939), Blyth (1951), Anderson (1955), Boll (1955), Stein (1959), based on the generalized Bayes approach. These results describe the structure of minimax, Bayes and regular (equivariant) decisions (estimators) for shift-type experiments. In particular, the minimax estimator for a shift experiment was shown to coincide with the generalized Bayes estimator for the uniform a priori distribution on the whole parameter space (this estimator is also called "Pitman estimator"). These results, together with the general asymptotic theorems, provide the tools to obtain assertions like the abovementioned Hajek–Le Cam theorem on the lower bounds of asymptotically minimax risk and Hajek's convolution theorem.

Further development of mathematical statistics arrived often at the consideration of models for which the limit experiment is *not* a Gaussian shift. A typical example is given by models for which the so-called ***mixed local asymptotical normality*** condition is true, see Jeganathan (1980), Swensen (1980), Davies (1985), etc. In this case the limit experiment is mixed Gaussian. More complicated models appear also for "non-regular" statistical problems, for example, if the density function of i.i.d. observations has discontinuities (a number of such models were considered in Ibragimov and Khasminskii (1981)). Other examples arise in problems of statistics of random processes, problems of sequential analysis, sequential experimental design, etc.

In order to obtain considerable asymptotical assertions for such models one has to extend, at least, the existing results for the shift experiments to the more general class of limit models. This is the problem that is studied in Chapter 3. We consider the class of experiments (with a finite-dimensional parametric space) called (γ, Γ)-*models*. (A shift experiment is a particular case of such models.) The characteristic feature of the experiments $\mathcal{E} = (\Omega, \mathcal{F};\ P_\alpha, \alpha \in R^d)$ from this class is a special form of the likelihood ratio

$$Z_\alpha(\omega) = \frac{dP_\alpha}{dP_0} = \frac{q(\gamma(\omega) - \Gamma(\omega))}{q(\gamma(\omega))}, \quad \alpha \in R^d,$$

where $q = q(x)$ is a density function on R^d and γ, Γ are random elements with values respectively in R^d and $\mathcal{M}^d$ (the set of all symmetrical positive $d \times d$-matrices). Note that $\Gamma(\omega)$ coincides with the unity $d \times d$-matrix for the shift experiments and the variables $\gamma(\omega)$ and $\Gamma(\omega)$ are independent for the so-called mixed models (in particular, for mixed normal models).

The important subclass of (γ, Γ)-models consists of experiments with the density $q = q(x)$ coinciding with the standard Gaussian density $\varphi = \varphi(x)$. For such models, called $(\gamma, \Gamma; \varphi)$-*models*, the likelihood ratio is of the form

$$Z_\alpha(\omega) = \exp\left\{(\gamma(\omega), \alpha) - \frac{1}{2}|\Gamma\alpha|^2\right\}.$$

The reason to highlight this class is that for rather general problems of statistics of random processes the corresponding statistical experiments are approximated (under some regularity conditions) by $(\gamma, \Gamma; \varphi)$-models.

Fortunately, *the main results concerning the shift experiments can be extended on arbitrary* (γ, Γ)-*models*. Furthermore, the conjunction of these assertions with the general theorems from Chapter 3 allows to establish asymptotical conclusions for the estimation problem for a wide class of statistical models.

Before discussing the results of the first three chapters from the point of view of their applicability to real statistical problems we mention one difficulty arising here.

We have already remarked that one cannot speak about convergence of original experiments $\mathbb{E}^n$, $n \to \infty$, but only about convergence of the localized experiments $\mathcal{E}^n(\theta_0)$ produced from the original ones with some localization point chosen in advance. Therefore, the asymptotical conclusions relate, not to the original parameter θ, but only to a local parameter α obtained from θ by reparametrization. Hence, these results, being reformulated in terms of the original parameter θ, have only *local* properties. For example, the asymptotically minimax risk for the localized experiments $\mathcal{E}^n(\theta_0)$ becomes the local asymptotically minimax risk for the original

$\mathbf{E}^n$ in an asymptotically small neighborhood of the point θ_0. Consequently, the approach developed so far in Chapters 2 and 3 does not lead to the construction of minimax (in an asymptotical sense) but only *locally* minimax decisions. In addition, usually the localization point θ_0 , unknown a priori, enters explicitly in the expression of these decisions.

To bypass this problem Le Cam (1960, 1969), proposed (under the LAN condition) the following idea: one can take, instead of a localization point θ_0 for the experiments $\mathbf{E}^n$, some "preliminary" estimator $\widetilde{\theta}_n$ such that the estimator sequence $(\widetilde{\theta}_n)$ is asymptotically consistent as $n \to \infty$. Thus, if $\widehat{\theta}_n = \widehat{\theta}_n(\theta_0)$ is such an estimator for a localization point θ_0 then one can try to take the new estimators $\overline{\theta}_n = \widehat{\theta}_n(\widetilde{\theta}_n)$ as *globally* minimax (i.e., independently of θ_0). It was shown in Le Cam (1960, 1986) that this procedure really "works" under some uniform variant of the LAN condition and some additional conditions on the pilot estimators $\widetilde{\theta}_n$. The results presented in Chapter 4 just claim the applicability of this approach for more general models under a uniform variant of the λ-convergence condition.

Chapter 5 is an illustration of the general results of the previous chapters. Indeed, these results are applied to one concrete, but sufficiently general model, a first order autoregressive model.

The special consideration of this model is of interest, first of all, because the corresponding observations are not independent (as for the classical statistics) and they are even strongly dependent for explosive models.

The contents of Chapter 5 are mostly well known, but a new result is included, one related to the asymptotical properties of least square estimators for an explosive autoregression and for a random number of observations.

The authors use the occasion to express their gratitude to D. Chibisov for his helpful remarks and critics which allowed them to improve the contents and the exposition of the book.

The book was written under support of Steklov Mathematical Institute and the Humboldt Research Awards Programmes (A. Shiryaev) and INTAS grants. We feel also very much indebted to Irina Spokoiny who helped during the preparation of the manuscript.

Chapter 1

Statistical Experiments and Their Comparison

1.1 Statistical Experiment (Main Definitions, the Problem of Comparison of Experiments)

1. The axiomatic system of the *Probability theory* proposed in 1933 by A. Kolmogorov is based on the concept of a *probability model* or a *probability space*

$$(\Omega, \mathcal{F}, P),$$

consisting of:

(1) $\Omega = \{\omega\}$, the set (space) of *issues (elementary events)*;
(2) $\mathcal{F}$, a σ-field of subsets in Ω called *events*;
(3) P, a σ-additive non-negative function of sets from $\mathcal{F}$ (i.e. a measure) normalized by the condition $P(\Omega) = 1$ (in other words, P is a *probability* on the measurable space $(\Omega, \mathcal{F})$).

The theory of *Mathematical Statistics* is based on the notion of a *statistical model* or, it is also said, of a *statistical experiment*

$$\mathcal{E} = (\Omega, \mathcal{F};\ P_\theta\,, \theta \in \Theta),$$

including the set of probability spaces $(\Omega, \mathcal{F}, P_\theta)$, $\theta \in \Theta$, with the family of measures $\mathcal{P} = \{P_\theta\,, \theta \in \Theta\}$ being "parameterized by a parameter θ which belongs to some set of possible "theories" Θ. The main problem in mathematical statistics is to determine (based on "observed data") the true "theory" θ, i.e. to choose such a probability model $(\Omega, \mathcal{F}, P_\theta)$ that corresponds "best of all" to the "observed data".

The questions of which "theory" P_θ is preferable, of what *quality* is the "decision" chosen, what is a "loss", etc. are the subject of the *statistical decision theory*. This theory gives us some logical framework which determines (for a fixed experiment $\mathcal{E}$) the concept of a "statistical decision" and proposes different definitions of "optimality". Whereas this theory relates to statistical decisions for a *fixed* experiment $\mathcal{E}$, the questions of *comparison*, *approximation*, *convergence*, etc., of different probabilistic-statistical models (experiments) are the subject of the *general theory of statistical experiments.*

However one formulate the problem of searching the "real theory" θ or the real distribution P_θ which determines the probabilistic-statistical mechanism of the model under consideration, the first question appearing here is the question of how much "statistical information" the considered experiment contains or, more precisely, how much "statistical information" with respect to θ is carried by the observed data. In spite of the various existing definitions of "information" (for example, the *Fisher information*, the *Shannon information*, the *Kullback information*) it seems impossible to give one universal notion of "information" contained in the observed data with respect to the unknown parameter θ that is suitable for a wide class of statistical problems simultaneously. That is why, instead of stating the question of "statistical information" in a considered experiment, it is natural to pose another question (having the meaning of comparison): "when will an experiment $\mathcal{E}$ be more informative than another, say $\mathcal{E}^*$?".

That such an approach is reasonable seems to be justified, for instance, by the following consideration. Let us assume that $\mathcal{E}$ is a Gaussian shift experiment (see Example 1.2 below). One may claim that about Gaussian shift experiments more or less "everything is known". Therefore, it is desirable to have a definition of when "an experiment $\mathcal{E}^*$ is less (or more) informative than some Gaussian shift experiment $\mathcal{E}$". From this we can conclude that "the statistical loss for the experiment $\mathcal{E}^*$ is more (or less) than for the Gaussian shift experiment $\mathcal{E}$" but the latter one can be usually found or, at least, estimated.

The investigation of the questions of comparison of statistical experiments was initiated by the paper Bohnenblust, Sharpley, Sherman (1949) followed by the papers Blackwell (1951, 1953) where the following definition was introduced: "$\mathcal{E}$ is more informative than $\mathcal{E}^*$" (in short, "$\mathcal{E} \overset{R}{\gg} \mathcal{E}^*$) if for any bounded loss function W and any decision ρ^* (see Section 1.4 below) in the experiment $\mathcal{E}^*$ there exists a decision ρ for experiment $\mathcal{E}$ such that

$$R_\theta(\mathcal{E}, W, \rho) \leq R_\theta(\mathcal{E}^*, W, \rho^*), \quad \theta \in \Theta.$$

Here we denote by $R_\theta(\mathcal{E}, W, \rho)$ and $R_\theta(\mathcal{E}^*, W, \rho^*)$ "the statistical risk" for the experiments $\mathcal{E}$ and $\mathcal{E}^*$ respectively. Unfortunately, such an approach to define that "$\mathcal{E}$ is more informative than $\mathcal{E}^*$" has an essential deficiency: the considered experiments may simply be non-comparable.

An important break in this direction occurred when Le Cam (1964), stated, instead of the question "Is $\mathcal{E}$ more informative than $\mathcal{E}^*$?" another question "How much do we loose if we use the experiment $\mathcal{E}^*$ instead of the experiment $\mathcal{E}$?", Such a setting suggested the introduction of the notion of a *deficiency* $\delta(\mathcal{E}, \mathcal{E}^*)$ in the experiment $\mathcal{E}$ with respect to the experiment $\mathcal{E}^*$. The meaning of this notion is as follows: how much does one fail to reconstruct (with the help of a *randomization*) the experiment $\mathcal{E}^*$ from the experiment $\mathcal{E}$. Although this value

$\delta(\mathcal{E},\mathcal{E}^*)$ is not easy to calculate or even estimate, the introduction of this notion was found to be very important and useful. First, it is ***always defined*** for any experiments $\mathcal{E} = (\Omega,\mathcal{F}; P_\theta, \theta \in \Theta)$ and $\mathcal{E}^* = (\Omega^*,\mathcal{F}^*; P^*_\theta, \theta \in \Theta)$ (with the same parameter space Θ but with different, in general, $(\Omega,\mathcal{F})$ and $(\Omega^*,\mathcal{F}^*)$). Second, one can compare with the help of this value the statistical risks $R_\theta(\mathcal{E},W,\rho)$ and $R_\theta(\mathcal{E}^*,W,\rho^*)$ by the inequalities of the form

$$R_\theta(\mathcal{E},W,\rho) \le R_\theta(\mathcal{E}^*,W,\rho^*) + \|W\|\delta(\mathcal{E},\mathcal{E}^*)$$

(for more details see Section 1.7).

In the case of $\delta(\mathcal{E},\mathcal{E}^*) = 0$ (i.e. if the experiment $\mathcal{E}^*$ can be obtained from the experiment $\mathcal{E}$ by a randomization) it is natural to say that "$\mathcal{E}$ is more informative than $\mathcal{E}^*$" (in this situation we use the notation $\mathcal{E} \overset{\delta}{\gg} \mathcal{E}^*$). The following fact is remarkable: the properties "$\mathcal{E} \overset{R}{\gg} \mathcal{E}^*$" and "$\mathcal{E} \overset{\delta}{\gg} \mathcal{E}^*$" are equivalent (see Section 1.7).

With the help of the notion of a deficiency Le Cam introduced another important (mutual) characteristic of the experiments $\mathcal{E}$ and $\mathcal{E}^*$, the so-called Δ-***distance*** (between $\mathcal{E}$ and $\mathcal{E}^*$):

$$\Delta(\mathcal{E},\mathcal{E}^*) = \max\{\delta(\mathcal{E},\mathcal{E}^*),\, \delta(\mathcal{E}^*,\mathcal{E})\}\,.$$

Besides the possibility of comparing experiments, the Δ-distance allows us to define the so-called ***strong convergence*** (or Δ-***convergence***) "$\mathcal{E}^n \xrightarrow{\Delta} \mathcal{E}$" of statistical experiments $\mathcal{E}^n$ to $\mathcal{E}$, $n \to \infty$. Together with the other two definitions of convergence (***weak*** or w-***convergence*** "$\mathcal{E}^n \xrightarrow{w} \mathcal{E}$", and also the λ-***convergence*** introduced below, "$\mathcal{E}^n \xrightarrow{\lambda} \mathcal{E}$") these three kinds of convergence permit us to carry out a detailed study of many important questions of the statistical decision theory in its asymptotical aspect, for example, "what are asymptotic (as $n \to \infty$) lower bounds of the risks for the experiments $\mathcal{E}^n, n \ge 1$" and "how to construct asymptotically efficient decisions for the experiments $\mathcal{E}^n$, $n \ge 1$, using the knowledge about efficient decisions for the experiment $\mathcal{E}$ which is the limit (in some sense) for $\mathcal{E}^n$, $n \ge 1$".

The first general results with respect to "asymptotic lower bounds" have been obtained by Hajek (1970, 1972) and Le Cam (1972) in the form of well-known "local asymptotic minimax theorems" and "convolution theorems" under the so-called ***local asymptotic normality*** (LAN) property.

Later, many results have been obtained in this direction connected mostly with the weakening of LAN-type conditions of the convergence of experiments and with more precise description of asymptotical lower bounds (see for more details the monographs Le Cam (1986), Strasser(1985), Ibragimov and Khasminskii (1981), Millar(1983)). An essential part of the present book is concerned with similar questions (see Chapters 2–4).

2. The main object under consideration below is a statistical experiment

$$\mathcal{E} = (\Omega, \mathcal{F};\ P_\theta\ , \theta \in \Theta)$$

where Θ is some parameter space describing possible "theories".

Let us cite some examples of statistical experiments.

Example 1.1 Let $(\Omega, \mathcal{F}) = (R, \mathcal{B}(R))$ be the real line R with the system of its Borel subsets $\mathcal{B}(R)$ and let $\mathcal{P} = \{P_\theta,\ \theta \in \Theta\}$ be the set of *all* probability measures on $(R, \mathcal{B}(R))$. Here the set Θ can be identified with the set of *all* distribution functions $F = F(x)$, $x \in R$. Then in this case a "point θ" is some function $F = F(x)$, $x \in R$.

Example 1.2 (*Gaussian shift experiment on the real line*) Let again $(\Omega, \mathcal{F}) = (R, \mathcal{B}(R))$. If

$$\mathcal{P} = \{P_\theta\ ,\ \theta \in R\}$$

is a family of Gaussian (normal) $\mathcal{N}(\theta, 1)$ distributions then $\mathcal{E}$ is said to be *a Gaussian shift experiment on the real line.*

Example 1.3 An experiment $\mathcal{E} = (\Omega, \mathcal{F}, \mathcal{P})$ with $(\Omega, \mathcal{F}) = (R, \mathcal{B}(R))$ and

$$\mathcal{P} = \{P_\theta\ , \quad \theta = (m, \sigma^2), \quad |m| < \infty, \quad 0 \le \sigma^2 < \infty\}$$

is *a Gaussian experiment on the real line*, where $P_\theta = P_{(m,\sigma^2)} = \mathcal{N}(m, \sigma^2)$.

Example 1.4 Let an experiment $\mathcal{E} = (\Omega, \mathcal{F}; P_\theta, \theta \in \Theta)$ fulfill the following condition:

(1) for each pair θ' and θ'' from Θ the measures $P_{\theta'}$ and $P_{\theta''}$ are equivalent, $(P_{\theta'} \sim P_{\theta''})$, i.e. mutually absolutely continuous $(P_{\theta'} \ll P_{\theta''}$ and $P_{\theta'} \gg P_{\theta''})$;

(2) there exists $\theta_0 \in \Theta$ such that the family of random variables $L = (L_\theta\ , \theta \in \Theta)$, with $L_\theta = \ln dP_\theta / dP_{\theta_0}$ being log of the Radon-Nikodym derivative of P_θ with respect to P_{θ_0}, is normally disctributed under the measure P_{θ_0}.

In this case the experiment $\mathcal{E}$ is called *a Gaussian shift experiment.*

Example 1.5 Consider an experiment $\mathcal{E} = (\Omega, \mathcal{F};\ P_\theta\ , \theta \in \Theta)$, where the set Θ consists only of two points θ_1 and θ_2. Such an experiment is usually called *binary.* (Sometimes for simplicity of notation we write P instead of P_{θ_1} and $\widetilde{P}$ instead of P_{θ_2}.)

Example 1.6 If all the measures P_θ in an experiment $\mathcal{E}_* = (\Omega, \mathcal{F}; P_\theta\ , \theta \in \Theta)$ coincide with one another, $(P_\theta = P_*\ , \theta \in \Theta)$ then it is natural to call $\mathcal{E}_*$ *a totally non-informative experiment.* On the contrary, if the measures P_θ are mutually singular $(P_{\theta'} \perp P_{\theta''}\ , \theta' \neq \theta'')$ then such an experiment is called *totally informative.*

Example 1.7 Let $(\Omega, \mathcal{F}) = (\mathbf{C}_T, \mathcal{B}(\mathbf{C}_T)$ be the measurable space $\mathbf{C}_T$ of continuous functions $x = (x_t)$, $0 \le t \le T$, with the σ-field $\mathcal{B}(\mathbf{C}_T)$ of its cylindric subsets. Assume that a standard Wiener process $W = (W_t(\omega))$, $0 \le t \le T$, with the values in $\mathbf{C}_T$ is defined on some probability space $(\Omega', \mathcal{F}', P')$ and let P^0 be its distribution ($P^0(B) = P'\{\omega : \omega \in B\}, B \in \mathcal{B}(\mathbf{C}_T)$). Also let $\theta = (\theta_t), 0 \le t \le T$, be a deterministic function such that

$$\int_0^T \theta_t^2 \, dt < \infty, \qquad \theta \in \Theta.$$

Put

$$X_t^\theta = \int_0^t \theta_s \, ds + W_t, \qquad 0 \le t \le T,$$

and let P^θ be the distribution of the process $X^\theta = (X_t^\theta)$, $0 \le t \le T$, in the space $(\mathbf{C}_T, \mathcal{B}(\mathbf{C}_T))$.

It is not difficult to see that the corresponding experiment $\mathcal{E} = (\mathbf{C}_T, \mathcal{B}(\mathbf{C}_T); P^\theta, \theta \in \Theta)$ is a Gaussian shift in the sense of the definition given in Example 1.4. Indeed, in the considered case $P^{\theta'} \sim P^{\theta''}$ for all $\theta', \theta'' \in \Theta$ and due to well-known formulas (see Liptser and Shiryaev (1974, Chapter 7))

$$Z_T^\theta = \frac{dP^\theta}{dP^0}(x) = \exp\left\{ \int_0^T \theta_t \, dx_t - \frac{1}{2} \int_0^T \theta_t^2 \, dt \right\}.$$

This evidently implies that the distribution (the law) $\mathcal{L}\left(\ln Z_T^\theta \mid P^0\right)$ of the family of random variables $\{\ln Z_T^\theta, \theta \in \Theta\}$ with respect to P^0 is Gaussian with

$$E_{P^0} \ln Z_T^\theta = -\frac{1}{2} \int_0^T \theta_t^2 \, dt, \qquad D_{P^0} \ln Z_T^\theta = \frac{1}{2} \int_0^T \theta_t^2 \, dt.$$

Example 1.8 Let $(\Omega, \mathcal{F}) = (\mathbf{D}, \mathcal{B}(\mathbf{D}))$ be the space of functions $x = (x_t)_{t \ge 0}$ without discontinuities of the second kind (continuous from the left and with right limits) $\mathcal{B}(\mathbf{D})$, the Borel σ-field, and let P^θ be a distribution of a *semimartingale* $X = (X_t)_{t \ge 0}$ with values in $\mathbf{D}$ and with the triple of predictable characteristics

$$\theta = (B(x), C(x), \nu(x))$$

(see Liptser and Shiryaev (1986), Jacod and Shiryaev (1987)) from some set Θ. The corresponding "semimartingale" experiments cover a wide class of probability-statistical models in the statistics of random processes.

From the point of view of the character of the data received by a statistician, the preceding examples are concerned with the case of *fixed* sample "size" ("volume") and statistical decisions are based on just these data. The following two examples deal with the situation when the data are obtained *sequentially* and a statistician can choose a time to stop the process of observation and then a decision is taken based on the data received in such a manner.

Example 1.9 *Filtered experiments with discrete time.* The statistical problems with *independent* observations (of "size" n) lead to the consideration of statistical experiments

$$\mathcal{E}^n = (\Omega^n, \mathcal{F}^n; P_\theta^n, \theta \in \Theta)$$

with $\Omega^n = \Omega^{(1)} \times \ldots \times \Omega^{(n)}$, $\mathcal{F}^n = \mathcal{F}^{(1)} \otimes \ldots \otimes \mathcal{F}^{(n)}$, $P_\theta^n = P_\theta^{(1)} \times \ldots \times P_\theta^{(n)}$, i.e. with experiments that are a *direct product* of the form

$$\mathcal{E}^n = \mathcal{E}^{(1)} \times \ldots \times \mathcal{E}^{(n)}$$

with $\mathcal{E}^{(i)} = (\Omega^{(i)}, \mathcal{F}^{(i)}; P_\theta^{(i)}, \theta \in \Theta)$, $i = 1, ..., n$.

For the case of *dependent* observations another framework is more convenient, the so-called *filtered experiment* framework, which is described as follows. Let $(\Omega, \mathcal{F}, \boldsymbol{\mathcal{F}})$ be a measurable space $(\Omega, \mathcal{F})$ endowed with a filtration of σ-fields $\boldsymbol{\mathcal{F}} = (\mathcal{F}_k)_{k \geq 0}$, $\mathcal{F}_0 \subset \mathcal{F}_1 \subset \ldots \subset \mathcal{F}_k \subset \ldots \subset \mathcal{F}$. The events from $\mathcal{F}_k$ are treated as the events observed before the moment k (inclusively). Usually $\mathcal{F}_0$ is a trivial σ-field, ($\mathcal{F}_0 = \{\emptyset, \Omega\}$). Sometimes more general situations are also of interest when $\mathcal{F}_0$ is non-trivial and it can be treated as a σ-field of "a priori data".

Let $\mathcal{P} = (P_\theta, \theta \in \Theta)$ be a family of probability measures on $(\Omega, \mathcal{F})$ which describes all the conceivable "theories" with respect to the probabilistic-statistical distributions of the observed data. Denote by

$$P_{\theta,k} = P_\theta \,|\, \mathcal{F}_k$$

the restriction of the measure P_θ on the σ-field $\mathcal{F}_k$, $k \geq 0$, $\theta \in \Theta$.

Let then $\tau = \tau(\omega)$ be a *stopping time*, i.e. a random variable with values $0, 1, 2, \ldots$ and such that

$$\{\omega : \tau(\omega) = k\} \in \mathcal{F}_k, \quad k \geq 0.$$

The event $\{\omega : \tau(\omega) = k\}$ is treated as the event for observations being stopped at the moment k. By $\mathcal{F}_\tau$ we denote the set of events observed before the moment τ. Formally, $A \in \mathcal{F}_\tau$ iff for each $k \geq 0$

$$A \cap \{\omega : \tau(\omega) = k\} \in \mathcal{F}_k.$$

If $\mathcal{E} = (\Omega, \mathcal{F}; P_\theta, \theta \in \Theta)$ is a statistical experiment and additionally a filtration $\boldsymbol{\mathcal{F}} = (\mathcal{F}_k)_{k \geq 0}$ is given on $(\Omega, \mathcal{F})$ then by

$$(\mathcal{E}, \boldsymbol{\mathcal{F}}) = (\Omega, \mathcal{F}, \boldsymbol{\mathcal{F}}; P_\theta, \theta \in \Theta)$$

we denote the statistical experiment with the filtration $\boldsymbol{\mathcal{F}}$. For simplicity it is also said that $(\mathcal{E}, \boldsymbol{\mathcal{F}})$ is *a filtered statistical experiment.*

Let P be a measure on $(\Omega, \mathcal{F})$ such that for any $\theta \in \Theta$ the measure P_θ is *locally* absolutely continuous with respect to P, i.e. $P_{\theta,k} \ll P_k$ with $P_k = P \,|\, \mathcal{F}_k$.

Put

$$p_{\theta,k}(\omega) = \frac{dP_{\theta,k}}{dP_k}(\omega), \quad k \ge 0,$$

$$g_{\theta,k}(\omega) = \frac{p_{\theta,k}(\omega)}{p_{\theta,k-1}(\omega)}, \quad k \ge 1,$$

with $p_{\theta,0} = 1$.

It is not difficult to see that it holds for the Radon-Nikodym derivatives P-a.s.:

$$\frac{dP_{\theta,k}}{dP_k}(\omega) = \prod_{i=1}^{k} g_{\theta,k}(\omega), \quad \theta \in \Theta, \quad k \ge 1.$$

If the σ-fields $\mathcal{F}_k$ are generated by random variables $\xi_0(\omega), \dots, \xi_k(\omega)$ then the functions $g_{\theta,k}(\omega)$ can be represented in the form $g_k(\theta, \xi_0(\omega), \dots, \xi_k(\omega))$. In particular, it follows for the case of *independent* observations

$$g_{\theta,k}(\omega) = g_k(\theta; \xi_k(\omega));$$

for the case of a *Markov chain* (with respect to the measure P)

$$g_{\theta,k}(\omega) = g_k(\theta; \xi_{k-1}(\omega), \xi_k(\omega));$$

for a *controlled Markov sequence*

$$g_{\theta,k}(\omega) = g_k(\theta; \xi_{k-1}(\omega), \xi_k(\omega), u_k(\omega));$$

with the "control" $u_k(\omega)$ being chosen by the preceding observations $\xi_0(\omega), \dots, \xi_{k-1}(\omega)$, i.e. $u_k(\omega) = u_k(\xi_0(\omega), \dots, \xi_{k-1}(\omega))$.

Example 1.10 *Filtered experiments in continuous time.* This notion is used for statistical experiments $\mathcal{E} = (\Omega, \mathcal{F}; P_\theta, \theta \in \Theta)$ where a process of observation is described by a filtration $\mathbf{\mathcal{F}} = (\mathcal{F}_t)_{t \ge 0}$ with $\mathcal{F}_s \subset \mathcal{F}_t \subset \mathcal{F}$, $s \le t$. For such experiments we also use the notation $(\mathcal{E}, \mathbf{\mathcal{F}}) = (\Omega, \mathcal{F}, \mathbf{\mathcal{F}}; P_\theta, \theta \in \Theta)$.

Example 1.11 *Statistical experiments corresponding to diffusion models.* This kind of statistical models is a natural generalization of that in Example 1.6. Assume that $W = (W_t(\omega'), t \le T)$ is a d-component *standard Wiener process* given on some probability space $(\Omega', \mathcal{F}', P')$. Denote by $\mathbf{C}_T$ the space $\mathbf{C}^d[0,T]$ of continuous vector-functions $x = (x_t, t \le T)$ on the interval $[0,T]$ with values in R^d; by $\mathcal{B}(\mathbf{C}_T)$ we denote the Borel σ-field in $\mathbf{C}_T$. Let then Θ be an open subset in R^d and let $f = f_t(x, \theta)$ be a Borel vector-function (in all the arguments (t, x, θ)) with values in R^d such that for each θ it is an *optional* functional, i.e. $f_t(x, \theta)$ for fixed $t \in [0,T]$ and $\theta \in \Theta$ is $\mathcal{B}_t(\mathbf{C}_T)$-measurable with the σ-sub-field $\mathcal{B}_t(\mathbf{C}_T)$ in $\mathcal{B}(\mathbf{C}_T)$ generated by $x_s, s \le t$.

Let us consider (for fixed $\theta \in \Theta$) a solution of the stochastic differential equation

$$dX_t = f_t(x, \theta)\, dt + dW_t, \quad X_0 = 0,$$

and assume that f satisfies all the necessary conditions for this equation to have a single strong solution (for more details see Liptser and Shiryaev (1974, Chapter IV, Section 4). Let P_W^T be the *Wiener measure* on $(\mathbf{C}_T, \mathcal{B}(\mathbf{C}_T))$, i.e. $P_W^T(B) = P'(W \in B)$, $B \in \mathcal{B}(\mathbf{C}_T)$, and let P_θ^T be the distribution of the process X that satisfies the equation considered. The collection $\mathcal{E}^T = (\Omega^T, \mathcal{F}^T; P_\theta^T, \theta \in \Theta)$ with $\Omega^T = \mathbf{C}_T, \mathcal{F}^T = \mathcal{B}(\mathbf{C}_T)$ is called a statistical experiment generated by a diffusion type process.

3. Let us give some definitions related to assumptions on the measurable spaces $(\Omega, \mathcal{F})$ as well as on the families of measures $\mathcal{P} = \{P_\theta, \theta \in \Theta\}$.

A. Dominated experiment. An experiment $\mathcal{E} = (\Omega, \mathcal{F}; P_\theta, \theta \in \Theta)$ is *dominated* if there is a σ-finite measure ν on $(\Omega, \mathcal{F})$ such that all the measures $P_\theta, \theta \in \Theta$, are absolutely continuous with respect to it, $(P_\theta \ll \nu, \theta \in \Theta)$. If such a (dominating) measure ν exits then there exists also a *probability* measure Q with the property $P_\theta \ll Q, \theta \in \Theta$. Moreover (Halmos and Savage (1949); see also Strasser (1985, p.94)), this measure Q can be chosen in the form

$$Q = \sum_{n=1}^{\infty} \alpha_n P_{\theta_n}$$

with

$$\theta_n \in \Theta, \qquad \alpha_n \geq 0, \qquad \sum_{n=1}^{\infty} \alpha_n = 1,$$

and $\mathcal{P} \sim Q$ (i.e. if $P_\theta(A) = 0$ for each $\theta \in \Theta$ then $Q(A) = 0$ and, vice versa, if $Q(A) = 0$ then $P_\theta(A) = 0$ for each $\theta \in \Theta$).

B. An experiment with a *finite* parameter set and its standard representation. Let $\mathcal{E}_k = (\Omega, \mathcal{F}; P_\theta, \theta \in \Theta_k)$ be an experiment with a *finite* parameter set Θ_k consisting of k points, $\Theta_k = \{\theta_1, \dots, \theta_k\}$.

Denote

$$\nu = \sum_{i=1}^{k} P_{\theta_i}$$

and let

$$f_i = \frac{dP_{\theta_i}}{d\nu}$$

be the density (i.e. Radon-Nikodym derivative) of the measure P_{θ_i} with respect to the measure ν, $i = 1, \dots, k$. Obviously $f_i = f_i(\omega) \geq 0$, $\sum_{i=1}^{k} f_i(\omega) = 1$ (ν-a.s.; without loss of generality one can assume even for *all* $\omega \in \Omega$). If we denote by Σ_k

the *simplex*

$$\Sigma_k = \left\{ f : f = (f_1, \dots, f_k),\ f_i \geq 0,\ \sum_{i=1}^{k} f_i = 1 \right\},$$

then $f(\omega) = (f_1(\omega), \dots, f_k(\omega))$ can be considered as a *random point* on Σ_k.

Introduce for Borel sets $B \in \mathcal{B}(\Sigma_k)$ on the simplex Σ_k the measure μ with

$$\mu(B) = \nu\{\omega : (f_1(\omega), \dots, f_k(\omega)) \in B\} \quad (= \nu(f^{-1}(B)).$$

The measure μ on $(\Sigma_k, \mathcal{B}(\Sigma_k))$ is called (due to Blackwell (1951)) *the standard* (sometimes *canonical*) *measure* of the experiment Σ_k. If we put for $B \in \mathcal{B}(\Sigma_k)$

$$\mu_i(B) = P_{\theta_i}\{\omega : \quad (f_1(\omega), \dots, f_k(\omega)) \in B\} \quad \left(= P_{\theta_i}(f^{-1}(B))\right), \quad i = 1, \dots, k$$

then by the formulae of change of variables in a Lebesgue integral (see Shiryaev (1980, p.211))

$$\begin{aligned} \int_B f_i\, \mu(df) &= \int_{f^{-1}(B)} f_i(\omega)\, \nu(d\omega) = \int_{f^{-1}(B)} \frac{dP_{\theta_i}}{d\nu}(\omega)\, \nu(d\omega) \\ &= P_{\theta_i}(f^{-1}(B)) = \mu_i(B), \quad i = 1, \dots, k. \end{aligned}$$

In other words, the probability measures μ_i can be reconstructed from the measure μ by integration. Since also $\mu = \mu_1 + \dots + \mu_k$, there is a one-to-one correspondence between μ and $(\mu_1, \dots, \mu_k)$.

Define the experiment

$$\mathcal{E}_k^* = (\Sigma_k, \mathcal{B}(\Sigma_k);\ \mu_i, i = 1, \dots, k).$$

This experiment is called *the standard representation* of the experiment $\mathcal{E}_k$ or its *standardization.* These terms are justified by the fact (see Section 1.7, Example 1.22) that the Δ-distance $\Delta(\mathcal{E}_k, \mathcal{E}_k^*) = 0$, i.e. the experiments $\mathcal{E}_k$ and $\mathcal{E}_k^*$ are equivalent in a clear sense.

C. Canonical experiments. An important role in many questions of mathematical statistics is played by experiments of the special form

$$\mathcal{E} = (X, \mathcal{X};\ P_\theta, \theta \in \Theta)$$

for which the measurable space $(X, \mathcal{X})$ is a *metric space* X endowed with a *metric* $d = d(x, y)$ and $\mathcal{X}$ is the σ-field of Borel subsets in X generated by open sets.

In probability theory, especially in the theory of weak convergence of distributions of random processes, *complete separable* metric spaces (the so-called *Polish spaces*) $(X, \mathcal{X})$ play an important role. Typical examples of such space are the spaces R, R^n, R^∞, the space $\mathbf{C}_T$ of continuous functions with the uniform metric $d(x, y) = \sup_{0 \leq t \leq T} |x_t - y_t|$, the space $\mathbf{D}$ with the *Skorohod metric* (see, for example, Billingsley (1968)). In order to highlight the importance for the statistical theory of experiments $\mathcal{E} = (X, \mathcal{X}; P_\theta, \theta \in \Theta)$ with a Polish space $(X, \mathcal{X})$ we shall call such experiments *canonical.*

1.2 Randomization of Statistical Experiments. I. Markov Kernels. Stochastic Operators

1. The notion of *randomization* is of great importance for the definition of "randomized decisions" as well as to introduce general notions like "$\mathcal{E}$ is more informative than $\mathcal{E}^*$", the deficiency $\delta(\mathcal{E}, \mathcal{E}^*)$, the Δ-distance between the experiments $\mathcal{E}$ and $\mathcal{E}^*$. In this section we consider two approaches to define the concept of "randomization". The first of them is classical. It is based on the notion of *a Markov* or *stochastic* kernel. The second one uses the notion of *a stochastic operator* but in this case experiments are assumed to be *dominated.* The general definition of randomization and its relation with the previous ones will be considered in Section 1.5. It is worth noting that for many situations it is sufficient to operate only with randomizations generated by Markov kernels. For instance, if experiments are dominated and are defined on a Polish space $(X, \mathcal{X})$ and if the question of when one experiment is more informative than another is considered (see Section 1.5). This assumption simplifies significantly all the considerations.

2. We begin with the definition and some example of Markov kernels.

Definition 1.1 Let $(\Omega, \mathcal{F})$ and $(\Omega^*, \mathcal{F}^*)$ be two measurable spaces. *A Markov* (or *stochastic*) *kernel* from $(\Omega, \mathcal{F})$ to $(\Omega^*, \mathcal{F}^*)$ is a function $M = M(\omega, A^*)$: $\Omega \times \mathcal{F}^* \to [0,1]$ such that

(1) $M(\cdot, A^*)$ is $\mathcal{F}$-measurable function for each $A^* \in \mathcal{F}^*$;
(2) $M(\omega, \cdot)$ is a probability measure on $(\Omega^*, \mathcal{F}^*)$ for each $\omega \in \Omega$.

The set of all such Markov kernels is denoted by $\mathcal{M}(\Omega, \mathcal{F}^*)$.

Example 1.12 Let $(\Omega^*, \mathcal{F}^*, P^*)$ be a probability space, $\Omega = \Omega^*$, σ-field $\mathcal{F} \subseteq \mathcal{F}^*$ (i.e. $\mathcal{F}$ is a σ-sub-field of $\mathcal{F}^*$). Then due to the well-known definition (see, for example Shiryaev (1980, p.243)) the *regular conditional distribution* $P^*(A^* \mid \mathcal{F})(\omega)$ (as a function on $\Omega \times \mathcal{F}^*$) is exactly the Markov kernel from $(\Omega, \mathcal{F})$ to $(\Omega, \mathcal{F}^*)$.

Let $(\Omega, \mathcal{F}), (\Omega^*, \mathcal{F}^*)$ be two measurable spaces and $M(\omega, A^*)$ be a Markov kernel. For each measure P on $(\Omega, \mathcal{F})$ one can define with the help of the Markov kernel M a measure MP on $(\Omega^*, \mathcal{F}^*)$ with

$$MP(A) = \int M(\omega, A^*)\, P(d\omega), \quad A^* \in \mathcal{F}^*. \tag{1.1}$$

Thus, an experiment $\mathcal{E} = (\Omega, \mathcal{F}; P_\theta, \theta \in \Theta)$ and a Markov kernel M define new experiment

$$M\mathcal{E} = (\Omega^*, \mathcal{F}^*; MP_\theta, \theta \in \Theta). \tag{1.2}$$

Definition 1.2 The experiment $M\mathcal{E}$ is called the *randomization* of the experiments $\mathcal{E}$ by means of the Markov (or stochastic) kernel M.

Definition 1.3 Let $\mathcal{E} = (\Omega, \mathcal{F}; P_\theta, \theta \in \Theta)$ and $\mathcal{E}^* = (\Omega^*, \mathcal{F}^*; P_\theta^*, \theta \in \Theta)$ be experiments with the *same* parametric set Θ. If for some Markov kernel M one has

$$P_\theta^* = MP_\theta, \quad \theta \in \Theta,$$

then the experiment $\mathcal{E}$ is called *exhaustive* (or Blackwell-sufficient) for the experiment $\mathcal{E}^*$. (For brevity we will write this property in the form "$\mathcal{E} \gg \mathcal{E}^*$" or "$\mathcal{E} \overset{M}{\gg} \mathcal{E}^*$". Sometimes, in order to emphasize the role of the corresponding Markov kernel M, one uses the notation "$\mathcal{E}^* = M\mathcal{E}$".)

Example 1.13 Let $\mathcal{E} = (\Omega, \mathcal{F}; P_\theta, \theta \in \Theta)$ be an experiment, let $(\Omega^*, \mathcal{F}^*)$ be a measurable space, and let $T = T(\omega)$ be a $\mathcal{F}/\mathcal{F}^*$-measurable mapping from $(\Omega, \mathcal{F})$ to $(\Omega^*, \mathcal{F}^*)$. Denote $T^{-1}A^* = \{\omega : T(\omega) \in A^*\}$, $A^* \in \mathcal{F}^*$ and $P_\theta^*(A^*) = P_\theta(T^{-1}A^*)$ $(= P_\theta \circ T^{-1}(A^*))$. Obviously

$$P_\theta^*(A^*) = P_\theta(T^{-1}A^*) = \int_\Omega I_{\{T^{-1}A^*\}}(\omega)\, P_\theta(d\omega) = \int_\Omega M(\omega, A^*)\, P_\theta(d\omega),$$

with $M(\omega, A^*) = I_{T^{-1}A^*}(\omega)$ being a Markov kernel. Therefore, the experiment $T\mathcal{E} = (\Omega^*, \mathcal{F}^*; P_\theta \circ T^{-1}, \theta \in \Theta)$ obtained from $\mathcal{E}$ by the mapping T can be considered as an example of a randomization of the experiment $\mathcal{E}$.

Example 1.14 If $\mathcal{E} = (\Omega, \mathcal{F}; P_\theta, \theta \in \Theta)$ is an experiment and $\mathcal{F}^0$ is a σ-subfield in $\mathcal{F}$ then, due to the classical definitions (see, for example, Strasser (1985, Section 25), Borovkov (1984)), $\mathcal{F}^0$ is *sufficient* for the family $\mathcal{P} = \{P_\theta, \theta \in \Theta\}$ (or $\mathcal{P}$-sufficient) if for each $A \in \mathcal{F}$ there is a $\mathcal{F}^0$-measurable function $f_A(\omega)$ such that for all $\theta \in \Theta$ the conditional probability of P_θ with respect to $\mathcal{F}^0$ obeys the equality

$$P_\theta(A \mid \mathcal{F}^0)(\omega) = f_A(\omega) \qquad (P_\theta\text{-a.s.}). \tag{1.3}$$

From the definition of conditional probability we get for any $B \in \mathcal{F}^0$ and $\theta \in \Theta$

$$P_\theta(A \cap B) = \int_B f_A(\omega)\, P_\theta^0(d\omega),$$

with $P_\theta^0 = P_\theta \mid \mathcal{F}^0$ being the restriction of the measure P_θ on $\mathcal{F}^0$. In particular,

$$P_\theta(A) = \int_\Omega f_A(\omega)\, P_\theta^0(d\omega). \tag{1.4}$$

Note that in (1.3) we do not assume, in general, $f_A(\omega)$ to be a measure (in A) on $(\Omega, \mathcal{F})$ for any $\omega \in \Omega$. This means that $f_A(\omega)$ is not assumed to be (as in Example 1.12) a variant of *regular* conditional distribution $P_\theta(A \mid \mathcal{F}^0)(\omega), \theta \in$

Θ. But if this property holds then (1.4) yields that the sub-experiment $\mathcal{E}^0 = (\Omega, \mathcal{F}^0; P_\theta^0, \theta \in \Theta)$ is exhaustive (Blackwell-sufficient) for the experiment $\mathcal{E}$.

Remark 1.1 The following condition is sufficient for the existence of a regular conditional distribution $f_A(\omega)$ such that $P_\theta(A \mid \mathcal{F}^0)(\omega) = f_A(\omega)$ (P_θ-a.s.) for each $\theta \in \Theta$: the family of measures $\mathcal{P} = \{P_\theta, \theta \in \Theta\}$ is *dominated* (by a probability measure Q) and for each $\theta \in \Theta$ there exists a regular conditional distribution $f_{\theta,A}(\omega)$. Under this assumption the regular conditional probability $Q(A \mid \mathcal{F}^0)(\omega)$ can be taken instead of $f_A(\omega)$.

Remark 1.2 If the experiment $\mathcal{E}$ is dominated and the sub-experiment $\mathcal{E}^0$ is exhaustive for the family $\mathcal{P}$ then σ-field $\mathcal{F}^0$ is sufficient (in the classical sense) for $\mathcal{P}$, Strasser (1985, p.103).

3. The notion of randomization introduced above and based on the use of Markov kernels is not sufficient for the *general theory of statistical experiments* (and the same holds for the statistical decision theory). This is related, in particular, to the fact that the set of all Markov kernels $\mathcal{M}(\Omega, \mathcal{F}^*)$ being *convex* is not *compact*, but compactness is a very desirable property for the theory on the whole.

The need to generalize the notion of randomization given before can be explained also by the situation of Example 1.12. Indeed, to justify the implication "$\mathcal{F}^0$ is sufficient for $\mathcal{F}$" $\Rightarrow$ "$\mathcal{E}^0$ is exhaustive for $\mathcal{E}$", assumptions about the existence of regular conditional probabilities are generally needed.

The next step in the definition of randomization will be now made under an additional assumption that the experiments $\mathcal{E}$ and $\mathcal{E}^*$ considered are *dominated*.

Definition 1.4 Let $(\Omega, \mathcal{F}, \nu)$ and $(\Omega^*, \mathcal{F}^*, \nu^*)$ be measurable spaces with σ-finite measures ν and ν^* respectively. Let $L(\nu) = L_1(\Omega, \mathcal{F}, \nu)$ and $L(\nu^*) = L_1(\Omega^*, \mathcal{F}^*, \nu^*)$. Any positive linear operator $M_1 : L(\nu) \to L(\nu^*)$ such that for each non-negative function $g = g(\omega)$, $\omega \in \Omega$,

$$\|M_1 g\|_{L(\nu^*)} = \|g\|_{L(\nu)},$$

is called a *stochastic operator*.

Let $(\mathcal{E}, \nu)$ be a dominated experiment and let $Z_\theta = \frac{dP_\theta}{d\nu}(\omega)$ be the density of the measure P_θ with respect to the measure ν, $\theta \in \Theta$. It is clear that $Z_\theta \in L(\nu)$ and the measure P_θ is completely defined by $Z_\theta(\omega)$ and the measure ν. Therefore the following definition is natural.

Definition 1.5 Let $(\mathcal{E}, \nu)$ and $(\mathcal{E}^*, \nu^*)$ be dominated experiments. We say that $(\mathcal{E}^*, \nu^*)$ is a *randomization* of the experiment $(\mathcal{E}, \nu)$ if there exists a stochastic operator M_1 such that

$$\frac{dP_\theta^*}{d\nu^*} = M_1 \left(\frac{dP_\theta}{d\nu} \right)$$

for each $\theta \in \Theta$. In this case we write $\mathcal{E}^* = M_1 \mathcal{E}$.

It is easy to show that in the dominated case any experiment $\mathcal{E}^*$ which is a randomization of $\mathcal{E}$ by a stochastic kernel can be obtained from $\mathcal{E}$ by a randomization in the sense of Definition 1.5. Thus, it is natural to ask when a randomization defined by a stochastic operator M_1 can be generated by some Markov kernel. One of the famous sufficient conditions is, for example, as follows: the experiments $\mathcal{E}$ and $\mathcal{E}^*$ are dominated, the space Ω^* is locally compact and with countable base, and $\mathcal{F}^*$ is its Borel σ-field (see Strasser (1985, Theorem 24.1)).

4. If $(\mathcal{E}, \nu)$ is a dominated experiment then the space $L(\nu)$ is evidently isomorphic to the subspace of the space $\Sigma(\Omega, \mathcal{F})$ of all *signed* σ-finite measures σ on $(\Omega, \mathcal{F})$ that are absolutely continuous with respect to ν. This subspace is denoted by $L(\mathcal{E})$ (or $L(\mathcal{E}, \nu)$) and it is called the *L-space of the* (dominated) *experiment* $\mathcal{E}$. The space $L(\mathcal{E})$ can be described also in the following way. Let $L_0(\mathcal{E})$ be the set of all measures $\sigma \in \Sigma(\Omega, \mathcal{F})$ represented as a finite combination $\sum a_i \sigma_i$ with $a_i \in R$ and $\sigma_i \in \Sigma(\Omega, \mathcal{F})$ and such that $\sigma_i \ll P_{\theta_i}$ for some $P_{\theta_i} \in \mathcal{P}$. Then the closure of $L_0(\mathcal{E})$ (with respect to the variation norm $\|\cdot\|$) is exactly the space $L(\mathcal{E})$. It is important to note that this description does not use the assumption for $\mathcal{E}$ to be dominated. Hence this construction of $L(\mathcal{E})$ as the closure of $L_0(\mathcal{E})$ is true for any experiment $\mathcal{E}$. As for the dominated case the space $L(\mathcal{E})$ is called the *L-space of the* (arbitrary) *experiment* $\mathcal{E}$. It is not very difficult to show that $L(\mathcal{E})$ permits also the following representation:

$$L(\mathcal{E}) = \{\sigma \in \Sigma(\Omega, \mathcal{F}) : \text{if } \mu \in \Sigma(\Omega, \mathcal{F}) \text{ and } \mu \perp P_\theta \text{ for all } \theta \in \Theta \text{ then } \mu \perp \sigma\}.$$

(Compare with Strasser (1985, Section 41).)

In Section 1.5 the *general* definition of randomization will be given and it will be shown that the above considered randomizations by Markov (stochastic) kernels and by stochastic operators are particular cases of it. In the next section we give a number of examples that explain the essence of the already introduced notions of randomization.

1.3 Examples of Comparison of Experiments with the Help of Randomization

1. Let $\mathcal{E} = (\Omega, \mathcal{F}; P_\theta, \theta \in \Theta)$ and $\mathcal{E}^* = (\Omega^*, \mathcal{F}^*; P_\theta^*, \theta \in \Theta)$ be experiments with the same parametric set Θ. We denote "$\mathcal{E} \gg \mathcal{E}^*$" if the experiment $\mathcal{E}^*$ is obtained by a randomization (in the sense of both definitions of Section 1.2) from the experiment $\mathcal{E}$.

Example 1.15 Let $\mathcal{E} = (R, \mathcal{B}(R); \mathcal{N}(\theta, \rho^2\sigma^2), \theta \in R)$ be a Gaussian shift experiment (i.e. $P_\theta = \mathcal{N}(\theta, \rho^2\sigma^2)$ with the mean θ and known σ^2 and $0 < \rho < 1$). Let also $\mathcal{E}^* = (R, \mathcal{B}(R); \mathcal{N}(\theta, \sigma^2), \theta \in R)$ with known σ^2. To simplify the exposition the following notation will be often used to specify experiments:

$$\mathcal{E} : \xi \sim \mathcal{N}(\theta, \rho^2\sigma^2), \ \theta \in R; \ \sigma^2 \text{ and } 0 < \rho < 1 \text{ are known.}$$

Under this notation the experiment $\mathcal{E}$ is treated as an observation of a random variable ξ given on some probability space and having the distribution $\mathcal{N}(\theta, \rho^2\sigma^2)$, θ being an unknown parameter which is to be estimated by the result of observation ξ. Similarly we write

$$\mathcal{E}^* : \xi^* \sim \mathcal{N}(\theta, \sigma^2), \ \theta \in R, \ \sigma^2 \text{ is known.}$$

It is clear that

$$\xi^* \stackrel{d}{=} \xi + \eta,$$

where $\eta \sim \mathcal{N}(0, (1-\rho^2)\sigma^2)$ and "$\stackrel{d}{=}$" means the equality by distribution of random variables. If $P_\theta = \mathcal{N}(\theta, \rho^2\sigma^2)$ and $P_\theta^* = \mathcal{N}(\theta, \sigma^2)$ then evidently

$$P_\theta^* = P_\theta * Q,$$

with the symbol "$*$" denoting the convolution of measures P_θ and $Q = \mathcal{N}(0, (1-\rho^2)\sigma^2)$. Rewriting this property in the form

$$P_\theta^*(B) = \int Q(B - x)\, P_\theta(dx), \quad B \in \mathcal{B}(R),$$

with $B - x = \{y : y + x \in B\}$ we see that the experiment $\mathcal{E}^*$ is obtained from $\mathcal{E}$ by a randomization ("$\mathcal{E} \gg \mathcal{E}^*$") defined by the Markov kernel $M(x, B) = Q(B - x)$.

Example 1.16 Using the notations from the preceding example let

$$\begin{aligned} &\mathcal{E} : \xi \sim \mathcal{N}(0, \rho^2\theta^2), && \theta \in R, \ 0 < \rho < 1 \text{ is known;} \\ &\mathcal{E}^* : \xi^* \sim \mathcal{N}(0, \theta^2), && \theta \in R. \end{aligned}$$

Then $\xi \stackrel{d}{=} \rho\xi^*$, $\xi^* \stackrel{d}{=} \xi/\rho$ and this implies (see Example 1.13 in Section 1.2) that the experiments $\mathcal{E}$ and $\mathcal{E}^*$ are *equivalent* ($\mathcal{E} \sim \mathcal{E}^*$) in the sense that $\mathcal{E} \gg \mathcal{E}^*$ and $\mathcal{E}^* \gg \mathcal{E}$.

Example 1.17 Let

$$\begin{aligned} &\mathcal{E} : \xi \sim \mathcal{N}(\theta_1, \rho^2\theta_2), && \theta_1 \in R, \quad \theta_2 > 0; \quad 0 < \rho < 1 \text{ is known,} \\ &\mathcal{E}^* : \xi^* \sim \mathcal{N}(\theta_1, \theta_2), && \theta_1 \in R, \quad \theta_2 > 0. \end{aligned}$$

It was shown in Hansen and Torgensen (1974) that these experiments $\mathcal{E}$ and $\mathcal{E}^*$ are *not comparable* in the sense that it is impossible to conclude that $\mathcal{E} \gg \mathcal{E}^*$ or $\mathcal{E}^* \gg \mathcal{E}$.

Example 1.18 Boll (1955), Hansen and Torgensen (1974), Torgensen (1991), Lehmann (1987). Let us consider two shift experiments

$$\mathcal{E} = (R, \mathcal{B}(R);\, P_\theta\,, \theta \in R), \qquad P_\theta(-\infty, x] = F(x-\theta);$$
$$\mathcal{E}^* = (R, \mathcal{B}(R);\, P_\theta^*\,, \theta \in R), \qquad P_\theta^*(-\infty, x] = F^*(x-\theta),$$

with known distribution functions F and F^*. It follows from the papers mentioned above that

$$\mathcal{E} \gg \mathcal{E}^* \iff \xi^* \stackrel{d}{=} \xi + \eta, \tag{1.5}$$

where ξ and ξ^* are the corresponding random variables generating the experiments $\mathcal{E}$ and $\mathcal{E}^*$, and η is an auxiliary random variable such that ξ and η are independent.

If $\varphi(t) = E_{P_0} e^{it\xi}$, $\varphi^*(t) = E_{P_0^*} e^{it\xi^*}$, then the property (1.5) can be rewritten in the following form:

$$\mathcal{E} \gg \mathcal{E}^* \iff \psi(t) \equiv \frac{\varphi^*(t)}{\varphi(t)} \text{ is a characteristic function.} \tag{1.6}$$

For example, if $F(x)$ has the density $f(x) = e^{-x} I(x > 0)$, and $F^*(x)$ has the density $f^*(x) = \frac{1}{2} e^{-|x|}$ then $\varphi(t) = (1-it)^{-1}$, $\varphi^*(t) = (1+it^2)^{-1}$, and hence the function $\psi(t) \equiv \frac{\varphi^*(t)}{\varphi(t)} = (1+it)^{-1}$ is obviously the characteristic function of the random variable $-\xi$. Therefore, taking $\eta \stackrel{d}{=} -\xi$ and not depending on ξ we state that $\xi^* \stackrel{d}{=} \xi + \eta$, and $\mathcal{E} \gg \mathcal{E}^*$ by (1.6).

Due to Lehmann (1988) the following unexpected result for shift experiments is a direct consequence of (1.6):

Theorem 1.1 *If ξ^* is normally distributed and $\mathcal{E} \gg \mathcal{E}^*$, then ξ is also normally distributed.*

Indeed, if $\mathcal{E} \gg \mathcal{E}^*$ then there exists a random variable η independent on ξ and such that $\xi^* \stackrel{d}{=} \xi + \eta$. It follows from the famous Cramer theorem that if a random variable ξ^* is normally distributed and it is a sum of two independent random variables then these variables also should be normally distributed.

Example 1.19 Lehmann (1988). Let, similarly to Example 1.18,

$$\mathcal{E}^* = (R, \mathcal{B}(R);\, P_\theta^*\,, \theta \in R), \qquad P_\theta^*(-\infty, x] = F^*(x-\theta),$$
$$\mathcal{E}_\lambda = (R, \mathcal{B}(R);\, P_{\theta,\lambda}\,, \theta \in R), \qquad P_{\theta,\lambda}(-\infty, x] = F^*\left(\frac{x-\theta}{\lambda}\right),$$

with a fixed constant λ from $(0,1)$. Then $\mathcal{E}_\lambda \gg \mathcal{E}^* \iff \psi(t) \equiv \frac{\varphi^*(t)}{\varphi^*(\lambda t)}$ is a characteristic function. Due to P. Lévy a characteristic function $\varphi^*(t)$ with the property that $\frac{\varphi^*(t)}{\varphi^*(\lambda t)}$ is again a characteristic function for any $0 < \lambda < 1$, corresponds to a distribution from the well-known class L, which is a subclass of the set of all infinitely divisible distributions.

For example, if $F^*(dx) = f^*(x)\,dx$ with $f^*(x) = \frac{1}{2}e^{-|x|}$ then $\varphi^*(t) = (1+t^2)^{-1}$ and the function

$$\psi(t) \equiv \frac{\varphi^*(t)}{\varphi^*(\lambda t)} = \frac{1+\lambda^2 t^2}{1+t^2} = \lambda^2 1 + (1-\lambda^2)\frac{1}{1+t^2}$$

is, for each $0 < \lambda < 1$, the characteristic function of the random variable η equal to zero with the probability λ^2 and having the distribution with the density $\frac{1}{2}e^{-|x|}$ with the probability $1-\lambda^2$. Therefore, for any fixed $\lambda \in (0,1)$ one has $\mathcal{E}_\lambda \gg \mathcal{E}^*$.

Example 1.20 Let $F^*(x-\theta)$ be the *uniform* distribution $U(\theta - 1/2, \theta + 1/2)$ concentrated on the interval $(\theta - 1/2, \theta + 1/2)$. If $F_\lambda(x-\theta) = F^*\left(\frac{x-\theta}{\lambda}\right)$ then contrary to the preceding example (where $\mathcal{E}_\lambda \gg \mathcal{E}^*$ for all $\lambda \in (0,1)$) here (due to Lehmann (1988)) $\mathcal{E}_\lambda \gg \mathcal{E}^*$ iff $\lambda = 1/k$ for some integer k.

Other examples of experiment comparison are given in Section 1.9.

1.4 The Main Components of the General Decision Theory

1. Let $\mathcal{E} = (\Omega, \mathcal{F};\, P_\theta, \theta \in \Theta)$ be a fixed statistical experiment. In order to formalize the questions of the type of which a probability measure P_θ (i.e. which a "theory θ") "best of all corresponds" to the observed data, what is the "quality" of a chosen decision with respect to θ, what is "the corresponding loss" etc., Wald (1950), invented the "Statistical Decision Theory". The main components of this theory (in the "Markov" version) are:

(1) *a decision space* $(\mathbf{A}, \mathcal{A})$, i.e. a measurable space of possible decisions (actions) with respect to θ or P_θ;

(2) *a loss function* $W = W_\theta(a)$ (denoted also by $W(\theta, a)$), $\theta \in \Theta$, $a \in \mathbf{A}$, which characterizes "the quality" of a decision a if the true value of the unknown parameter is θ (usually W is assumed to be $\mathcal{A}$-measurable for each $\theta \in \Theta$ and to satisfy the condition $\inf_a W_\theta(a) > -\infty$);

(3) Markov kernels $\rho = \rho(\omega, A)$ (from $(\Omega, \mathcal{F})$ to $(\mathbf{A}, \mathcal{A})$) called *randomized* (or *Markov*) *decisions*. In the case that for each $\omega \in \Omega$ the measure $\rho(\omega, \cdot)$ is concentrated at one point $\rho(\omega) \in \mathbf{A}$, the decision $\rho = \rho(\omega)$, $\omega \in \Omega$, is said to be *non-randomized*. (The decision $\rho(\omega)$ is assumed to be a $\mathcal{F}/\mathcal{A}$-measurable function.)

The set of all Markov kernels is denoted by $\mathfrak{R}(\Omega, \mathbf{A})$. By $\mathfrak{R}_0(\Omega, \mathbf{A})$ we denote the set of all non-randomized decisions.

For any decision $\rho \in \mathfrak{R}(\Omega, \mathbf{A})$ and any loss function W one can define *the loss* (at the point θ for given ω)

$$W_\theta(\rho) = \int_{\mathbf{A}} W_\theta(a)\,\rho(\omega, da)$$

and *the mean loss*, or *risk* (at the point θ)

$$R_\theta(W,\rho) = E_\theta\, W_\theta(\rho).$$

If $\rho \in \mathfrak{R}_0(\Omega, \mathbf{A})$ then, obviously,

$$R_\theta(W,\rho) = E_\theta\, W_\theta(\rho(\omega)).$$

If $\pi = \pi(d\theta)$ is an a priori distribution on the parameter space Θ endowed with a σ-field $\mathcal{B}(\Theta)$ then one can connect with the risk $R_\theta(\rho, W)$ the π-risk

$$R_\pi(W,\rho) = \int_\Theta R_\theta(W,\rho)\,\pi(d\theta).$$

The value

$$\mathcal{R}_\pi(\mathcal{E}, W) = \inf_{\rho\in\mathfrak{R}(\Omega,\mathbf{A})} R_\pi(\rho, W)$$

is called a *π-Bayes risk* for the loss function W in the experiment $\mathcal{E}$.

The value

$$\mathcal{R}(\mathcal{E}, W) = \inf_{\rho\in\mathfrak{R}(\Omega,\mathbf{A})} \sup_{\theta\in\Theta} R_\theta(W,\rho)$$

is *the minimax risk* in the experiment $\mathcal{E}$ for the loss function W.

The function

$$P_\theta\rho(A) = E_\theta\, \rho(\omega, A), \quad A \in \mathcal{A},$$

is called *the operating characteristic* of a decision ρ. Under this notation the risk $R_\theta(W,\rho)$ can be written in the form

$$R_\theta(W,\rho) = \int_{\mathbf{A}} W_\theta(a)\, P_\theta\rho(da).$$

2. Another approach to define randomized decisions is also often used in statistical problems. This approach is reduced to the following: *a randomized decision for an experiment $\mathcal{E}$ is a non-randomized decision, but for an experiment $\widetilde{\mathcal{E}}$ which is an "extension" of $\mathcal{E}$.* (Such an approach is found to be very useful for a number of situations and we will use it in Section 3.3.)

Let ω be an "observation" in an experiment $\mathcal{E} = (\Omega, \mathcal{F};\, P_\theta\,, \theta \in \Theta)$. From the above considerations it follows that any non-randomized decision is determined by a $\mathcal{F}/\mathcal{A}$-measurable function $\rho = \rho(\omega)$ but a randomized decision is defined by a Markov kernel $\rho = \rho(\omega, A)$.

Now we suppose that $(\Omega', \mathcal{F}', P')$ is some new probability space and $\omega' \in \Omega'$ is treated as the result of an auxiliary "randomizing machine" (like throwing a coin). Then for each $\mathcal{F}\otimes\mathcal{F}'/\mathcal{A}$-measurable function $\widetilde{\rho} = \widetilde{\rho}(\omega, \omega')$ one can construct the Markov kernel

$$\rho(\omega, A) = P'\left\{\omega' : \widetilde{\rho}(\omega,\omega') \in A\right\}, \quad \omega \in \Omega,\ A \in \mathcal{A}.$$

Therefore, a non-randomized decision $\widetilde{\rho} = \widetilde{\rho}(\omega, \omega')$ for the experiment $\mathcal{E} \times \mathcal{E}' = (\Omega \times \Omega', \mathcal{F} \otimes \mathcal{F}'; P_\theta \times P', \theta \in \Theta)$ (called *a direct extension* or *extension* of the experiment $\mathcal{E}$) generates some randomized decision $\rho = \rho(\omega, A)$ for the original experiment $\mathcal{E}$.

It is possible to claim the inverse result (see, for example Wald and Wolfowits (1951), van der Vaart (1991)).

Lemma 1.1 *Let $(\mathbf{A}, \mathcal{A})$ be a complete separable metric space with the Borel σ-field $\mathcal{A}$ and let $\rho = \rho(\omega, A)$ be a Markov kernel from $(\Omega, \mathcal{F})$ to $(\mathbf{A}, \mathcal{A})$. Then one can construct on the measurable space $(\Omega \times \Omega', \mathcal{F} \otimes \mathcal{F}')$ with $\Omega' = [0,1]$, and $\mathcal{F}'$ being the Borel σ-field on Ω', a $\mathcal{F} \otimes \mathcal{F}'/\mathcal{A}$-measurable function $\widetilde{\rho} = \widetilde{\rho}(\omega, \omega')$ such that*

$$\rho(\omega, A) = P'\left\{\omega' : \widetilde{\rho}(\omega, \omega') \in A\right\}, \quad \omega \in \Omega,\ A \in \mathcal{A}$$

with the Lebesgue measure P' on $(\Omega', \mathcal{F}')$.

Remark 1.3 Note that the operating characteristic $P_\theta \rho$ of a (randomized) decision ρ in the experiment $\mathcal{E}$ coincides with the operating characteristic $(P_\theta \times P')\rho$ of a (non-randomized) decision $\widetilde{\rho}$ in the experiment $\mathcal{E} \times \mathcal{E}'$:

$$P_\theta \rho = (P_\theta \times P')\widetilde{\rho}.$$

(In other words, the distribution $\mathcal{L}(\widetilde{\rho} \mid P_\theta \times P')$ of the decision $\widetilde{\rho}$ with respect to $P_\theta \times P'$, i.e. the measure $(P_\theta \times P')\widetilde{\rho}$, coincides with the distribution $\mathcal{L}(\rho \mid P_\theta)$ of the decision ρ with respect to P_θ, i.e. with the measure $P_\theta \rho$, compare with Lemma 1.3 below.)

3. The idea of randomization is also often used in the theory of controlled Markov processes, for sequential experimental design, for the problem of stochastic approximation and generally, for problems related to recursive and sequential decision procedures. In this situation the distribution of the result ω_k of "a randomized machine" at the k-th step can depend on the preceding observation.

Similar and even more general models are described by the following definition.

Definition 1.6 Let $\mathcal{E} = (\Omega, \mathcal{F}; P_\theta, \theta \in \Theta)$ be an experiment, let $(\Omega', \mathcal{F}')$ be a measurable space, and let $M = M(\omega, B)$ be a Markov kernel from $(\Omega, \mathcal{F})$ to $(\Omega', \mathcal{F}')$. Define the experiment $\widetilde{\mathcal{E}} = (\widetilde{\Omega}, \widetilde{\mathcal{F}}; \widetilde{P}_\theta, \theta \in \Theta)$ with $\widetilde{\Omega} = \Omega \times \Omega'$, $\widetilde{\mathcal{F}} = \mathcal{F} \otimes \mathcal{F}'$ and with the measures $\widetilde{P}_\theta$ being the semi-direct product of the measures P_θ and the Markov kernel M, i.e.

$$\widetilde{P}_\theta(A \times B) = \int_A M(\omega, B)\, P_\theta(d\omega), \quad A \in \mathcal{F},\ B \in \mathcal{F}',\ \theta \in \Theta.$$

This experiment $\widetilde{\mathcal{E}}$ is called *a Markov* (or *stochastic*) *extension* of the experiment $\mathcal{E}$ by the Markov kernel M.

Obviously, if the Markov kernel $M = M(\omega, B)$ does not depend on ω ($M = M(B)$) then the Markov extension coincides with the direct extension considered above.

Let us give some simple properties of Markov extensions that follow directly from the definition.

Lemma 1.2 *Let an experiment $\widetilde{\mathcal{E}}$ be the Markov extension of an experiment $\mathcal{E}$ by a Markov kernel M. Then*

(i) The projection (marginal) of the experiment $\widetilde{\mathcal{E}}$ on $(\Omega, \mathcal{F})$ coincides with $\mathcal{E}$ (i.e. the image of $\widetilde{\mathcal{E}}$ for the mapping $T_1 : (\omega, \omega') \to \omega$ from $\Omega \times \Omega'$ to Ω coincides with $\mathcal{E}$).

(ii) The projection (marginal) of $\widetilde{\mathcal{E}}$ on $(\Omega', \mathcal{F}')$ (i.e. the image of $\widetilde{\mathcal{E}}$ for the mapping $T_2 : (\omega, \omega') \to \omega'$ from $\Omega \times \Omega'$ to Ω') coincides with the experiment $\mathcal{E}' = (\Omega', \mathcal{F}'; P'_\theta \in \Theta)$ which is a Markov randomization of $\mathcal{E}$ by the Markov kernel M, $\mathcal{E}' = M\mathcal{E}$.

(iii) The experiment $\widetilde{\mathcal{E}}$ is equivalent to the experiment $\mathcal{E}$ and it is the randomization of $\mathcal{E}$ by the Markov kernel $\widetilde{M}$ (from $(\Omega, \mathcal{F})$ to $(\Omega \times \Omega', \mathcal{F} \otimes \mathcal{F}')$) defined as follows

$$\widetilde{M}(\omega, A \times B) = I_{\{\omega \in A\}} M(\omega, B), \quad \omega \in \Omega,\ A \in \mathcal{F},\ B \in \mathcal{F}'. \tag{1.7}$$

(iv) If the experiment $\mathcal{E}$ is dominated (by a measure P) and $Z_\theta(\omega) = dP_\theta/dP(\omega)$ then the experiment $\widetilde{\mathcal{E}}$ is dominated by the measure $\widetilde{P} = P \times M$ which is the semi-direct product of the measure P and the Markov kernel M, and ($\widetilde{P}$-a.s.)

$$\frac{d\widetilde{P}_\theta}{d\widetilde{P}}(\omega, \omega') = Z_\theta(\omega).$$

Proof. For any $A \in \mathcal{F}$ one has

$$\widetilde{P}_\theta \circ T_1^{-1}(A) = \widetilde{P}_\theta(A \times \Omega') = \int_A M(\omega, \Omega')\, P_\theta(d\omega) = P_\theta(A), \quad \theta \in \Theta,$$

that implies the first assertion of the lemma.

Similarly for any $B \in \mathcal{F}'$

$$\widetilde{P}_\theta \circ T_2^{-1}(A) = \widetilde{P}_\theta(\Omega \times B) = \int_\Omega M(\omega, B)\, P_\theta(d\omega) = P_\theta^*(B), \quad \theta \in \Theta,$$

and (ii) follows.

Let now $\widetilde{M}$ be defined by (1.7). Then

$$\begin{aligned}
\int_\Omega \widetilde{M}(\omega, A \times B)\, P_\theta(d\omega) &= \int_\Omega I_{\{\omega \in A\}} M(\omega, B)\, P_\theta(d\omega) \\
&= \int_A M(\omega, B)\, P_\theta(d\omega) \\
&= \widetilde{P}_\theta(A \times B), \qquad A \in \mathcal{F},\ B \in \mathcal{F}',\ \theta \in \Theta.
\end{aligned}$$

Thus $\widetilde{\mathcal{E}}$ is the Markov randomization of $\mathcal{E}$ by the kernel $\widetilde{M}$. Since from (i) and Example 1.13, Section 1.2, $\mathcal{E}$ is a randomization of $\widetilde{\mathcal{E}}$, the experiments $\mathcal{E}$ and $\widetilde{\mathcal{E}}$ are equivalent.

The last assertion of the lemma is evident. □

4. Now we show how the notion of a Markov decision for an experiment $\mathcal{E}$ can be connected with the notion of a Markov extension of $\mathcal{E}$.

Lemma 1.3 *Let $\mathcal{E} = (\Omega, \mathcal{F}; P_\theta, \theta \in \Theta)$ be an experiment, let $(\mathbf{A}, \mathcal{A})$ be a decision space and let $\rho = \rho(\omega, A)$ be a Markov decision, $\rho \in \mathcal{R}(\mathcal{E}, \mathbf{A})$. Then there exists a Markov extension $\widetilde{\mathcal{E}}$ of the experiment $\mathcal{E}$ and a non-randomized decision $\widetilde{\rho} = \widetilde{\rho}(\widetilde{\omega})$ for $\widetilde{\mathcal{E}}$ such that*

$$\mathcal{L}(\widetilde{\rho} \mid \widetilde{P}_\theta) = \mathcal{L}(\rho \mid P_\theta), \quad \theta \in \Theta,$$

i.e. the distributions of the decisions $\widetilde{\rho}$ and ρ under the measures $\widetilde{P}_\theta$ and P_θ respectively coincide for each $\theta \in \Theta$.

Proof. Define $\widetilde{\mathcal{E}}$ as the extension of $\mathcal{E}$ by a Markov kernel ρ, i.e. let $\widetilde{\mathcal{E}} = (\Omega \times \mathbf{A}, \mathcal{F} \otimes \mathcal{A}; \widetilde{P}_\theta, \theta \in \Theta)$ with the measures $\widetilde{P}_\theta$ being the semi-direct product of the measures P_θ and the Markov kernel ρ. Then define the non-randomized decision $\widetilde{\rho}$ for $\widetilde{\mathcal{E}}$ as follows

$$\widetilde{\rho}(\omega, a) = a, \quad \omega \in \Omega, \, a \in \mathbf{A}.$$

For $A \in \mathcal{A}$, $\theta \in \Theta$ one has

$$\widetilde{P}_\theta \, \widetilde{\rho}(A) = \widetilde{P}_\theta(\widetilde{\rho}^{-1}(A)) = \widetilde{P}_\theta(\Omega \times A) = \int_\Omega \rho(\omega, A) \, P_\theta(d\omega) = P_\theta \, \rho(A),$$

which is the assertion of the lemma. □

5. Similarly to the situation in the theory of statistical experiments where the notion of randomization based on Markov kernels is not sufficient, it is necessary for the statistical decision theory to address to a more general than the above mentioned notion of (generalized) decision.

The approach here is as follows. Taking into account that the definition of a quality of decisions and their comparison is based (in the Markov case) on the risk function

$$R_\theta(W, \rho) = \int_\Theta \int_{\mathbf{A}} W_\theta(a) \, \rho(\omega, da) \, P_\theta(d\omega), \tag{1.8}$$

it is natural to introduce the notion of a generalized decision in the following way.

Let ν be a signed measure on $(\Omega, \mathcal{F})$ with a finite variation norm $\|\nu\| \equiv \sup \sum |\nu(A_i)|$ where sup is taken over all the finite partitions of the space $\mathbf{A}$ on measurable sets A_i.

Now we suppose that $(\mathbf{A}, \mathcal{A})$ is a *topological* space with the Baire σ-field $\mathcal{A}$ ($= \mathcal{B}_0(\mathbf{A})$) which is the minimal σ-field of subset in $\mathbf{A}$ such that all the continuous functions on $\mathbf{A}$ are $\mathcal{A}$-measurable. (Note that $\mathcal{B}_0(\mathbf{A}) \subseteq \mathcal{B}(\mathbf{A})$ with the σ-field $\mathcal{B}(\mathbf{A})$ generated by all the open sets; if $\mathbf{A}$ is a *metric* space then $\mathcal{B}_0(\mathbf{A}) = \mathcal{B}(\mathbf{A})$.) If $f = f(a)$ is a continuous function on $\mathbf{A}$ ($f \in C(\mathbf{A})$), then we denote $\|f\| = \sup|f(a)|$. If $\|f\| < \infty$ then we also say that $f \in C_b(\mathbf{A})$.

Let ρ be a Markov kernel. We define analogously to (1.8) the *bilinear* functional

$$\beta_\rho(f,\nu) = \int_\Omega \int_{\mathbf{A}} f(a)\,\rho(\omega, da)\,\nu(d\omega). \tag{1.9}$$

It is obvious that the functional $\beta_\rho(f,\nu)$ defined by the Markov kernel $\rho = \rho(\omega, A)$ has the following properties:

(*a*) $\beta_\rho(f,\nu)$ is positive, i.e. $\beta_\rho(f,\nu) \geq 0$ if $f \geq 0$, $\nu \geq 0$;

(*b*) $|\beta_\rho(f,\nu)| \leq \|f\|\|\nu\|$;

(*c*) $\beta_\rho(1,\nu) = \|\nu\|$ if $\nu \geq 0$.

Let now $\mathcal{E} = (\Omega, \mathcal{F}; P_\theta, \theta \in \Theta)$ be a fixed experiment and let $L(\mathcal{E})$ be its L-space (Section 1.4). The bilinear functional $\beta_\rho(f,\nu)$ is defined for all $f \in C_b(\mathbf{A})$ and wittingly for all $\nu \in L(\mathcal{E})$ and the properties (*a*), (*b*), (*c*) obviously hold. This remark explains the following definition.

Definition 1.7 Let $\mathcal{E}$ be a statistical experiment, let $(\mathbf{A}, \mathcal{A})$ be a topological space (of decisions) with the Baire σ-field $\mathcal{A}$. Any real-valued bilinear functional $\beta(f,\nu)$ defined on $C_b(\mathbf{A}) \times L(\mathcal{E})$ and satisfying the properties (*a*), (*b*), (*c*) above is *a generalized decision* (or *generalized decision rule*, or *β-decision*).

Together with the class $\mathfrak{R}(\mathcal{E}, \mathbf{A})$ of decisions defined by Markov kernels we denote by $\mathfrak{B}(\mathcal{E}, \mathbf{A})$ the class of all generalized decisions.

The following theorem collects a number of important assertions about the class $\mathfrak{B}(\mathcal{E}, \mathbf{A})$ (see Strasser (1985, Section 42)).

Theorem 1.2

(1) The set $\mathfrak{B}(\mathcal{E}, \mathbf{A})$ is convex and compact for the topology of pointwise convergence in $C_b(\mathbf{A}) \times L(\mathcal{E})$ ("weak topology").

(2) The set $\mathfrak{R}(\mathcal{E}, \mathbf{A})$ is dense in $\mathfrak{B}(\mathcal{E}, \mathbf{A})$ for the weak topology.

(3) If an experiment $\mathcal{E}$ is dominated and $\mathbf{A}$ is a locally compact countable space, then for every $\beta \in \mathfrak{B}(\mathcal{E}, \mathbf{A})$ there exists a kernel ρ such that

$$\beta(f,\nu) = \beta_\rho(f,\nu) \qquad \left(\equiv \int_\Omega \int_{\mathbf{A}} f(a)\,\rho(\omega, da)\,\nu(d\omega)\right).$$

6. If $\rho \in \mathfrak{B}(\mathcal{E}, \mathbf{A})$, then the risk $R_\theta(W,\rho)$ defined by (1.7) is (see (1.9)) exactly $\beta_\rho(W_\theta, P_\theta)$. This leads to the following natural definition of the risk $R_\theta(W,\beta)$ for β-decisions.

Definition 1.8 Let $W_\theta(\cdot) \in C_b(\mathbf{A})$. The ***risk*** $R_\theta(W,\beta)$ (or $R_\theta(\mathcal{E},W,\beta)$) of a β-decision at the point θ in the experiment $\mathcal{E}$ is defined as the value $\beta(W_\theta, P_\theta)$ of the corresponding functional

$$R_\theta(W,\beta) = \beta(W_\theta, P_\theta).$$

7. Bilinear functionals defining β-decisions were determined for bounded and continuous f and $\nu \in L(\mathcal{E})$. Now we extend this definition to a wider class of functions f.

Definition 1.9 Let $f = f(a)$ be a real-valued function bounded from below and *lower semicontinuous* (i.e. all the sets of the form $\{a : f(a) > x\}$ are open). For each β-decision put

$$\beta(f,\nu) = \sup\{\beta(\varphi,\nu),\ \varphi \le f,\ \varphi \in C_b(\mathbf{A})\}$$

where $\beta(\varphi,\nu)$ is the corresponding bilinear functional determining the β-decision.

Definition 1.10 Let $\mathcal{E}$ be a statistical experiment and let $W_\theta(\cdot)$ be a lower semicontinuous function bounded from below. The ***risk*** $R_\theta(W,\beta)$ (or $R_\theta(\mathcal{E},W,\beta)$) of a β-decision at the point θ in the experiment $\mathcal{E}$ is defined as $R_\theta(W,\beta) = \beta(W_\theta, P_\theta)$.

Note that if an experiment $\mathcal{E}$ is dominated and $\mathbf{A}$ is a compact metric space, then for each β-decision there is a Markov kernel ρ such that $\beta(f,\nu) = \beta\rho(f,\nu)$ for each lower semicontinuous and bounded from below function f and any non-negative measure ν from $L(\mathcal{E})$ (see Strasser (1985, Section 43)).

1.5 Randomization of Statistical Experiments. II. General case

1. If one states the question of the *comparison* of two experiments $\mathcal{E}$ and $\mathcal{E}^*$ (of the type "$\mathcal{E} \gg \mathcal{E}^*$") based on the notion of randomization then the considerations from the previous two sections permit us to do it only for the following two cases:

(1) if the randomization is defined by a Markov (stochastic) kernel, or
(2) if all the considered experiments are dominated (in this case randomizations are determined by stochastic operators).

Now we pass to the general case using the idea of considering *bilinear functionals*. Let $\mathcal{E} = (\Omega,\mathcal{F};\ P_\theta, \theta \in \Theta)$ be a statistical experiment, let $(\Omega^*,\mathcal{F}^*)$ be a measurable space and let $L_\infty(\Omega^*,\mathcal{F}^*)$ be the Banach space of real-valued bounded function $f = f(\omega^*)$. If $M = M(\omega, A^*)$ is a Markov kernel (from $(\Omega,\mathcal{F})$ to $(\Omega^*,\mathcal{F}^*)$) then the corresponding bilinear operator $B_M(f,\nu)$ is defined for $f \in L_\infty(\Omega^*,\mathcal{F}^*)$ and $\nu \in L(\mathcal{E})$ by the formula

$$B_M(f,\nu) = \int_\Omega \int_{\Omega^*} f(\omega^*)\, M(\omega, d\omega^*)\, \nu(d\omega).$$

If, in particular, $f(\omega^*) = I_{A^*}(\omega^*), A^* \in \mathcal{F}^*$, then

$$B_M(I_{A^*}, P_\theta) = \int_\Omega M(\omega, A^*)\, P_\theta(d\omega) \quad \Big(= MP_\theta(A^*)\Big). \tag{1.10}$$

The bilinear functional $B_M(f,\nu)$ defined by the Markov kernel M satisfies the following properties (compare with Section 1.4):

(a) $B_M(f,\nu) \geq 0$ if $f \geq 0$, $\nu \geq 0$ i.e. $B_M(\cdot,\cdot)$ is positive;

(b) $|B_M(f,\nu)| \leq \|f\|\,\|\nu\|$ with $\|f\| = \sup_{\omega^*} |f(\omega^*)|$;

(c) $B_M(1,\nu) = \|\nu\|$ if $\nu \geq 0$.

Therefore, the following definition seems natural.

Definition 1.11 *A generalized randomization* of an experiment $\mathcal{E}$ in $(\Omega^*, \mathcal{F}^*)$ is a *bilinear functional* $B(f,\nu)$ defined on $L_\infty(\Omega^*, \mathcal{F}^*) \times L(\mathcal{E})$ and possessing the properties $(a), (b), (c)$. The set of all such functionals is denoted by $\mathfrak{B}(\mathcal{E}, \mathcal{F}^*)$.

Let us put in this definition $\nu = P_\theta$ and let BP_θ be an element of the space $L^*_\infty(\Omega^*, \mathcal{F}^*)$ (the dual space for $L_\infty(\Omega^*, \mathcal{F}^*)$) such that for any $f \in L_\infty(\Omega^*, \mathcal{F}^*)$

$$(BP_\theta)f = B(f, P_\theta).$$

It is well known that BP_θ is a *finite-additive* probability $(BP_\theta \in \Sigma^f(\Omega^*, \mathcal{F}^*))$ and $(BP_\theta)f$ is the integral of f with respect to BP_θ (see Hewitt and Ross (1963)). Hence, one can say that a generalized randomization defined by a bilinear functional B for an experiment $\mathcal{E}$ determines *a generalized experiment*

$$B\mathcal{E} = (\Omega^*, \mathcal{F}^*;\ BP_\theta, \theta \in \Theta).$$

It is natural to call $B\mathcal{E}$ the (generalized) experiment obtained from $\mathcal{E}$ with the help of the (generalized) randomization B.

Definition 1.12 Let $\mathcal{E}$ and $\mathcal{E}^*$ be experiments and let for some bilinear functional B the measure $BP_\theta \in \Sigma(\Omega^*, \mathcal{F}^*)$, $\theta \in \Theta$ and

$$P^*_\theta = BP_\theta \quad \text{for all} \quad \theta \in \Theta.$$

In this situation we say that the experiment $\mathcal{E}^*$ is obtained from $\mathcal{E}$ by the generalized randomization (in short, "$\mathcal{E} \gg \mathcal{E}^*$").

If $\mathcal{E}$ and $\mathcal{E}^*$ are two experiments then the set $\mathfrak{B}(\mathcal{E}, \mathcal{F}^*)$ of all generalized randomization from $\mathcal{E}$ to $(\Omega^*, \mathcal{F}^*)$ is denoted by $\mathfrak{B}(\mathcal{E}, \mathcal{E}^*)$.

Similarly to the previous section we introduce the *weak* topology (of topology of pointwise convergence) on the set $\mathfrak{B}(\mathcal{E}, \mathcal{F}^*)$. With respect to this topology the set $B(\mathcal{E}, \mathcal{F}^*)$ is compact and evidently convex.

2. It is natural to ask whether the notion of generalized randomization by a bilinear functional agrees with the definitions of randomization by a Markov kernel (if it exists) and by a stochastic operator (if the experiments $\mathcal{E}$ and $\mathcal{E}^*$ are dominated).

A. If M is a Markov kernel and $\mathcal{E}^* = M\mathcal{E}$ then from (1.10) and (1.10) it follows that $\mathcal{E}^* = B_M\mathcal{E}$. In other words, the randomization with the help of Markov kernels is a particular case of the generalized randomization. The following theorem describes some situations when the inverse result holds; Strasser (1985, Section 55).

Theorem 1.3 *(1) Let $(\Omega, \mathcal{F})$ and $(\Omega^*, \mathcal{F}^*)$ be locally compact separable metric spaces and let an experiment $\mathcal{E}^*$ be obtained by a generalized randomization B from $\mathcal{E}$ $(\mathcal{E}^* = B\mathcal{E})$. Then there exists a Markov kernel M such that $\mathcal{E}^* = M\mathcal{E}$.*

(2) Let $\mathcal{E}$ be a dominated experiment and let $(\Omega^*, \mathcal{F}^*)$ be a locally compact separable space. If an experiment $\mathcal{E}^*$ is obtained from $\mathcal{E}$ by a generalized randomization B $(\mathcal{E}^* = B\mathcal{E})$ then there exists a Markov kernel M such that $\mathcal{E}^* = M\mathcal{E}$.

B. Let us consider now the situation when $\mathcal{E}$ and $\mathcal{E}^*$ are dominated experiments (with dominating measures ν and ν^* respectively) and $L(\mathcal{E}) = L_1(\Omega, \mathcal{F}, \nu)$, $L(\mathcal{E}^*) = L_1(\Omega^*, \mathcal{F}^*, \nu^*)$. Let M_1 be a stochastic operator and $\mathcal{E}^* = M_1\mathcal{E}$. For $f \in L(\mathcal{E}^*)$ and any measure $\sigma \in \Sigma(\Omega, \mathcal{F})$ such that $\sigma \ll \nu$ put

$$B_{M_1}(f, \sigma) = (M_1\sigma)f.$$

It is clear that $B_{M_1} \in \mathcal{B}(\mathcal{E}, \mathcal{F}^*)$. Hence, from $\mathcal{E}^* = M_1\mathcal{E}$ it follows $\mathcal{E}^* = B_{M_1}\mathcal{E}$, i.e. the randomization with the help of stochastic operators is again a particular case of the generalized randomization. But now the inverse result also is true, Strasser (1985, Section 55).

Theorem 1.4 *If $\mathcal{E}$ and $\mathcal{E}^*$ are dominated experiments and $\mathcal{E}^* = B\mathcal{E}$ then there exists a stochastic operator M_1 such that $\mathcal{E}^* = M_1\mathcal{E}$.*

C. The result of this theorem suggests the idea that in the general case (not assuming the experiments $\mathcal{E}$ and $\mathcal{E}^*$ to be dominated) it is also possible to define any generalized randomization by a suitable stochastic operator (from $L(\mathcal{E})$ to $L(\mathcal{E}^*)$).

Definition 1.13 Any positive linear operator $\overline{M} : L(\mathcal{E}) \to L(\mathcal{E}^*)$ such that $\|\overline{M}\sigma\| = \|\sigma\|$, $\sigma \in L(\mathcal{E})$, $\sigma \geq 0$, is *a stochastic operator* (from $L(\mathcal{E})$ to $L(\mathcal{E}^*)$).

The set of all such operators is denoted by $\overline{\mathcal{M}}(\mathcal{E}, \mathcal{F}^*)$.

If $P_\theta^* = \overline{M}P_\theta, \theta \in \Theta$, then the experiment $\mathcal{E}^*$ is said to be obtained from $\mathcal{E}$ by a stochastic operator $\overline{M}$ $(\mathcal{E}^* = \overline{M}\mathcal{E})$.

Note that any stochastic operator $\overline{M}$ defines the generalized randomization $(B_{\overline{M}})$ if we put by definition

$$B_{\overline{M}}(f, \sigma) = (\overline{M}\sigma)f.$$

It is worthwhile now to recall that a generalized randomization by a bilinear functional B defines, for each measure P_θ, an element $BP_\theta \in \Sigma^f(\Omega^*, \mathcal{F}^*)$ which is a finite–additive measure. But a stochastic operator $\overline{M}$ transforms all the measures P_θ into σ-additive measures $\overline{M}P_\theta$. The following interesting result claims that in some quite definite sense any generalized randomization can be considered as defined by a stochastic operator.

Theorem 1.5 *(1) Let B be a generalized randomization via a bilinear functional $B(f,\nu)$ defined on $L_\infty(\Omega^*, \mathcal{F}^*) \times L(\mathcal{E})$. Then there exists a stochastic operator $\overline{M}$ (from $L(\mathcal{E})$ to $L(\mathcal{E}^*)$) such that for any $\theta \in \Theta$, the following inequality between two variation norms holds:*

$$||\overline{M}P_\theta - P_\theta^*|| \leq ||BP_\theta - P_\theta^*||.$$

(2) If an experiment $\mathcal{E}^*$ is obtained by a generalized randomization B from $\mathcal{E}$ ($\mathcal{E}^* = B\mathcal{E}$) then there exists a stochastic operator $\overline{M}$ such that $\mathcal{E}^* = \overline{M}\mathcal{E}$.

Proof. It is rather simple (compare with Strasser (1985, Section 55).) Indeed, let π be an arbitrary probability measure from $L(\mathcal{E}^*)$. If $\sigma \in \Sigma^f(\Omega^*, \mathcal{F}^*)$ is a finite-additive measure on $(\Omega^*, \mathcal{F}^*)$ then we define an operator $T : \Sigma^f(\Omega^*, \mathcal{F}^*) \to L(\mathcal{E}^*)$ by the formula

$$T\sigma = \sigma_1 + \sigma_2(\Omega^*)\pi$$

where σ_1 and σ_2 are defined from the (unique) decomposition $\sigma = \sigma_1 + \sigma_2$ of the measure σ as the sum of two orthogonal ($\sigma_1 \perp \sigma_2$) components with $\sigma_1 \in L(\mathcal{E}^*)$ and $\sigma_2 \in L(\mathcal{E}^*)^\perp$. If we denote by $|\sigma|$ the measure which is the total variation of σ and $||\sigma|| = |\sigma|(\Omega^*)$ then

$$\begin{aligned} ||T\sigma|| &= |T\sigma|(\Omega^*) = |\sigma_1 + \sigma_2(\Omega^*)\pi|(\Omega^*) \leq |\sigma_1|(\Omega^*) + |\sigma_2|(\Omega^*)\pi(\Omega^*) = \\ &= |\sigma_1 + \sigma_2|(\Omega^*) = |\sigma|(\Omega^*) = ||\sigma||. \end{aligned}$$

Since for non-negative functions σ one has $T\sigma(\Omega^*) = \sigma(\Omega^*)$ then $||T|| = 1$. It is clear also that the operator T maps σ-additive measures into σ-additive measures.

This implies that for $\overline{M} = T \circ B$

$$\begin{aligned} ||\overline{M}P_\theta - P_\theta^*|| &= ||(T \circ B)P_\theta - P_\theta^*|| = ||(T \circ B)P_\theta - TP_\theta^*|| = \\ &= ||T(BP_\theta - P_\theta^*)|| \leq ||BP_\theta - P_\theta^*||, \qquad \theta \in \Theta. \end{aligned}$$

Hence, (i) is proved. The statement (ii) obviously follows from the first one. □

1.6 Comparison of Experiments via Decision Problems

1. Let $\mathcal{E}$ and $\mathcal{E}^*$ be two statistical experiments. As we have seen, these experiments can be compared via the notion of randomization. However, there exists another way to compare experiments based on the comparison of the corresponding risk functions. The relations between both definitions are of great interest. Fortunately it turned out that these definitions are really equivalent in some quite clear sense. We begin with some definitions.

Definition 1.14 An experiment $\mathcal{E}$ is *more informative* than an experiment $\mathcal{E}^*$ for a statistical decision problem defined (see Section 1.5) by a topological decision space $(\mathbf{A}, \mathcal{A})$ with the Baire σ-field $\mathcal{A} = \mathcal{B}_0(\mathcal{A})$ and a continuous loss function $W = W_\theta(a)$, $0 \leq W \leq 1$, if for every β^*-decision for the experiment $\mathcal{E}^*$ there exists a β-decision for the experiment $\mathcal{E}$ such that

$$R_\theta(W, \beta) \leq R_\theta(W, \beta^*), \qquad \theta \in \Theta. \tag{1.11}$$

(This property of "preferability" we will denote as "$\mathcal{E} \overset{(\mathbf{A},W)}{\gg} \mathcal{E}^*$" or "$\mathcal{E}^* \overset{(\mathbf{A},W)}{\ll} \mathcal{E}$".)

Definition 1.15 An experiment $\mathcal{E}$ is *preferable* to an experiment $\mathcal{E}^*$ "($\mathcal{E} \overset{R}{\gg} \mathcal{E}^*$") if "$\mathcal{E} \overset{(\mathbf{A},W)}{\gg} \mathcal{E}^*$" for *any* decision space $(\mathbf{A}, \mathcal{A})$ and *any continuous loss function* W with $0 \leq W \leq 1$.

The notation "$\mathcal{E} \overset{R}{\gg} \mathcal{E}^*$" stresses the fact that the experiment comparison is based on the comparison of *risk* functions; compare with Definition 1.12 in Section 1.5 where the notation "$\mathcal{E} \gg \mathcal{E}^*$" was used to emphasize that an experiment $\mathcal{E}^*$ was obtained by a *randomization* from an experiment $\mathcal{E}$.

This definition requiring the property (1.11) for too large sets of loss functions and admissible decisions seems to be very frightening. (Sometimes the comparison of experiments can be done under some additional constraints on the class of functions W, on the space $(\mathbf{A}, \mathcal{A})$ and the structure of decisions.)

But the following important theorem claims equivalence of the conditions "$\mathcal{E} \overset{R}{\gg} \mathcal{E}^*$" and "$\mathcal{E} \gg \mathcal{E}^*$".

Theorem 1.6 *The following implications are true:*

$$\mathcal{E} \gg \mathcal{E}^* \iff \mathcal{E} \overset{R}{\gg} \mathcal{E}^*. \tag{1.12}$$

The proof of this property (and particular implications) was given for various assumptions in the papers Blackwell (1951, 1953), Sherman(1951), Stein(1951)) among others. A more general result than (1.12), using the notion of a deficiency $\delta(\mathcal{E}, \mathcal{E}^*)$, was stated by Le Cam and will be given in Section 1.7 below where also the discussion of the result (1.12) will be continued.

2. Now we show how the result (1.12) can be used to construct an example where the property "$\mathcal{E} \gg \mathcal{E}^*$" is not satisfied (i.e. $\mathcal{E}^*$ can not be obtained from $\mathcal{E}$ by a randomization). For this we continue to consider Example 1.20 from Section 1.3 and show that for $\lambda \neq 1/k$ with any integer k the property "$\mathcal{E}_\lambda \gg \mathcal{E}^*$" is not satisfied. Our exposition follows Lehmann (1988).

In accordance with (1.12) it is sufficient to find W and β^* such that the property (1.11) does not hold for such λ. Put

$$a(\theta) = \begin{cases} 1, & \theta < -1/2, \\ 1/2, & |\theta| \le 1/2, \\ 0, & \theta > 1/2. \end{cases}$$

Let ω and ω^* be observations in experiments

$$\mathcal{E}_{\lambda_0} = (R, \mathcal{B}(R);\, P_{\theta,\lambda_0}\,,\, \theta \in (-\infty,\infty)) \quad \text{and} \quad \mathcal{E}^* = (R, \mathcal{B}(R);\, P^*_\theta\,,\, \theta \in (-\infty,\infty)),$$

with

$$P_{\theta,\lambda_0} \sim U(\theta - \frac{\lambda_0}{2}, \theta + \frac{\lambda_0}{2}), \quad P^*_\theta \sim U(\theta - \frac{1}{2}, \theta + \frac{1}{2}).$$

Note that $P^*_\theta(-\infty, 0) = a(\theta)$. Hence for the non-randomized decision

$$\rho^*(\omega^*) = I_{(-\infty,0)}(\omega^*)$$

it follows

$$E^*_\theta\, \rho^*(\omega^*) = a(\theta).$$

In other words, $\rho^*(\omega^*)$ is an unbiased estimator of $a(\theta)$. It is not difficult also to see that $\mathbf{D}^*_\theta\, \rho^*(\omega^*) = 0$ for $|\theta| > 1$.

Now we consider the loss function

$$W(\theta, \rho^*) = (a(\theta) - \rho^*)^2.$$

Then it holds for the risk of ρ^*:

$$R_\theta(\mathcal{E}^*, W, \rho^*) = \mathbf{D}^*_\theta\, \rho^*(\omega^*) = 0, \qquad |\theta| > 1.$$

For the proof that the property "$\mathcal{E}_{\lambda_0} \overset{R}{\gg} \mathcal{E}^*$" for $\lambda_0 \ne 1/k$ is not true, we shall show that $R_\theta(\mathcal{E}_{\lambda_0}, W, \beta) > 0$ if we state that for any decision β for the experiment $\mathcal{E}_{\lambda_0}$ with $\lambda_0 \ne 1/k$ and $|\theta| > 1$.

Let $\rho = \rho(\omega)$ be a non-randomized unbiased estimator of the function $a(\theta)$ in the experiment $\mathcal{E}_{\lambda_0}$ and assume that $\mathbf{D}_{\theta,\lambda_0}\rho(\omega) = 0$ if $|\theta| > 1$. Then it is easy to see that the function $\rho = \rho(\omega)$ must be constant for $\omega > 1 - \lambda_0/2$ and $\omega < -1 + \lambda_0/2$. In the first case $\rho(\omega) = 0$ and in the second one $\rho(\omega) = 1$.

The unbiasedness property $E^*_\theta \rho^*(\omega^*) = a(\theta)$ implies

$$\frac{1}{\lambda_0} \int_{\theta-\lambda_0/2}^{\theta+\lambda_0/2} \rho(\omega)\, d\omega = a(\theta), \qquad -\infty < \theta < \infty.$$

It easily implies that (almost surely with respect to the Lebesgue measure)

$$\rho(\omega - \lambda_0/2) - \rho(\omega + \lambda_0/2) = \lambda_0 I(|\omega| < 1/2). \tag{1.13}$$

Since $\rho(\omega) = 0$ for $\omega > 1 - \lambda_0/2$ and $\rho(\omega) = 1$ for $\omega < -1 + \lambda_0/2$ it follows from (1.13) that the functions $\rho(\omega \pm j\lambda_0)$, $j = 0, 1, \ldots$ can change their values only by jumps of the variable λ_0. This yields that λ_0 must be of the form $\lambda_0 = 1/k$ with some integer k and therefore, for any non-randomized decision $\rho = \rho(\omega)$, the covariance $\mathbf{D}_{\theta,\lambda_0}\rho(\omega) = 0$, $|\theta| > 1$, only if $\lambda_0 = 1/k$.

Considering additionally randomized (Markov) decisions brings anything new. Indeed, if $\widetilde{\rho}(\omega, \omega')$ is the function from Lemma 1.1 of Section 1.4 such that $E_{\theta,\lambda_0} \times E'\widetilde{\rho}(\omega, \omega') = a(\theta)$ and $E_{\theta,\lambda_0} \times E'(\widetilde{\rho}(\omega, \omega') - a(\theta))^2 = 0$ for $|\theta| > 1$ then $\rho(\omega) = E'\widetilde{\rho}(\omega, \omega')$ will be an unbiased estimator for $a(\theta)$ with $\mathbf{D}_{\theta,\lambda_0}\rho(\omega) = 0$ for $|\theta| > 1$. It implies that a transition to randomized Markov decision cannot decrease the risk value. From the general results of Sections 1.4 and 1.5 one can also derive that the same is true for generalized randomizations.

1.7 Deficiency $\delta(\mathcal{E}, \mathcal{E}^*)$ and Δ-Distance $\Delta(\mathcal{E}, \mathcal{E}^*)$

1. As we have already noted above, one cannot expect two experiments to be generally comparable, i.e. that one is obtained by a randomization from the other. On the contrary, usually two different experiments are *uncomparable*.

The most significant step in the problem of comparison of experiments was made after by Le Cam who introduced (in December, 1958, see Le Cam (1964; 1985, p.234)) the notions of *deficiency* and Δ-*distance* based on the idea of defining a numerical measure which permits to evaluate how much is one experiment impossible to be reconstructed from another with the help of randomizations.

Definition 1.16 *A deficiency* of an experiment $\mathcal{E}$ (or in the experiment $\mathcal{E}$) with respect to an experiment $\mathcal{E}^*$ is the value

$$\delta(\mathcal{E}, \mathcal{E}^*) = \inf_B \sup_\theta \|BP_\theta - P_\theta^*\| \tag{1.14}$$

where inf is taken over the class $\mathfrak{B}(\mathcal{E}, \mathcal{F}^*)$ of all randomizations B and the norm is treated as the total variation in $L_\infty^*(\Omega^*, \mathcal{F}^*)$, see Jacod and Shiryaev (1987, Section V.4a).

It follows directly from this definition (see Theorems 1.8 and 1.10 below) that if $\delta(\mathcal{E}, \mathcal{E}^*) = 0$ then the experiment $\mathcal{E}^*$ is a randomization of $\mathcal{E}$. But the opposite assertion is evident and, hence, $\delta(\mathcal{E}, \mathcal{E}^*) = 0 \Longleftrightarrow \mathcal{E} \gg \mathcal{E}^*$.

Recall that due to Theorem 1.5 from Section 1.5 it is sufficient to take inf in (1.14) only over the class of randomizations $B_{\overline{M}}$ defined by stochastic operators $\overline{M} \in \overline{\mathcal{M}}(\mathcal{E}, \mathcal{F}^*)$:

$$\delta(\mathcal{E}, \mathcal{E}^*) = \inf_{\overline{M} \in \overline{\mathcal{M}}(\mathcal{E}, \mathcal{F}^*)} \sup_\theta \|\overline{M}P_\theta - P_\theta^*\|. \tag{1.15}$$

The next natural question here would be whether can one take inf only over the class of stochastic operators defined by Markov (stochastic) kernels.

The following result generalizes Theorem 1.3 from Section 1.5 (Strasser (1985, Section 55)).

Theorem 1.7 *Let any of the following two conditions hold:*

(i) $(\Omega,\mathcal{F})$ *and* $(\Omega^*,\mathcal{F}^*)$ *are locally compact separable metric spaces;*
(ii) $\mathcal{E}$ *is a dominated experiment and* $(\Omega^*,\mathcal{F}^*)$ *is a locally compact metric space.*

Then

$$\delta(\mathcal{E},\mathcal{E}^*) = \inf_{M\in\mathcal{M}(\mathcal{E},\mathcal{F}^*)} \sup_{\theta} \|MP_\theta - P_\theta^*\|. \tag{1.16}$$

2.

Definition 1.17 For any two experiments $\mathcal{E}$ and $\mathcal{E}^*$ with the same parametric set Θ denote

$$\Delta(\mathcal{E},\mathcal{E}^*) = \max\left\{\delta(\mathcal{E},\mathcal{E}^*),\delta(\mathcal{E}^*,\mathcal{E})\right\}. \tag{1.17}$$

The value $\Delta(\mathcal{E},\mathcal{E}^*)$ is called Δ-*distance* between $\mathcal{E}$ and $\mathcal{E}^*$.

Remark 1.4 Strictly speaking, the definition (1.17) implies only that $\Delta(\cdot,\cdot)$ is a *pseudo-metric* in the set of experiments (with the same parametric space);

$$\Delta(\mathcal{E},\mathcal{E}^*) \le \Delta(\mathcal{E},\mathcal{E}^{**}) + \Delta(\mathcal{E}^{**},\mathcal{E}^*), \tag{1.18}$$

$$\Delta(\mathcal{E},\mathcal{E}) = 0. \tag{1.19}$$

In order to be a *metric*, the distance $\Delta(\cdot,\cdot)$ must obey the property that $\Delta(\mathcal{E},\mathcal{E}^*) = 0$ implies $\mathcal{E} = \mathcal{E}^*$. Of course, this is not true. Nevertheless, equivalent experiments with $\mathcal{E} \sim \mathcal{E}^*$ (i.e. such that $\Delta(\mathcal{E},\mathcal{E}^*) = 0$) can be collected in *equivalence* classes (types). If $\dot{\mathcal{E}}$ and $\dot{\mathcal{E}}^*$ are two such types then the value $\Delta(\dot{\mathcal{E}},\dot{\mathcal{E}}^*)$ (treated as $\Delta(\mathcal{E},\mathcal{E}^*)$ for any $\mathcal{E} \in \dot{\mathcal{E}}$ and $\dot{\mathcal{E}}^* \in \mathcal{E}^*$) becomes a *real metric* between types of (equivalent) experiments. In what follows, whenever we speak of $\Delta(\mathcal{E},\mathcal{E}^*)$ as the distance between experiments $\mathcal{E}$ and $\mathcal{E}^*$ we understand just the above meaning.

3. We have already noted that if $\delta(\mathcal{E},\mathcal{E}^*) = 0$ then "$\mathcal{E} \gg \mathcal{E}^*$", i.e. the experiment $\mathcal{E}^*$ is a randomization of $\mathcal{E}$ and thus "the experiment $\mathcal{E}^*$ is reconstructed from $\mathcal{E}$" or "$\mathcal{E}$ is more informative than $\mathcal{E}^*$". (It would be more correct to say that "$\mathcal{E}$ is not less informative than $\mathcal{E}^*$" but for the simplicity one says usually "more" instead of "not less".)

Now we present a number of examples of experiments having these properties.

Example 1.21 Let $\mathcal{E}_* = (\Omega,\mathcal{F};\ P_\theta,\theta\in\Theta)$ be a totally non-informative experiment (see Example 1.4) with $P_\theta = P_*$ for all $\theta \in \Theta$. Such experiment is obtained by (Markov) randomization from any other experiment $\mathcal{E}$ (with the same parameter set Θ) by the Markov kernel $M(\omega, B) = P_*(B)$, $B \in \mathcal{F}$. In other words,

$$\delta(\mathcal{E}, \mathcal{E}^*) = 0.$$

Let now $\mathcal{E}^* = (\Omega^*, \mathcal{F}^*; P_\theta^*, \theta \in \Theta)$ be a totally informative experiment (see Example 1.4 in Section 1.1) with singular measures P_θ for different values of θ, i.e.

$$P_\theta^* \perp P_{\theta'}^*, \qquad \theta \neq \theta'.$$

Assume the set Θ to be not more than *countable*, $\Theta = \{\theta_1, \theta_2, \dots\}$. Let $\Omega^* = \sum \Omega_i^*$ with $\Omega_i^* \bigcap \Omega_j^* = \emptyset$ for $i \neq j$ and $P_{\theta_i}^*(\Omega_j^*) = 0$ for $i \neq j$, $P_{\theta_i}^*(\Omega_i^*) = 1$. (Such sets Ω_i^*, $i \geq 1$, exist by virtue of the singularity property of measures $P_{\theta_i}^*$, $i \geq 1$). If $\mathcal{E} = (\Omega, \mathcal{F}; P, \theta \in \Theta))$ is an experiment with the same parameter set $\Theta = \{\theta_1, \theta_2, \dots\}$ then, denoting for $A \in \mathcal{F}$ and $\omega^* \in \Omega^*$,

$$M(\omega^*, A) = P_{\theta_i}(A) \quad \text{if} \quad \omega^* \in \Omega_i^*,$$

we get for any $i \geq 1$

$$P_{\theta_i}(A) = \int_{\Omega_i^*} P_{\theta_i}(A)\, P_{\theta_i}^*(d\omega^*) = \int_\Omega M(\omega^*, A)\, P_{\theta_i}^*(d\omega^*), \quad \theta \in \Theta, \quad A \in \mathcal{F}.$$

Hence, for the case of a *countable* set Θ any experiment $\mathcal{E}$ is obtained by a *Markov randomization* from the *totally informative experiment* $\mathcal{E}^*$; therefore

$$\delta(\mathcal{E}, \mathcal{E}^*) = 0.$$

It follows from Theorem 1.8 below that this assertion remains true for *any* parameter set Θ.

Example 1.22 Let $\mathcal{E}_k$ be an experiment with a *finite* parametric set $\Theta = \{\theta_1, \dots, \theta_k\}$ and let $\mathcal{E}_k^*$ be its standardization (see for more details Section 1.3). Since by the definition $\mathcal{E}_k^* = f\mathcal{E}_k$ (Example 1.13, Section 1.2), it follows $\delta(\mathcal{E}_k, \mathcal{E}_k^*) = 0$. The inverse assertion also holds, i.e. $\delta(\mathcal{E}_k^*, \mathcal{E}_k) = 0$. It follows from the fact that the σ-field $\mathcal{F}^f$ is sufficient for $\mathcal{F}$. This can be established quite easily if we suppose that the space $(\Omega, \mathcal{F})$ is Borel (see, for example, Shiryaev(1980, Section II.2)). Then there exist regular conditional distributions and with their help one can construct the corresponding Markov kernel in the following manner. Let $A \in \mathcal{F}$. Then $P_\theta(A) = E_\theta\, E_\theta\left[I(A) \mid \mathcal{F}^f\right]$ and if $Q = \frac{1}{k}\sum_{i=1}^k P_{\theta_i}$ then by the chain rule for conditional expectations (see Shiryaev (1980, Section II.7)) we find ($\oplus$ being the symbol of the *pseudo-inversion*, i.e. $a^\oplus = a^{-1}$ if $a > 0$ and $a^\oplus = 0$ if $a = 0$) that for $\theta \in \Theta$ (P_θ-a.s.)

$$\begin{aligned} E_\theta\Big[I(A) \mid \mathcal{F}^f\Big] &= \left[E_Q\left(\frac{dP_\theta}{dQ} \mid \mathcal{F}^f\right)\right]^\oplus E_Q\left[I(A)\frac{dP_\theta}{dQ} \mid \mathcal{F}^f\right] \\ &= f_\theta^\oplus\, f_\theta\, E_Q\Big[I(A) \mid \mathcal{F}^f\Big]. \end{aligned}$$

Since $P_\theta(f_\theta = 0) = E_Q\, I_{\{f_\theta=0\}}\, f_\theta = 0$. it also holds $f_\theta^{\oplus} f_\theta = 1$ (P_θ-a.s.). Thus $E_\theta\left[I(A) \mid \mathcal{F}^f\right] = E_Q\left[I(A) \mid \mathcal{F}^f\right]$ (P_θ-a.s.) and

$$P_\theta(A) = E_\theta\, E_\theta \left[I(A) \mid \mathcal{F}^f\right] = \int_{\Sigma_k} M(x, A)\, \mu(dx),$$

where $M(x, A) = Q(A \mid f = x)$ is a Markov kernel provided that the space $(\Omega, \mathcal{F})$ is Borel. Therefore, in the studied case the experiment $\mathcal{E}_k$ is obtained from $\mathcal{E}_k^*$ by the randomization defined by the Markov kernel M and, hence, $\delta(\mathcal{E}_k^*, \mathcal{E}_k) = 0$. This property remains valid also in the general case. For this situation the randomization will be determined by a stochastic operator (see Strasser (1985, Theorem 24.13)).

The above consideration implies that $\Delta(\mathcal{E}_k^*, \mathcal{E}_k) = 0$, i.e. the experiment $\mathcal{E}_k$ and its standardization $\mathcal{E}_k^*$ are equivalent.

4. Let $\mathcal{E} = (\Omega, \mathcal{F};\, \mathbf{P}_\theta, \theta \in \Theta)$ be an experiment and let S be some subset of Θ, $S \subseteq \Theta$. We denote by $\mathcal{E}_S$ "the restriction of the experiment $\mathcal{E}$ on S", i.e.

$$\mathcal{E}_S = (\Omega, \mathcal{F};\, P_\theta, \theta \in S).$$

The following important result is due to Le Cam (1964, 1986).

Theorem 1.8 *(1) For any two experiments $\mathcal{E}$ and $\mathcal{E}^*$ with the same parametric set Θ*

$$\delta(\mathcal{E},\mathcal{E}^*) = \sup\, \delta(\mathcal{E}_\sigma, \mathcal{E}_\sigma^*), \qquad \Delta(\mathcal{E},\mathcal{E}^*) = \sup\, \Delta(\mathcal{E}_\sigma, \mathcal{E}_\sigma^*) \tag{1.20}$$

where sup *is taken over all the finite subsets σ of Θ.*

(2) There exists an operator $K \in \mathfrak{B}(\mathcal{E}, \mathcal{F}^*)$ such that

$$\delta(\mathcal{E},\mathcal{E}^*) = \sup_\theta \|K P_\theta - P_\theta^*\|. \tag{1.21}$$

Proof. Assertion (1.20) can be derived directly from the representation for $\delta(\mathcal{E},\mathcal{E}^*)$ and $\Delta(\mathcal{E},\mathcal{E}^*)$ given in Section 1.6. Assertion (1.21) is a consequence of the equivalence of of the statements (A) and (D) in Theorem 1.10 below. □

The importance of the property (1.20) can be explained, for example, by the fact that when calculating or estimating the deficiency $\delta(\mathcal{E},\mathcal{E}^*)$, the parametric set Θ can be substituted by its finite or countable subset σ (taking sup over all such subsets) thus giving us the possibility to use the properties of dominateness for such experiments $\mathcal{E}_\sigma, \mathcal{E}_\sigma^*$.

5. Let $\mathcal{E} = (\Omega, \mathcal{F};\, P_\theta, \theta \in \Theta)$ and $\mathcal{E} = (\Omega^*, \mathcal{F}^*;\, P_\theta^*, \theta \in \Theta)$ be two experiments with a finite parametric set $\Theta = \{\theta_1, \dots, \theta_k\}$ and with the standard measures μ and μ^* respectively (see Section 1.3). Besides the variation distance $\|\mu - \mu^*\|$ defined as

$$\|\mu - \mu^*\| = \sup_{|h(x)| \le 1} \left| \int_{\Sigma_k} h(x)\, d(\mu - \mu^*) \right| \tag{1.22}$$

we introduce a distance, Dudley (1968),

$$\|\mu - \mu^*\|_D = \sup \left| \int_{\Sigma_k} h(x)\, d(\mu - \mu^*) \right| \tag{1.23}$$

where sup is taken over all the functions $h = h(x)$, $x \in \Sigma_k$, such that $|h(x)| \le 1$ and h satisfies the Lipschitz condition ($|h(x) - h(y)| \le |x - y|$).

Let also $\mathcal{L}(P, P^*)$ be the Lévy–Prokhorov distance between measures P and P^* (on R^k):

$$\mathcal{L}(P, P^*) = \inf\left\{ \varepsilon : \ P^*(A) \le P(A^\varepsilon) + \varepsilon \ \text{ and } \ P(A) \le P^*(A^\varepsilon) + \varepsilon, \quad A \in \mathcal{B}(R^k) \right\}$$

with $A^\varepsilon = \{x : \rho(x, A) \le \varepsilon\}$, $\rho(x, A) = \inf\{\rho(x, y) : y \in A\}$, and $\rho(x, y) = \max_{1 \le i \le k} |x_i - y_i|$.

Theorem 1.9 *Let $\mathcal{E}$ and $\mathcal{E}^*$ be two statistical experiments with the standard measures μ and μ^* on $(\Sigma_k, \mathcal{B}(\Sigma_k))$ respectively. Then there exists a non-negative function $\gamma = \gamma(a)$, $\gamma(a) \downarrow 0$ as $a \downarrow 0$ such that*

$$\Delta(\mathcal{E}, \mathcal{E}^*) \le \|\mu - \mu^*\|_D \le \gamma(\Delta(\mathcal{E}, \mathcal{E}^*)); \tag{1.24}$$

$$\frac{1}{4k} \Delta(\mathcal{E}, \mathcal{E}^*) \le \mathcal{L}(\frac{1}{k}\mu, \frac{1}{k}\mu^*) \le \left[\frac{3}{2k} \gamma(\Delta(\mathcal{E}, \mathcal{E}^*)) \right]^{1/2}. \tag{1.25}$$

Moreover,

$$\Delta(\mathcal{E}, \mathcal{E}^*) = 0 \quad \Longleftrightarrow \quad \mu = \mu^*. \tag{1.26}$$

Proof. (1.24) is given in Pollard (1987). The inequality (1.25) follows from (1.24) and from the inequalities (see Shiryaev (1980, Section III.3))

$$\|P - P^*\| \le 4\mathcal{L}(P, P^*), \qquad \mathcal{L}(P, P^*) \le \left(3/2 \|P - P^*\|_D \right)^{1/2}. \tag{1.27}$$

The assertion (1.26) follows directly from (1.24). □

6. One of the main results connected with the notion of deficiency and Δ-distance consists in the fact that with their help one can *compare* the values of risks in various experiments. One of the possible results in this direction has been already given in Section 1.6, see Theorem 1.6. Its generalization is the following important theorem, Le Cam (1964; 1986, Chapter 2, Theorem 2).

Theorem 1.10 *(The randomization criterion)* *Let $\mathcal{E} = (\Omega, \mathcal{F}; P_\theta, \theta \in \Theta)$ and $\mathcal{E}^* = (\Omega^*, \mathcal{F}^*; P_\theta^*, \theta \in \Theta)$ be two experiments with the same parametric set Θ. Let $(\mathbf{A}, \mathcal{A})$ denote an arbitrary topological decision space $\mathbf{A}$ with the Baire σ-field $\mathcal{A}$ and let $\mathcal{W} = \{W = W_\theta(a),\ a \in \mathbf{A}, ,\theta \in \Theta\}$ be a collection of loss functions which are lower continuous in a for each θ with the norm $\|W_\theta\| = \sup_a |W_\theta(a)|$, $\theta \in \Theta$. Then for every $\varepsilon > 0$ the following assertions are equivalent:*

(A) *there exists a randomization* $B \in \mathfrak{B}(\mathcal{E},\mathcal{E}^*)$ *such that*

$$\sup_\theta \|BP_\theta - P^*_\theta\| \le \varepsilon; \tag{1.28}$$

(B) *for any space* $(\mathbf{A},\mathcal{A})$, *any loss function* $W \in \mathcal{W}$ *and any decision* $\beta^* \in \mathfrak{B}(\mathcal{E}^*,\mathbf{A})$ *there exists a decision* $\beta \in \mathfrak{B}(\mathcal{E},\mathbf{A})$ *such that*

$$R_\theta(\mathcal{E},W,\beta) \le R_\theta(\mathcal{E}^*,W,\beta^*) + \varepsilon\|W_\theta\|, \quad \theta \in \Theta; \tag{1.29}$$

(C) *for any space* $(\mathbf{A},\mathcal{A})$, *any loss function* $W \in \mathcal{W}$ *and any probability measure* $\pi = \pi(d\theta)$ *with a finite support* $(\pi \in \Pi^0)$ *the corresponding* π*-Bayes risks satisfy the inequality*

$$\mathcal{R}_\pi(\mathcal{E},W) \le \mathcal{R}_\pi(\mathcal{E}^*,W) + \varepsilon \int \|W_\theta\|\,\pi(d\theta);$$

(D) $\quad \delta(\mathcal{E},\mathcal{E}^*) \le \varepsilon.$

Taking into account the importance of this theorem we give the sketch of the proof. The exposition follows Le Cam (1964, 1986) and Strasser (1985).

Proof. $(A) \Rightarrow (C)$. Due to Definition 1.9 from Section 1.4 one can suppose here (and further) that the loss function is continuous in a for each θ and $0 \le W_\theta \le 1$. We have to check that for any $\pi \in \Pi^0$

$$\inf_{\beta\in\mathfrak{B}(\mathcal{E},\mathbf{A})} \int R_\theta(\mathcal{E},W,\beta)\,\pi(d\theta) \le \inf_{\beta\in\mathfrak{B}(\mathcal{E}^*,\mathbf{A})} \int R_\theta(\mathcal{E}^*,W,\beta^*)\,\pi(d\theta) + \varepsilon.$$

Because the set $\mathfrak{R}(\mathcal{E}^*,\mathbf{A})$ is dense in $\mathfrak{B}(\mathcal{E}^*,\mathbf{A})$ (Theorem 1.2, Section 1.4) one can take inf in the right-hand side of (C) over the set $\mathfrak{R}(\mathcal{E}^*,\mathbf{A})$ of decisions generated by Markov kernels ρ^*. Let us choose some ρ^* and construct the bilinear functional β^*

$$\beta^*(f,\nu^*) = \beta^*_{\rho^*}(f,\nu^*) = \iint f(a)\,\rho^*(\omega^*,da)\,\nu^*(d\omega^*).$$

Let BP_θ be the measure ocurring in (A). From Theorem 1.5, Section 1.5, this measure can be assumed σ-finite. Put

$$\beta(f,\nu) = \beta^*(f,B\nu).$$

According to Definition 1.7 in Section 1.4 and (A) the risk of β satisfies

$$\begin{aligned} R_\theta(\mathcal{E},W,\beta) &\equiv \beta(W_\theta,P_\theta) \\ &= \beta^*(W_\theta,BP_\theta) \\ &= \iint W_\theta(a)\,\rho^*(\omega^*,da)\,(BP_\theta)(d\omega^*) \\ &\le \iint W_\theta(a)\,\rho^*(\omega^*,da)\,P^*_\theta(d\omega^*) + \varepsilon. \end{aligned}$$

Hence, for $\pi \in \Pi^0$

$$\int R_\theta(\mathcal{E}, W, \beta)\pi(d\theta) \le \int R_\theta(\mathcal{E}^*, W, \beta^*)\,\pi(d\theta) + \varepsilon$$

and

$$\inf_{\beta\in\mathfrak{B}(\mathcal{E},\mathbf{A})} \int R_\theta(\mathcal{E}, W, \beta)\,\pi(d\theta) \le \inf_{\beta\in\mathfrak{B}(\mathcal{E}^*,\mathbf{A})} \int R_\theta(\mathcal{E}^*, W, \beta^*)\,\pi(d\theta) + \varepsilon$$

which proves $(A) \Rightarrow (C)$.

$(C) \Rightarrow (A)$. The proof of this implication is the most difficult part of the whole proof. It is necessary to prove the existence of a bilinear operator $B \in \mathfrak{B}(\mathcal{E}, \mathcal{E}^*)$ such that the corresponding measures BP_θ satisfy the property

(A) $\quad \|BP_\theta - P_\theta^*\| \le \varepsilon$ for all $\theta \in \Theta$.

In turn, this fact is equivalent to

(A^*) $\quad \inf_{B\in\mathfrak{B}(\mathcal{E}^*,\mathbf{A})} \int \|BP_\theta - P_\theta^*\|\,\pi(d\theta) \le \varepsilon \qquad$ for any $\pi \in \Pi^0$.

(For a proof of the equivalence of (A) and (A^*) see, for example, Strasser (Section 45.7, p.239)).

It follows from (C) and the remark about the density of $\mathfrak{R}(\mathcal{E}^*, \mathbf{A})$ in $\mathfrak{B}(\mathcal{E}^*, \mathbf{A})$ that for any $\pi \in \Pi^0$

$$\inf_{\beta\in\mathfrak{B}(\mathcal{E},\mathbf{A})} \int R_\theta(\mathcal{E}, W, \beta)\,\pi(d\theta) \le \inf_{\rho^*\in\mathfrak{R}(\mathcal{E}^*,\mathbf{A})} \int R_\theta(\mathcal{E}^*, W, \beta^*_{\rho^*})\,\pi(d\theta) + \varepsilon \qquad (1.30)$$

where

$$R_\theta(\mathcal{E}^*, W, \beta^*_{\rho^*}) = \beta^*_{\rho^*}(W_\theta, P_\theta^*) = \iint W_\theta(a)\,\rho^*(\omega^*, da)\,P_\theta^*(d\omega^*). \qquad (1.31)$$

Since the property (C) is assumed to hold for any space $(\mathbf{A}, \mathcal{A})$, we may take as one of them the "standard" space

$$\mathbf{A} = [-1, 1]^\Theta = \{a = (a_\theta\,, \theta \in \Theta),\ |a_\theta| \le 1\}$$

which is the space of functions $a = (a_\theta\,, \theta \in \Theta)$ defined on Θ with values in $[-1, 1]$. A loss function $W = W_\theta(a)$ is also chosen to be ("standard")

$$W_\theta(a) = a_\theta.$$

Therefore

$$\int_{\mathbf{A}} W_\theta(a)\,\rho^*(\omega^*, da) = \int_{\mathbf{A}} a_\theta\,\rho^*(\omega^*, da).$$

Denote $\chi^*(\omega^*) = \{\chi_\theta^*(\omega^*),\ \theta \in \Theta\}$ with

$$\chi_\theta^*(\omega^*) = \int_{\mathbf{A}} a_\theta\,\rho^*(\omega^*, da).$$

For each $\omega^* \in \Omega^*$ we have $\chi^*(\omega^*) \in \mathbf{A} = [-1,1]^\Theta$. Define the non-randomized decision $\chi^*(\omega^*, da) = \delta_{\chi^*(\omega^*)}(da)$. Obviously

$$\int_{\mathbf{A}} W_\theta(a)\,\chi^*(\omega^*, da) = W_\theta(\chi^*(\omega^*)) = \chi_\theta^*(\omega^*) = \int_{\mathbf{A}} W_\theta(a)\,\rho^*(\omega^*, da).$$

In other words, for the considered "standard" decision problem it is sufficient to take inf in the right-hand side of (1.30) only over the class of all non-randomized decisions.

The next step of the proof is to verify that for $\rho^*(\omega^*, da) = \delta_{\rho^*(\omega^*)}(da)$ with $\rho^*(\omega^*) \in \mathbf{A} = [-1,1]^\Theta$ and for any $\pi \in \Pi^0$ there is a randomization $B = B(\pi, \rho^*) \in \mathfrak{B}(\mathcal{E},\mathcal{E}^*)$ such that

$$\int \left[\int \rho_\theta^*(\omega^*)\,(BP_\theta)(d\omega^*)\right] \pi(d\theta) \le \int \left[\int \rho_\theta^*(\omega^*)\,P_\theta^*(d\omega^*)\right] \pi(d\theta) + \varepsilon,$$

i.e. that

$$R_\pi(B\mathcal{E}, W, \rho^*) \le R_\pi(\mathcal{E}^*, W, \rho^*) + \varepsilon. \tag{1.32}$$

The set $\mathfrak{R}_0(\mathcal{E}^*, \mathbf{A})$ of simple non-randomized Markov decisions $\rho_0^*(\omega^*)$ of the form

$$\rho_0^*(\omega^*) = \sum_{j=1}^n y_j\, I_{C_j^*}(\omega^*), \qquad \omega^* \in \Omega^*, \tag{1.33}$$

with $y_j \in \mathbf{A} = [-1,1]^\Theta$, $\sum_{j=1}^n C_j^* = \Omega^*$, $n < \infty$ is dense in $\mathfrak{R}(\mathcal{E}^*, \mathbf{A})$, Strasser (1985, Lemma 50.4). Hence one can at once suppose $\rho^*(\omega^*) = \rho_0^*(\omega^*)$ in (1.32) so that this simple decision ρ_0^* is defined by $(y_1, \dots, y_n;\ C_1^*, \dots, C_n^*)$ as in (1.33).

Fixing the collection $(y_1, \dots, y_n)$ of elements from $\mathbf{A} = [-1,1]^\Theta$ we form a new decision space $\mathbf{A}_y$ with its elements being the points x_i from the convex hull of the set $y_1, \dots, y_n$, i.e. the points of the form

$$x_i = \sum_{j=1}^n \alpha_{ij}\, y_j, \quad \sum_{j=1}^n \alpha_{ij} = 1, \quad \alpha_{ij} \ge 0.$$

The randomization B looked for is constructed in the following way. Let $\delta > 0$. Because of (C) there exists (for the decision space $\mathbf{A}_y$) a simple Markov decision $\rho_0(\omega)$ with

$$\rho_0(\omega) = \sum_{i=1}^m x_i\, I_{C_i}(\omega), \quad x_i = \sum_{j=1}^n \alpha_{ij}\, y_j \in \mathbf{A}_y, \quad 1 \le i \le n, \quad \sum_{j=1}^m C_i = \Omega,$$

such that

$$R_\pi(\mathcal{E}, W, \rho_0) \le R_\pi(\mathcal{E}^*, W, \rho_0^*) + \varepsilon + \delta. \tag{1.34}$$

Let points ω_j^* be such that $\rho_0^*(\omega_j^*) = y_j$, $1 \le j \le n$. Define a randomization $B_0 \in \mathfrak{B}(\mathcal{E}, \mathcal{E}^*)$, letting

$$B_0(f,\nu) = \sum_{i=1}^{m}\sum_{j=1}^{n} \alpha_{ij}\, f(\omega_j^*)\,\nu(C_i), \qquad f \in \mathcal{L}_\infty(\Omega^*, \mathcal{F}^*), \quad \nu \in L(\mathcal{E}).$$

It is clear that each measure $B_0 P_\theta$ is concentrated at the points $\omega_1^*, \dots, \omega_n^*$ and

$$(B_0 P_\theta)\{\omega_j^*\} = \sum_{i=1}^{m} \alpha_{ij}\, P_\theta(C_i).$$

We shall show that

$$R_\pi(\mathcal{E}, W, \rho_0) = R_\pi(B_0\mathcal{E}, W, \rho_0^*). \tag{1.35}$$

Indeed, it is easy to see that

$$\int \rho_{0,\theta}(\omega)\, P_\theta(d\omega) = \sum_{i=1}^{m} x_i\, P_\theta(C_i) \tag{1.36}$$

and

$$\int \rho_{0,\theta}^*(\omega^*)\,(B_0 P_\theta)(d\omega^*) = \sum_{i=1}^{m}\sum_{j=1}^{n} \alpha_{ij}\, y_j\, P_\theta(C_i). \tag{1.37}$$

But $x_i = \sum_{j=1}^{n} \alpha_{ij}\, y_j$, hence the right-hand sides of (1.36) and (1.37) coincide and, therefore, (1.35) is true. Together with (1.34) it implies the inequality

$$R_\pi(B_0\mathcal{E}, W, \rho_0^*) \le R_\pi(\mathcal{E}^*, W, \rho_0^*) + \varepsilon + \delta. \tag{1.38}$$

Since $\delta > 0$ is arbitrary and $\rho_0^* = \rho^*$, it holds

$$\inf_{B \in \mathfrak{B}(\mathcal{E},\mathcal{E}^*)} \Big[R_\pi(B\mathcal{E}, W, \rho^*) - R_\pi(\mathcal{E}^*, W, \rho^*) \Big] \le \varepsilon.$$

Now the existence of a randomization $B = B(\pi, \rho^*)$ follows from the fact that inf (in B) is attained on the compact set $\mathfrak{B}(\mathcal{E}, \mathcal{E}^*)$.

Thus, (1.32) is proved. If one denotes

$$\Phi(B, \rho^*, \pi) = \iint \rho_\theta^*(\omega^*)\,\big[BP_\theta - P_\theta^*\big](d\omega^*)\,\pi(d\theta)$$
$$\Big(= R_\pi(B\mathcal{E}, W, \rho_0^*) - R_\pi(\mathcal{E}^*, W, \rho^*) \Big)$$

with $B \in \mathfrak{B}(\mathcal{E}, \mathcal{E}^*)$, $\pi \in \Pi^0$, then we have

$$\sup_{\rho^* \in \mathfrak{R}(\mathcal{E}^*, \mathbf{A})} \; \inf_{B \in \mathfrak{B}(\mathcal{E},\mathcal{E}^*)} \Phi(B, \rho^*, \pi) \le \varepsilon. \tag{1.39}$$

The function' $\Phi(B, \rho^*, \pi)$ is convex in B, continuous and concave in ρ^*. The set $\mathfrak{B}(\mathcal{E}, \mathcal{E}^*)$ is convex and compact and $\mathfrak{R}(\mathcal{E}^*, \mathbf{A})$ is convex. Thus the classical minimax theorem applies, (see e.g. Strasser (1985, Theorem 45.8)), and we conclude from (1.39) that

$$\inf_{B\in\mathfrak{B}(\mathcal{E},\mathcal{E}^*)} \sup_{\rho^*\in\mathfrak{R}(\mathcal{E}^*,\mathbf{A})} \Phi(B,\rho^*,\pi) \le \varepsilon.$$

The function (in B) $\sup_{\rho^*\in\mathfrak{R}(\mathcal{E}^*,\mathbf{A})} \Phi(B,\rho^*,\pi)$ is lower semicontinuous on the compact set $\mathfrak{B}(\mathcal{E},\mathcal{E}^*)$, hence inf (in B) is attained, say at the point B_π. Therefore

$$\sup_{\rho^*\in\mathfrak{R}(\mathcal{E}^*,\mathbf{A})} \Phi(B_\pi,\rho^*,\pi) \le \varepsilon. \tag{1.40}$$

Now we show that

$$\sup_{\rho^*\in\mathfrak{R}(\mathcal{E}^*,\mathbf{A})} \Phi(B_\pi,\rho^*,\pi) = \int \|B_\pi P_\theta - P^*_\theta\|\,\pi(d\theta). \tag{1.41}$$

It is clear directly from the definition of $\Phi(B,\rho^*,\pi)$, that

$$\sup_{\rho^*\in\mathfrak{R}(\mathcal{E}^*,\mathbf{A})} \Phi(B_\pi,\rho^*,\pi) \le \int \|B_\pi P_\theta - P^*_\theta\|\,\pi(d\theta). \tag{1.42}$$

On the other hand, the definition of $\|B_\pi P_\theta - P^*_\theta\|$ implies that for every $\delta > 0$ and every θ, the there exists a function $h_\theta(\omega^*)$ such that

$$\int h_\theta(\omega^*)\,(B_\pi P_\theta - P^*_\theta)(d\omega^*) \ge \|B_\pi P_\theta - P^*_\theta\| - \delta. \tag{1.43}$$

The function $h = \{h_\theta(\omega^*),\ \theta\in\Theta\}$ belongs to $\mathbf{A} = [-1,1]^\Theta$; thus by (1.43)

$$\sup_{\rho^*\in\mathfrak{R}(\mathcal{E}^*,\mathbf{A})} \Phi(B_\pi,\rho^*,\pi) \ge \Phi(B_\pi,h,\pi) \ge \int \|B_\pi P_\theta - P^*_\theta\|\,\pi(d\theta) - \delta.$$

Since $\delta > 0$ was chosen arbitrarily, the last assertion and (1.42) yield the desired equality (1.41) and hence, by (1.41) and (1.40) we have $(C) \Rightarrow (A^*)$. Since also the assertions (A) and (A^*) are equivalent, we get $(C) \Rightarrow (A)$.

$(A) \iff (D)$. The implication $(\Rightarrow)$ is obvious. To prove the inverse one we suppose that

$$\delta(\mathcal{E},\mathcal{E}^*) = \inf_{B\in\mathfrak{B}(\mathcal{E},\mathcal{E}^*)} \sup_\theta \|BP_\theta - P^*_\theta\| \le \varepsilon.$$

Let us fix $\delta > 0$ and choose B_δ in such a way that

$$\sup_\theta \|B_\delta P_\theta - P^*_\theta\| \le \varepsilon + \delta.$$

This implies for any $\pi \in \Pi^0$,

$$\int \|B_\delta P_\theta - P^*_\theta\|\,\pi(d\theta) \le \varepsilon + \delta,$$

and hence

$$\inf_{B\in\mathfrak{B}(\mathcal{E}^*,\mathbf{A})} \int \|BP_\theta - P_\theta^*\| \, \pi(d\theta) \le \varepsilon,$$

which yields (A^*). Since $(A) \Longleftrightarrow (A^*)$ we obtain $(D) \Longleftrightarrow (A)$.

To complete the proof of the theorem it remains only to show that $(B) \Longleftrightarrow (C)$. (This assertion is called "the Blackwell Theorem"). The implication $(B) \Rightarrow (C)$ is obvious. The assertion $(C) \Rightarrow (B)$ follows from the general results of functional analysis (see, for example, Strasser (1985, Section 45.7)). □

7. The problem of finding or, at least, estimating the deficiency $\delta(\mathcal{E},\mathcal{E}^*)$ and Δ-distance $\Delta(\mathcal{E},\mathcal{E}^*)$ is a very important task. This has been illustrated by Theorem 1.10 which allows to compare, for example, the risks $\mathcal{R}_\pi(\mathcal{E},W)$ and $\mathcal{R}_\pi(\mathcal{E}^*,W)$. If $W \in \mathcal{W}$ then this theorem implies for every $\pi \in \Pi^0$

$$|\mathcal{R}_\pi(\mathcal{E},W) - \mathcal{R}_\pi(\mathcal{E}^*,W)| \le \Delta(\mathcal{E},\mathcal{E}^*)\|W\|. \tag{1.44}$$

Therefore, the knowledge of the value $\Delta(\mathcal{E},\mathcal{E}^*)$ permits us to compare π-Bayes risks in the experiments $\mathcal{E}$ and $\mathcal{E}^*$ for any bounded loss function $W \in \mathcal{W}$.

Remark 1.5 Let experiments $\mathcal{E}$ and $\mathcal{E}^*$ be defined on a common measurable space $(\Omega,\mathcal{F})$. It immediately follows from the definition of $\Delta(\mathcal{E},\mathcal{E}^*)$ that

$$\Delta(\mathcal{E},\mathcal{E}^*) \le \sup_\theta \|P_\theta - P_\theta^*\|.$$

In some situation this estimate could be very rough. For instance, let $\theta \in [0,1]$, $P_\theta \sim U[\theta,\theta+1]$ be the uniform distribution on $[\theta,\theta+1]$ and let $P_\theta^* \sim U[\theta+2,\theta+3]$ be the uniform distribution on $[\theta+2,\theta+3]$. Then for every $\theta \in [0,1]$ one has $\|P_\theta - P_\theta^*\| = 2$ but at the same time obviously $\Delta(\mathcal{E},\mathcal{E}^*) = 0$.

The inequality (1.44) shows how one can estimate the distance between Bayes risks if the Δ-distance between the corresponding experiments is known. In the following section the inverse (in some sense) assertion will be stated, showing how the Δ-distance can be determined via the corresponding risks.

1.8 On Calculation of Deficiency $\delta(\mathcal{E},\mathcal{E}^*)$ and Δ-Distance with the Help of Bayes Risks

1. We start with the case of binary experiments

$$\mathcal{E} = (\Omega,\mathcal{F};\, P_1,P_2) \qquad \text{and} \qquad \mathcal{E}^* = (\Omega^*,\mathcal{F}^*;\, P_1^*,P_2^*).$$

Let $\mathbf{A} = \{a_1,a_2\}$, where a decision "a_i" is treated as the hypothesis that the true distribution is P_i (in the experiment $\mathcal{E}$) and P_i^* (in the experiment $\mathcal{E}^*$). Due to Theorem 1.2 in Section 1.4 for any decision problem with such (binary) experiments and such **A** one can restrict oneself to the case of decisions defined by Markov kernels.

Let $K(\omega,\cdot)$ be a Markov kernel for the experiment $\mathcal{E}$ from $(\Omega,\mathcal{F})$ to **A**. Since $K(\omega,a_2) = 1 - K(\omega,a_1)$, the kernel K can be defined by the function $\varkappa(\omega) = K(\omega,a_2)$ which is treated as the probability to choose the decision a_2 if the observation is ω.

Let $W(\theta,a)$ be a loss function, $\theta = 1,2$, and $a = a_1, a_2$. We will assume $W(i,a_i) = 0$ and denote $W_{12} = W(1,a_2)$, $W_{21} = W(2,a_1)$.

If $R_1(W,\varkappa)$ and $R_2(W,\varkappa)$ are the risks for $\theta = 1,2$, respectively, then

$$R_1(W,\varkappa) = W_{12}\, E_1\, \varkappa, \quad R_2(W,\varkappa) = W_{21}\, E_2(1-\varkappa),$$

with $E_1\,\varkappa$ and $E_1\,(1-\varkappa)$ being *the errors of the first and the second type.*

If for $0 \le \pi \le 1$ we define

$$R_\pi(W,\varkappa) = \pi\, W_{12}\, E_1\, \varkappa + (1-\pi)\, W_{21}\, E_2\,(1-\varkappa),$$

then the corresponding π-Bayes risk is

$$\mathcal{R}_\pi(\mathcal{E},W) = \inf_{\varkappa} R_\pi\,(W,\varkappa).$$

In the particular case of $W_{12} = W_{21} = 1$ this π-Bayes risk is denoted by

$$g_\pi(\mathcal{E}) = \inf_{\varkappa}\left[\pi\, E_1\, \varkappa + (1-\pi)\, E_2\,(1-\varkappa)\right].$$

If we put $Q = \pi\, P_1 + (1-\pi)\, P_2$ and $z_1 = \dfrac{dP_1}{dQ}$, $z_2 = \dfrac{dP_2}{dQ}$ then it is easy to see that

$$g_\pi(\mathcal{E}) = \inf_{\varkappa}\left[\pi\, E_1\, \varkappa + (1-\pi)\, E_2\,(1-\varkappa)\right] = (1-\pi) + \inf_{\varkappa} E_Q\, \varkappa\,(\pi\, z_1 - (1-\pi)\, z_2).$$

This representation readily yields that inf is attained at

$$\widetilde{\varkappa} = I\left(Z > \frac{\pi}{1-\pi}\right)$$

with $Z = \dfrac{z_2}{z_1}$ being exactly the *Lebesgue derivative* dP_1/dP_2 of the measure P_2 with respect to P_1 (see Shiryaev (1984, Section VII.6)) or, which is the same, the Radon–Nikodym derivative (density) of the *absolutely continuous component* of the measure P_2 with respect to P_1. Hence

$$g_\pi(\mathcal{E}) = \frac{1}{2} - \frac{1}{2} E_Q\, |\pi\, z_1 - (1-\pi)\, z_2|.$$

In particular, for $\pi = 1/2$

$$g_{1/2}(\mathcal{E}) = \frac{1}{2}\left[1 - \frac{1}{2} E_Q |z_1 - z_2|\right] = \frac{1}{2}\left[1 - \frac{1}{2}\|P_1 - P_2\|\right].$$

The next result states the relation of deficiency $\delta(\mathcal{E},\mathcal{E}^*)$ and Δ-distance $\Delta(\mathcal{E},\mathcal{E}^*)$ with the π-Bayes risks $g_\pi(\mathcal{E})$ and $g_\pi(\mathcal{E}^*)$.

Theorem 1.11 *(Le Cam (1964), Strasser (1985)) The following assertions hold:*

$$\delta(\mathcal{E},\mathcal{E}^*) = \sup_{0\le\pi\le1} \left[g_\pi(\mathcal{E}) - g_\pi(\mathcal{E}^*)\right], \tag{1.45}$$

$$\Delta(\mathcal{E},\mathcal{E}^*) = \sup_{0\le\pi\le1} |g_\pi(\mathcal{E}) - g_\pi(\mathcal{E}^*)|. \tag{1.46}$$

Proof. The first assertion follows directly from Theorem 1.10, Section 1.7. In fact, if $\delta(\mathcal{E},\mathcal{E}^*) \le \varepsilon$ then this theorem implies for the risks $\mathcal{R}_\pi(\mathcal{E},W) = g_\pi(\mathcal{E})$ and $\mathcal{R}_\pi(\mathcal{E}^*,W) = g_\pi(\mathcal{E}^*)$ and for any $0\le\pi\le1$ the inequality $g_\pi(\mathcal{E}) - g_\pi(\mathcal{E}^*) \le \varepsilon$ and hence $\sup_\pi \left[g_\pi(\mathcal{E}) - g_\pi(\mathcal{E}^*)\right] \le \varepsilon$. Therefore

$$\delta(\mathcal{E},\mathcal{E}^*) \le \varepsilon \quad \Longrightarrow \quad \sup_{0\le\pi\le1} \left[g_\pi(\mathcal{E}) - g_\pi(\mathcal{E}^*)\right] \le \varepsilon. \tag{1.47}$$

Next, it is easy to see that for any loss function W such that $0\le W\le 1$

$$g_\pi(\mathcal{E}) - g_\pi\mathcal{E}^*) \le \varepsilon \quad \Longrightarrow \quad \mathcal{R}_\pi(\mathcal{E},W) - \mathcal{R}_\pi(\mathcal{E}^*,W) \le \varepsilon.$$

Now, again by Theorem 1.10 of Section 1.7, the inequalities $\mathcal{R}_\pi(\mathcal{E},W) - \mathcal{R}_\pi(\mathcal{E}^*,W) \le \varepsilon$ for all $0\le\pi\le1$ and any loss function W with $0\le W\le 1$ imply $\delta(\mathcal{E},\mathcal{E}^*)\le\varepsilon$. Hence the inverse implication in (1.47) is also true yielding (1.45), and the statement (1.46) evidently follows from (1.45). □

2. In order to state in the general case a result similar to Theorem 1.11 it is useful to note at the beginning the following. For the binary case which is considered above with $W_{11} = W_{22} = 0$, the risk $\mathcal{R}_\pi(\mathcal{E},W)$ fulfills

$$\begin{aligned}\mathcal{R}_\pi(\mathcal{E},W) &= \inf_{\varkappa}\left[\pi\, W_{12} E_1\, \varkappa + (1-\pi)\, W_{21}\, E_2\,(1-\varkappa)\right] \\ &= E_Q \inf_{\varkappa}\left[\pi\, z_1\, W_{12}\, \varkappa + (1-\pi)\, z_2\, W_{21}\,(1-\varkappa)\right].\end{aligned}$$

If we denote $C = 1 - W$ (so that $C_{11} = C_{22} = 1$) then it is easily seen that

$$\mathcal{R}_\pi(\mathcal{E},W) = R_\pi(\mathcal{E},1-C) = 1 - \nu_\pi(\mathcal{E},C) \tag{1.48}$$

with ($\pi_1 = \pi$, $\pi_2 = 1-\pi$) and

$$\begin{aligned}\nu_\pi(\mathcal{E},C) &= E_Q \sup_{\varkappa}\left[\pi\, z_1\,(1-C_{12})\,\varkappa + (1-\pi)\, z_2\,(1-C_{21})\,(1-\varkappa)\right] \\ &= E_Q \max\left[\pi_1\, z_1 + \pi_2\, C_{21}\, z_2\,,\; \pi_1\, C_{12}\, z_1 + \pi_2\, z_2\right] \\ &= E_Q \max\left[\pi_1\, C_{11}\, z_1 + \pi_2\, C_{21}\, z_2\,,\; \pi_1\, C_{12}\, z_1 + \pi_2\, C_{22}\, z_2\right].\end{aligned} \tag{1.49}$$

The proof of Theorem 1.11 leads to the following representation for $\delta(\mathcal{E},\mathcal{E}^*)$:

$$\delta(\mathcal{E},\mathcal{E}^*) = \sup_{(\pi,C)} \left[\nu_\pi(\mathcal{E},C) - \nu_\pi(\mathcal{E}^*,C)\right] \tag{1.50}$$

with sup over all $0 \le \pi \le 1$ and $0 \le C \le 1$ $(C = C_{ij}, i,j = 1,2)$.

A similar representation can be given for the case of arbitrary (not only binary) experiments. For this we note that the value $\nu_\pi(\mathcal{E},C)$ is just the π-Bayes *gain*, i.e.

$$\nu_\pi(\mathcal{E},C) = \sup_{\rho\in\mathfrak{R}(\mathcal{E},\mathbf{A})} \nu_\pi(C,\rho) \tag{1.51}$$

with $\nu_\pi(C,\rho) = 1-R_\pi(W,\rho)$ for $W = 1-C$ and with a distribution π concentrated at a finite set.

The formula (1.49) shows that the gain $\nu_\pi(\mathcal{E},C)$ can be rewritten in the form

$$\nu_\pi(\mathcal{E},C) = E_Q \max_{j=1,2} \sum_{i=1}^{2} \pi_i\, C_{ij}\, z_i. \tag{1.52}$$

Because the set $\mathfrak{R}_0(\mathcal{E},\mathbf{A})$ of simple non-randomized Markov decisions is dense in $\mathfrak{R}(\mathcal{E},\mathbf{A})$ we obtain in the general case for $\nu_\pi(\mathcal{E},C)$ the following representation similar to (1.52):

$$\nu_\pi(\mathcal{E},C) = E_Q \sup_j \sum_\theta \pi_\theta\, C_{\theta j}\, z_\theta \tag{1.53}$$

where the measure π has a finite support, $C_{\theta j} \in [0,1]$, the probability measure Q is such that $P_\theta \ll Q$ for all θ with $\pi(\{\theta\}) \neq 0$, and $z_\theta = dP_\theta/dQ$.

Then the same arguments as for the case of binary experiments carry us to the following result generalizing Theorem 1.11.

Theorem 1.12 *(Le Cam (1964), Strasser (1985, Section 15)) For any two experiments $\mathcal{E}$ and $\mathcal{E}^*$*

$$\begin{aligned} \delta(\mathcal{E},\mathcal{E}^*) &= \sup_{(\pi,C)} \left[\nu_\pi(\mathcal{E},C) - \nu_\pi(\mathcal{E}^*,C)\right], \\ \Delta(\mathcal{E},\mathcal{E}^*) &= \sup_{(\pi,C)} \left|\nu_\pi(\mathcal{E},C) - \nu_\pi(\mathcal{E}^*,C)\right| \end{aligned}$$

with π and C as in (1.53).

1.9 Some Explicit Formulas for Deficiency and Δ-Distance

1. Consider two experiments $\mathcal{E}$ and $\mathcal{E}_*$ corresponding to *linear* Gaussian (normal) models (using the notations of Section 1.3):

$$\mathcal{E} : X = A'\beta + \xi \quad \text{and} \quad \mathcal{E}_* : X_* = A_*'\beta + \xi_*. \tag{1.54}$$

Here $A = A(k \times n)$ is a matrix with k rows and n columns, $\xi = \xi(n \times 1)$ is a Gaussian column-vector with independent components ξ_i, $i = 1, \dots, n$ of the form $\xi_i \sim \mathcal{N}(0, \sigma^2)$, $0 < \sigma^2 < \infty$, $X = X(n \times 1)$; similarly, $A_* = A_*(k \times n_*)$, $\xi_* = \xi_*(n_* \times 1)$ is also a column-vector with independent components $\xi_{*i} \sim \mathcal{N}(0, \sigma^2)$, and $X_* = X_*(n_* \times 1)$. Finally, the vector $\beta = \beta(k \times 1)$ is assumed to be unknown.

In the case when σ^2 is known, $\theta = \beta$ is considered as unknown parameter. If σ^2 is unknown then $\theta = (\beta, \sigma^2)$.

Theorem 1.13 *(Luschgy (1992); Lehmann (1988))*
(1) *Let σ^2 be known. Then the following assertions are equivalent:*

(i) $\delta(\mathcal{E}_*, \mathcal{E}) = 0$;
(ii) *the matrix $AA' - A_*A_*'$ is non-negative.*

(2) *Let σ^2 be unknown. Then* (i) is equivalent to the following condition:

(ii') the matrix $AA' - A_*A_*'$ is non-negative and $n - n_* \geq \operatorname{rank}(AA' - A_*A_*')$.

Theorem 1.14 *(Luschgy (1992); Lehmann (1988)) Let σ^2 be known, the matrix $AA' - A_*A_*'$ be non-negative (thus $\delta(\mathcal{E}, \mathcal{E}_*) = 0$) and the matrix A_*A_*' be non-singular. Then*

$$\delta(\mathcal{E}_*, \mathcal{E}) = \|\mathcal{N}(0, (A_*A_*')^{-1}) - \mathcal{N}(0, (AA')^{-1})\|, \tag{1.55}$$

or, equivalently,

$$\delta(\mathcal{E}_*, \mathcal{E}) = E\left|\left[\frac{\det(AA')}{\det(A_*A_*')}\right]^{\frac{1}{2}} \exp\left\{-\frac{1}{2}\eta'(AA' - A_*A_*')\eta\right\} - 1\right|, \tag{1.56}$$

with $\eta \sim \mathcal{N}(0, (A_*A_*')^{-1})$.

We omit the proof noting only that it is based on the following results (Boll (1955), Torgensen (1972)) for "translation" (or "shift") experiments: Let $\Theta = X$ be a locally compact topological group, and $P_\theta(A) = P(A\theta^{-1})$ with some probability measure P. (The experiment $\mathcal{E}_P = (X, \mathcal{X}; P_\theta, \theta \in \Theta = X)$ is called *a shift experiment.*) If $\mathcal{E}_P$ and $\mathcal{E}_Q$ are two shift experiments corresponding to the measures P and Q respectively, and if the experiment $\mathcal{E}_P$ is dominated then

$$\delta(\mathcal{E}_P, \mathcal{E}_Q) = \inf_\mu \|\mu * P - Q\|$$

where the symbol "$*$" denotes the convolution of measures. Moreover, the inf is attained at some ("least preferable") distribution μ_0.

2. Now we consider the following analog (in continuous time) of the linear Gaussian models (1.54). Define

$$\begin{aligned} \mathcal{E}\ &: X_t = \theta(t) + \xi_t, \qquad 0 \leq t \leq T, \\ \mathcal{E}^* &: X_t^* = \theta(t) + \xi_t^*, \qquad 0 \leq t \leq T, \end{aligned} \tag{1.57}$$

with some Gaussian processes $\xi = (\xi_t)$ and $\xi^* = (\xi_t^*)$ with distributions P and P^* on R^T, having zero means and covariance operators $K = K(s,t)$ and $K^* = K^*(s,t)$ respectively. Also we suppose that ξ, ξ^* and $\theta = (\theta(t))_{t\le T}$ belong to some separable Banach space $B \subseteq R^T$ with the Borel σ-field $\mathcal{B}(B)$.

Theorem 1.15 *(Luschgy (1992))* 1) *The following assertions are equivalent:*

(i) $\delta(\mathcal{E},\mathcal{E}^*) = 0$;

(ii) $K^* \ge K$ *(i.e.* $K^* - K$ *is a non-negative operator);*

(iii) $P(A) \ge P^*(A)$ *for every symmetric convex set* $A \in \mathcal{B}(B)$.

2) *If* $\delta(\mathcal{E},\mathcal{E}^*) = 0$ *(or, equivalently, if* (i) *or* (iii) *is true), then*

$$\delta(\mathcal{E}^*,\mathcal{E}) = \|P - P^*\|. \tag{1.58}$$

Remark 1.6 Suppose for the model (1.57) that P is a Gaussian measure and P^* is *an arbitrary* measure on $(B,\mathcal{B}(B))$. Due to Luschgy (1992), the deficiency $\delta(\mathcal{E},\mathcal{E}^*) = 0$ iff $P^* = P * R$ with some probability measure R on $(B,\mathcal{B}(B))$. This property yields the measure P to be "less dispersed" than the measure P^* in the sense that $P(A) \ge P^*(A)$ for each symmetric convex set A. In fact, $P(A+x) \le P(A)$ for all $x \in B$ hence

$$P^*(A) = \int P(A-x)\,dR \le P(A).$$

Example 1.23 (Luschgy (1992); Kramkov (1994)) Let $P^* = P_T$ be a Wiener measure on $\mathbf{C}[0,T]$ with $K^*(s,t) = \min(s,t)$ and let $P = P_T^A$ be a measure corresponding to a diffusion process $\xi = (\xi_t)_{t\le T}$ with

$$d\xi_t = A(t)\,\xi_t\,dt + dW_t\,, \quad \xi_0 = 0, \tag{1.59}$$

where $W = (W_t)_{t\le T}$ is a standard Wiener process and $A = A(t)$ is a continuously differentiable function on $[0,T]$.

The solution $\xi = (\xi_t)_{t\le T}$ of the linear stochastic equation (1.59) admits the following representation

$$\xi_t = \psi(t)\int_0^t \psi^{-1}(s)\,ds \tag{1.60}$$

with

$$\psi(t) = \exp\left\{\int_0^t A(s)\,ds\right\}.$$

It easily follows from (1.60) for the covariance operator K of ξ

$$K(s,t) = \psi(t)\,\psi(s)\int_0^{\min(s,t)} \psi^{-2}(u)\,du.$$

Starting from this representation Luschgy (1992) stated that $K^* \geq K$ iff $A'(t) + A^2(t) \geq 0$ on $[0,T]$ and $A(T) \leq 0$. Thus the following result does hold.

Theorem 1.16 *Let* $\theta = (\theta(t))_{t\leq T} \in \mathbf{C}[0,T]$ *and*

$$\begin{aligned} &\mathcal{E}_T^0 : X_t = \theta(t) + \xi_t\,, \quad \xi_t = W_t\,, && 0 \leq t \leq T; \\ &\mathcal{E}_T^A : X_t = \theta(t) + \xi_t\,, \quad d\xi_t = A(t)\,\xi_t\,dt + dW_t, && 0 \leq t \leq T. \end{aligned} \tag{1.61}$$

Then

$$\delta(\mathcal{E}_T^A, \mathcal{E}_T^0) = 0 \iff A'(t) + A^2(t) \geq 0,\ t \in [0,T] \text{ and } A(T) \leq 0; \tag{1.62}$$

$$\delta(\mathcal{E}_T^0, \mathcal{E}_T^A) = 0 \iff A'(t) + A^2(t) \leq 0,\ t \in [0,T] \text{ and } A(T) \geq 0. \tag{1.63}$$

This result, in particular, implies the following interesting consequences:

(*a*) if $A(t) \leq 0$ and $A(t) \uparrow$ then $\delta(\mathcal{E}_T^A, \mathcal{E}_T^0) = 0$;

(*b*) if $\delta(\mathcal{E}_T^A, \mathcal{E}_T^0) = 0$ then $A(T) \leq 0$;

(c) if $\delta(\mathcal{E}_T^0, \mathcal{E}_T^A) = 0$ then $A(t) \downarrow$ and $A(T) \geq 0$.

From Theorem 1.15, if $\delta(\mathcal{E}_T^A, \mathcal{E}_T^0) = 0$ then

$$\delta(\mathcal{E}_T^0, \mathcal{E}_T^A) = \|P_T^0 - P_T^A\|.$$

Due to Theorem V.4.21 in Jacod and Shiryaev (1987) the following two-side inequalities hold for the variation distance $\|P_T^0 - P_T^A\|$:

$$2\left[1 - \sqrt{E^0 \exp(h_T)}\right] \leq \|P_T^0 - P_T^A\| \leq 4\sqrt{E^0 h_T}$$

where

$$h_T = \frac{1}{8}\int_0^T A^2(s)\, W_s^2\, ds$$

is the *Hellinger process* of order $\frac{1}{2}$, Jacod and Shiryaev (1987, Chapter V). Since $E^0\, h_T = \frac{1}{8}\int_0^T A^2(s)\, s\, ds$, we have for $\delta(\mathcal{E}_T^0, \mathcal{E}_T^A)$ (under the assumption $\delta(\mathcal{E}_T^A, \mathcal{E}_T^0) = 0$) the following lower estimate

$$\delta(\mathcal{E}_T^0, \mathcal{E}_T^A) \leq 2\left[\int_0^T A^2(s)\, s\, ds\right]^{1/2}.$$

In particular, for $A(s) = -1$ (Gaussian-Markov case)

$$\delta(\mathcal{E}_T^0, \mathcal{E}_T^A) \leq T.$$

For this case Luschgy (1992) carried out the detailed analysis of the value $\delta(\mathcal{E}_T^0, \mathcal{E}_T^A)$ under the assumption that T is small stating the following asymptotic relation:

$$\delta(\mathcal{E}_T^0, \mathcal{E}_T^A) = \left(\frac{2}{\pi e}\right)^{1/2} T + O(T^2) \qquad (\approx 0.48\,T + O(T^2)).$$

Remark 1.7 The fact that under the assumption $A'(t) + A^2(t) \geq 0,\ t \leq T$, and $A(T) \leq 0$ the experiment $\mathcal{E}_T^A$ is more informative than $\mathcal{E}_T^0$, has very interesting interpretations for applied statistics.

Indeed, estimating an unknown "trend" $\theta = (\theta(t))_{t \leq T}$ in a Gaussian noise (as in (1.61)) it is natural to state the question of which noise is "noisier" and hence, the corresponding loss of estimation is larger. Due to (1.62) $\delta(\mathcal{E}_T^A, \mathcal{E}_T^0) = 0$ and from Theorem 1.10, Section 1.7, for any loss function $W = W_\theta(a) \in \mathcal{W}$ and any decision ρ^0 in the experiment $\mathcal{E}_T^0$ there is a decision ρ^A in the experiment $\mathcal{E}_T^A$ such that $R_\theta(W, \rho^A) \leq R_\theta(W, \rho^0),\ \theta \in \Theta$. This just means that a "Wiener" noise is noisier than, for example, a Gaussian-Markov one (i.e. if $A(t) = -1$).

The accurate value or some estimates for $\delta(\mathcal{E}_T^0, \mathcal{E}_T^A)$ is of great interest because, as usual, the solution of decision problems in the "Wiener" white noise case is simpler than, for example, in the "Gaussian-Markov" one. Thus, knowledge about the value $\delta(\mathcal{E}_T^0, \mathcal{E}_T^A)$, again due to Theorem 1.10 from Section 1.7, delivers some information about the risks in the experiment $\mathcal{E}_T^A$, if the risks for the experiments $\mathcal{E}_T^0$ have been already calculated.

Let us consider another example with $T < 1$ and $A(t) = \frac{1}{t-1}$ that corresponds to the case when $\xi = (\xi_t)_{t \leq 1}$ is a Brownian bridge,

$$d\xi_t = -\frac{\xi_t}{t-1}\,dt + dW_t, \quad \xi_0 = 0,\ 0 \leq t \leq 1.$$

Here $K(s,t) = \min(s,t) - s\,t,\ A'(t) + A^2(t) = 0,\ A(t) < 0$ for each $t < 1$. Thus $\delta(\mathcal{E}_T^A, \mathcal{E}_T^0) = 0$.

It is worth noting that the value $\delta(\mathcal{E}_T^0, \mathcal{E}_T^A)$ can be found *exactly* in this case. In fact, for the model (1.61)

$$\delta(\mathcal{E}_T^0, \mathcal{E}_T^A) = \|P_T^0 - P_T^A\| = E^0|Z_T - 1|,$$

with $Z_T(W) = \frac{dP_T^A}{dP_T^0}(W)$ defined by the formula (Liptser and Shiryaev (1974, Chapter 7)):

$$Z_T(W) = \exp\left\{\int_0^T A(s)\,W_s\,dW_s - \frac{1}{2}\int_0^T A^2(s)\,W_s^2\,ds\right\}.$$

Applying the Itô formula to $\frac{1}{2}A^2(t)\,W_t^2$ we obtain

$$Z_T(W) = \psi^{-1/2}(T)\exp\left\{\frac{1}{2}A(T)\,W_T^2 - \frac{1}{2}\int_0^T [A'(s) + A^2(s)]\,W_s^2\,ds\right\}.$$

Since in the case under consideration $A' + A^2 = 0$, it follows

$$Z_T(W) = \psi^{-1/2}(T)\exp\left\{-\frac{1}{2}\frac{W_T^2}{1-T}\right\} = \sqrt{1-T}\exp\left\{-\frac{W_T^2}{2(1-T)}\right\},$$

and, obviously,

$$E^0\,|Z_T - 1| = \|\mathcal{N}(0,T) - \mathcal{N}(0,T(1-T))\|,$$

where $\mathcal{N}(0,S)$ denotes the normal distribution with zero mean and the covariance S. Therefore

$$\begin{aligned}\delta(\mathcal{E}_T^0, \mathcal{E}_T^A) &= \|\mathcal{N}(0,T) - \mathcal{N}(0,T(1-T))\| \\ &= \left(\frac{\pi\, e}{2}\right)^{-1/2} T + (2\,\pi\, e)^{-1/2}T^2 + o(T^2), \quad T \to 0,\end{aligned}$$

where the last equality is a particular case of the following result, Luschgy (1992, Lemma 2.7)): if $0 < \sigma^2 < \tau^2$ then

$$\|\mathcal{N}(0,\sigma^2) - \mathcal{N}(0,\tau^2)\| = 4\left\{\Phi\left(\left[\frac{\tau^2\ln(\tau^2/\sigma^2)}{\tau^2-\sigma^2}\right]^{1/2}\right) - \Phi\left(\left[\frac{\sigma^2\ln(\tau^2/\sigma^2)}{\tau^2-\sigma^2}\right]^{1/2}\right)\right\}.$$

The assertion of Theorem 1.16 can be proved directly without referring to Theorem 1.15. We give this proof due to Kramkov (1994) under the assumption that the unknown parameter θ is an element of the set

$$\Theta = \left\{\theta = (\theta_t)_{t\le T} : \ \theta_t = \int_0^t \theta'_s\, ds, \ t \le T, \ \int_0^T (\theta'_s)^2 ds < \infty\right\}.$$

Let $P^0_{T,\theta}$ and $P^A_{T,\theta}$ be the measures of the processes $X = (X_t)_{t\le T}$ in the experiments $\mathcal{E}_T^0$ and $\mathcal{E}_T^A$ defined by (1.61). Let also $h_T^{0,\alpha\beta}$ and $h_T^{A,\alpha\beta}$ be the Hellinger processes (of the order 1/2) corresponding to the pairs of measures $(P^0_{T,\alpha}, P^0_{T,\beta})$ and $(P^A_{T,\alpha}, P^A_{T,\beta})$ respectively; $\alpha, \beta \in \Theta$.

Lemma 1.4 *It holds*

$$h_T^{0,\alpha\beta} = \frac{1}{8}\int_0^T (\alpha'_s - \beta'_s)^2\, ds, \tag{1.64}$$

$$\begin{aligned} h_T^{A,\alpha\beta} = \frac{1}{8}\bigg\{&\int_0^T (\alpha'_s - \beta'_s)^2 ds + \int_0^T (\alpha_s - \beta_s)^2\,(A'(s) + A^2(s))\, ds \\ &- (\alpha_T - \beta_T)^2\, A(T)\bigg\}. \end{aligned} \tag{1.65}$$

Proof. The result (1.64) is well known, see e.g. Jacod and Shiryaev (1987, IV, §1c). To prove (1.65) note that by (1.61)

$$dX_t = \theta'_T\, d_t + A(T)\,(X_t - \theta_t)\,dt + dW_t.$$

Then, (see again Jacod and Shiryaev (1987, IV, §1c))

$$\begin{aligned}
8h_T^{A,\alpha\beta} &= \int_0^T \left[(\alpha'_s - \beta'_s) - A(s)(\alpha_s - \beta_s)\right]^2 ds \\
&= \int_0^T (\alpha'_s - \beta'_s)^2\, ds + \int_0^T A^2(s)\,(\alpha_s - \beta_s)^2\, ds - \int_0^T A(s)\, d(\alpha_s - \beta_s)^2 \\
&= \int_0^T (\alpha'_s - \beta'_s)^2\, ds + \int_0^T \left[A^2(s) + A'(s)\right](\alpha_s - \beta_s)^2\, ds \\
&\quad - A(T)(\alpha_T - \beta_T)^2
\end{aligned}$$

and (1.65) follows. □

Note that by (1.64) and (1.65) the Hellinger processes $h^{0,\alpha\beta}$ and $h^{A,\alpha\beta}$ are *deterministic* (non-random) in the case under consideration.

Next we turn to the proof of (1.62). If $\delta(\mathcal{E}_T^A, \mathcal{E}_T^0) = 0$ then it is easy to obtain for the Hellinger integrals (of order 1/2)

$$\begin{aligned}
H_T^{0,\alpha\beta} &= \int \left(dP_{T,\alpha}^0\right)^{1/2} \left(dP_{T,\beta}^0\right)^{1/2} \\
H_T^{A,\alpha\beta} &= \int \left(dP_{T,\alpha}^A\right)^{1/2} \left(dP_{T,\beta}^A\right)^{1/2}
\end{aligned}$$

the following inequality:

$$H_T^{0,\alpha\beta} \geq H_T^{A,\alpha\beta}.$$

Since also $H_T^{0,\alpha\beta} = \exp\left\{-h_T^{0,\alpha\beta}\right\}$, $H_T^{A,\alpha\beta} = \exp\left\{-h_T^{A,\alpha\beta}\right\}$, it follows

$$h_T^{0,\alpha\beta} \leq h_T^{A,\alpha\beta},$$

and by (1.64) and (1.65) (taking into account that $\alpha, \beta \in \Theta$) we obtain $A'(t) + A^2(t) \geq 0$, $t \in [0, T]$ and $A(T) \leq 0$.

Now we prove the implication $\Leftarrow$ in (1.62). For this we construct a new experiment $\tilde{\mathcal{E}}_T^A$ such that $\tilde{\mathcal{E}}_T^A \sim \mathcal{E}_T^A$ and $\tilde{\mathcal{E}}_T^A \gg \mathcal{E}_T^0$. Of course, this yields $\mathcal{E}_T^A \gg \mathcal{E}_T^0$, i.e. $\delta(\mathcal{E}_T^A, \mathcal{E}_T^0) = 0$).

Put (under the assumption $A'(t) + A^2(t) \geq 0$ and $A(T) \leq 0$) for all $\theta \in \Theta$

$$\begin{aligned}
\mathcal{E}_T^{(1)} : X_t &= \int_0^t \theta_s \left[A'(s) + A^2(s)\right]^{1/2} ds + W_t, \quad t \leq T; \\
\mathcal{E}_T^{(2)} : X_T &\sim \mathcal{N}\left(\theta_T \sqrt{-A(T)}, 1\right)
\end{aligned} \tag{1.66}$$

and define the experiment $\tilde{\mathcal{E}}_T^A$ as the direct product of the experiments $\mathcal{E}_T^0$, $\mathcal{E}_T^{(1)}$ and $\mathcal{E}_T^{(2)}$:

$$\tilde{\mathcal{E}}_T^A = \mathcal{E}_T^0 \times \mathcal{E}_T^{(1)} \times \mathcal{E}_T^{(2)}. \tag{1.67}$$

If $\widetilde{P}^A_{T,\theta}$ are the measures in this experiment then obviously for any finite set $\theta_1,\dots,\theta_N \in \Theta$ and any numbers $a_1,\dots,a_N$ such that $a_i \ge 0$ and $\sum_{i=1}^N a_i = 1$ the Hellinger integral (or Hellinger transformation, see Section 1.11 below) satisfies the condition

$$H(a_1,\dots,a_N;\widetilde{\mathcal{E}}^A_T) = \int \prod_i \left(d\widetilde{P}^A_{T,\theta_i}\right)^{a_i} = \exp\left\{-2\sum_{i,j=1}^N a_i\, a_j\, h_T^{A,\theta_i\theta_j}\right\},$$

that coincides with $H(a_1,\dots,a_N;\mathcal{E}^A_T)$. This yields the experiment $\widetilde{\mathcal{E}}^A_T$ to be equivalent to the experiment $\mathcal{E}^A_T$ ($\widetilde{\mathcal{E}}^A_T \sim \mathcal{E}^A_T$).

From the definition (1.67) evidently $\mathcal{E}^0_T \ll \widetilde{\mathcal{E}}^A_T$. Thus, if $A'(t) + A^2(t) \ge 0,\ t \in [0,T]$ and $A(T) \le 0$, then $\mathcal{E}^0_T \ll \mathcal{E}^A_T$, i.e. the experiment $\mathcal{E}^0_T$ can be obtained by a randomization from the experiment $\widetilde{\mathcal{E}}^A_T$. Moreover (Kramkov (1994)), in this case it is possible to construct explicitly the corresponding Markov kernel $M(X;B)$ such that

$$P^0_{T,\theta} = \int_{\mathbf{C}[0,T]} M(X;B)\, P^A_{T,\theta}(dX)$$

with $X \in \mathbf{C}[0,T]$, $B \in \mathcal{B}(\mathbf{C}[0,T])$.

1.10 A Particular Case of Comparison of Experiments (Due to Blackwell, Shannon, Fisher)

1. As we have already noted before, it is difficult (or maybe impossible) to propose an universal definition of "information" contained in an experiment which permits us "to serve" all various statistical problems simultaneously. From this point of view it is of interest to compare the different famous notions of information in the Statistical and Information theories.

We shall suppose in this section that all the considered experiments are defined on the real line

$$\mathcal{E} = (R, \mathcal{B}(R);\ P_\theta, \theta \in \Theta \subseteq R),$$

with distributions P_θ having densities $p_\theta(x)$ with respect to the Lebesgue measure on the real line.

Definition 1.18 (Compare with Definition 1.3, Section 1.2) An experiment $\mathcal{E}$ is *exhaustive* (or *Blackwell-sufficient*) for an experiment $\mathcal{E}^*$ ("$\mathcal{E} \gg \mathcal{E}^* - \mathrm{Bl}$") if there exists a Markov kernel such that $\mathcal{E}^*$ is the Markov randomization of $\mathcal{E}$ using this kernel.

Definition 1.19 Let $\mathcal{E}$ be an experiment and let $\pi = \pi(d\theta)$ be a priori distribution on Θ (such that $\pi(\Theta) = 1$). *The Shannon information* contained in the experiment $\mathcal{E}$ with respect to the unknown parameter θ with the prior distribution π is the value

$$J(\mathcal{E},\pi) = \iint \log \frac{p_\theta(x)}{p(x)} \, p_\theta(x)\, dx\, \pi(d\theta)$$

with $p(x) = \int p_\theta(x)\, \pi(d\theta)$.

An experiment $\mathcal{E}$ is *more informative in Shannon sense* than the experiment $\mathcal{E}^*$ ("$\mathcal{E} \gg \mathcal{E}^*$ – Sh") if for *any* prior distribution π

$$J(\mathcal{E},\pi) \geq J(\mathcal{E}^*,\pi).$$

Definition 1.20 Let $p_\theta(x)$ have derivatives $\partial p_\theta(x)/\partial\theta$. The value

$$I(\mathcal{E},\theta) = E_\theta \left[\frac{\partial}{\partial\theta} \log p_\theta(x)\right]^2$$

is called *the Fisher information* in the experiment $\mathcal{E}$ at the point θ. The experiment $\mathcal{E}$ is said to be *more informative in Fisher sense* than an experiment $\mathcal{E}^*$ ("$\mathcal{E} \gg \mathcal{E}^*$ – Fsh") if for all $\theta \in \Theta$

$$I(\mathcal{E},\theta) \geq I(\mathcal{E}^*,\theta).$$

Stone (1961) stated that under some "regularity" conditions on the density function $p_\theta(x)$

$$\text{"}\mathcal{E} \gg \mathcal{E}^* - \text{Sh"} \quad \Rightarrow \quad \text{"}\mathcal{E} \gg \mathcal{E}^* - \text{Fsh"}.$$

Lindsey (1972) showed that always

$$\text{"}\mathcal{E} \gg \mathcal{E}^* - \text{Bl"} \quad \Rightarrow \quad \text{"}\mathcal{E} \gg \mathcal{E}^* - \text{Sh"}.$$

Therefore (in the "regular" cases)

$$\text{"}\mathcal{E} \gg \mathcal{E}^* - \text{Bl"} \quad \Rightarrow \quad \text{"}\mathcal{E} \gg \mathcal{E}^* - \text{Sh"} \quad \Rightarrow \quad \text{"}\mathcal{E} \gg \mathcal{E}^* - \text{Fsh"},$$

and, hence, the Fisher information "serves" a wider class of statistical applications based on comparison of experiments. However, it should be mentioned that there exist situations when "$\mathcal{E} \gg \mathcal{E}^*$ – Sh" but the Fisher information $I(\mathcal{E},\theta)$ and $I(\mathcal{E}^*,\theta)$ are simply undefined.

1.11 The Hellinger and Mellin Transformations

1. These transformations play in the theory of statistical experiments with a finite parameter set a role similar to that of the Laplace transformation in various considerations in the probability theory. Moreover, all these transformations are closely connected to each other.

Definition 1.21 Let $(\Sigma_k, \mathcal{B}(\Sigma_k))$ be the simplex

$$\Sigma_k = \left\{ x = (x_1, \dots, x_k) : x_i \geq 0,\ i = 1, \dots, k,\ \sum_{i=1}^{k} x_i = 1 \right\}$$

with the Borel σ-field $\mathcal{B}(\Sigma_k)$ and let η be a measure on $(\Sigma_k, \mathcal{B}(\Sigma_k))$ such that $\eta(\Sigma_k) < \infty$ and

$$\int_{\Sigma_k} x_i \, \eta(dx) = 1, \quad i = 1, \dots, k. \tag{1.68}$$

(Measures with this property are usually called *standard*; we denote the set of all such measures by $\mathcal{H}_k$.)

The Hellinger transformation of the measure η is the function

$$H(a; \eta) = \int_{\Sigma_k} x_1^{a_1} \dots x_k^{a_k} \, \eta(dx), \tag{1.69}$$

defined for $a = (a_1, \dots, a_k) \in \Sigma_k$, i.e. $a_i \geq 0$ and $\sum a_i = 1$, $i = 1, \dots, k$.

The next theorem summaries the main properties of the Hellinger transformations (see, for example, Strasser (1985, Section 5)).

Theorem 1.17

(1) If η_1 and η_2 are two standard measures such that for all $a \in \Sigma_k$

$$H(a; \eta_1) = H(a; \eta_2)$$

then $\eta_1 = \eta_2$.

(2) The set of standard measures on $(\Sigma_k, \mathcal{B}(\Sigma_k))$ is compact for the weak topology (of the weak convergence).

(3) A sequence of standard measures (η_n) on $(\Sigma_k, \mathcal{B}(\Sigma_k))$ weakly converges to a measure η iff the following pointwise convergence takes place

$$\lim_n H(a; \eta_n) = H(a; \eta), \quad a \in \Sigma_k.$$

Definition 1.22 Let μ be a probability measure on $(R_+^k, \mathcal{B}(R_+^k))$ with

$$R_+^k = \{ x = (x_1, \dots, x_k) : x_i \geq 0,\ i = 1, \dots, k \}$$

and with the Borel σ-field $\mathcal{B}(R_+^k)$ such that

$$\int_{R_+^k} x_i \, \mu(dx) \leq 1, \quad i = 1, \dots, k. \tag{1.70}$$

(The set of all such measures is denoted by $\mathcal{M}_k$.)

The Mellin transformation of the measure μ is a function

$$M(a;\eta) = \int_{R_+^k} x_1^{a_1} \dots x_k^{a_k} \, \mu(dx), \tag{1.71}$$

on the set

$$\Delta_k = \left\{ a = (a_1, \dots, a_k) : \; 0 \le a_i < 1, \; i = 1, \dots, k, \; \sum_{i=1}^{k} a_i < 1 \right\}.$$

The next theorem (see, for example, Strasser (1985, Section 5)) is an analog of the previous one.

Theorem 1.18

(1) If μ_1 and μ_2 are two standard measures such that

$$M(a;\mu_1) = M(a;\mu_2)$$

for all $a \in \Delta_k$ then $\mu_1 = \mu_2$.

(2) The set $\mathcal{M}_k$ of all standard measures on $(R_+^k, \mathcal{B}(R_+^k))$ is compact for the weak topology.

(3) A sequence of standard measures μ_n on $(R_+^k, \mathcal{B}(R_+^k))$ weakly converges to a measure $\mu \in \mathcal{M}_k$ iff the pointwise convergence

$$\lim_n M(a;\mu_n) = M(a;\mu), \quad a \in \Delta_k,$$

holds.

2. Let us consider applications of the Hellinger and Mellin transformations to measures related to a statistical experiment

$$\mathcal{E} = (\Omega, \mathcal{F}; P_0, P_1, \dots, P_k).$$

The Hellinger transformation gives us an analytic tool to study the properties of the likelihood

$$f_i = \frac{dP_i}{dQ}, \quad i = 0, 1, \dots, k$$

with respect to the measure $Q = P_0 + P_1 + \dots + P_k$.

The Mellin transformation is very useful for the investigation of likelihood ratios (Radon–Nikodym derivatives)

$$Z_i = \frac{dP_i}{dP_0}, \quad i = 1, \dots, k$$

with respect to the measure P_0.

Let us consider these statements in more detail. Let η be a measure on the simplex $(\Sigma_{k+1}, \mathcal{B}(\Sigma_{k+1}))$, defined as follows:

$$\eta(B) = Q\{\omega : \; (f_0(\omega), \dots, f_k(\omega)) \in B\} = Q(f^{-1}(B)),$$

for $B \in \mathcal{B}(\Sigma_{k+1})$ and $f = (f_0, \dots, f_k)$. Since

$$\int_{\Sigma_{k+1}} x_i\, \eta(dx) = \int \frac{dP_i}{dQ}\, dQ = 1, \quad i = 0, 1, \dots, k,$$

η is a standard measure, $\eta \in \mathcal{H}_{k+1}$.

Taking into account that the measures $P_0, P_1, \dots, P_k$ are related to the experiment $\mathcal{E}$, we write $H(a; \mathcal{E})$ instead of $H(a; \eta)$. This expression is called *the Hellinger transformation of the experiment* $\mathcal{E}$.

It is clear that by the chain rule for Lebesgue integrals

$$H(a; \mathcal{E}) = \int_\Omega \left(\frac{dP_0}{dQ}\right)^{a_0} \dots \left(\frac{dP_k}{dQ}\right)^{a_k} dQ = \int_\Omega f_1^{a_1} \dots f_k^{a_k}\, dQ, \tag{1.72}$$

with $a = (a_0, a_1, \dots, a_k)$, $a_i \geq 0$, $i = 0, 1, \dots, k$ and $\sum a_i = 1$.

The transformation $H(a; \mathcal{E})$ formally depends on the measure Q and this dependence should be show explicitly, for example, denoting this transformation as $H_Q(a; \mathcal{E})$. But really if Q' is another measure such that $P_i \ll Q'$, $i = 0, 1, \dots, n$ then $H'_Q(a; \mathcal{E}) = H_Q(a; \mathcal{E})$. In fact, it is sufficient to prove that if $Q \ll Q'$ then

$$\int_\Omega \prod_{i=0}^{k} \left(\frac{dP_i}{dQ'}\right)^{a_i} dQ' = \int_\Omega \prod_{i=0}^{k} \left(\frac{dP_i}{dQ}\right)^{a_i} dQ.$$

Let $Z = dQ/dQ'$. Then $dP_i/dQ' = (dP_i/dQ)\, Z$ and, since $a_0 + a_1 + \dots + a_k = 1$, we have

$$\begin{aligned}
\int_\Omega \prod_{i=0}^{k} \left(\frac{dP_i}{dQ}\right)^{a_i} dQ &= \int_\Omega \prod_{i=0}^{k} \left(\frac{dP_i}{dQ}\right)^{a_i} Z\, dQ' \\
&= \int_\Omega \prod_{i=0}^{k} \left(\frac{dP_i}{dQ} Z\right)^{a_i} dQ' = \int_\Omega \prod_{i=0}^{k} \left(\frac{dP_i}{dQ'}\right)^{a_i} dQ'.
\end{aligned}$$

Hence the right-hand side of (1.72) really does not depend on the measure Q, if only the property $P_i \ll Q$, $i = 0, 1, \dots, n$ does hold.

The definition of the Hellinger transformation (1.72) is often rewritten in the following *symbolic* form

$$H(a; \mathcal{E}) = \int_\Omega (dP_0)^{a_0} \dots (dP_k)^{a_k}\,. \tag{1.73}$$

In particular, for a *binary* experiment $\mathcal{E} = (\Omega, \mathcal{F}; P_0, P_1)$ putting $a_0 = \alpha$, $a_1 = 1 - \alpha$ and denoting the value $H((\alpha, 1-\alpha); \mathcal{E})$ by $H(\alpha; P_0, P_1)$ or simply $H(\alpha)$ we obtain

$$H(\alpha) = \int_\Omega (dP_0)^\alpha\, (dP_1)^{1-\alpha}\,. \tag{1.74}$$

If we introduce a probability measure $Q = \frac{1}{2}(P_0 + P_1)$, and denote by E_Q the integration with respect to this measure then

$$H(\alpha) = E_Q \zeta_0^{\alpha} \zeta_1^{1-\alpha} \tag{1.75}$$

with

$$\zeta_0 = \frac{dP_0}{dQ}, \qquad \zeta_1 = \frac{dP_1}{dQ}.$$

The value $H(\alpha; P_0, P_1)$ is called *the Hellinger integral of the order* α, $0 \le \alpha \le 1$. In the case of $\alpha = 1/2$ the value

$$H\left(\frac{1}{2}; P_0, P_1\right) = \int_\Omega \sqrt{dP_0}\sqrt{dP_1}$$

is often called *an affinity* between the measures P_0 and P_1 and denoted by $H(P_0, P_1)$.

The affinity $H(P_0, P_1)$ is closely related with another important notion, the *Kakutani-Hellinger* distance $\rho(P_0, P_1)$ between the measures P_0 and P_1, which is defined as a (non-negative) number such that

$$\rho^2(P_0, P_1) = \frac{1}{2}\int_\Omega \left(\sqrt{dP_0} - \sqrt{P_1}\right)^2 \tag{1.76}$$

where, similarly to (1.74), the integral is treated as $E_Q\left(\sqrt{\zeta_0} - \sqrt{\zeta_1}\right)^2$. In other words, by definition

$$\rho^2(P_0, P_1) = \frac{1}{2} E_Q \left(\sqrt{\zeta_0} - \sqrt{\zeta_1}\right)^2. \tag{1.77}$$

Comparing (1.75) and (1.76) we conclude that

$$\rho^2(P_0, P_1) = 1 - H(P_0, P_1). \tag{1.78}$$

It is not difficult to verify that the Kakutani-Hellinger distance really satisfies the axioms of "distance" on the set of all probability measures and hence, defines a metric on this set which is closely connected to the *variation* distance $\mathrm{Var}(P_0, P_1)$ between the measures P_0 and P_1. Indeed, if μ is a finite signed measure on $(\Omega, \mathcal{F})$, then the *total variation* $||\mu||$ is defined by

$$||\mu|| = \sup\left\{\int_\Omega \varphi(\omega)\, \mu(d\omega)\right\} \tag{1.79}$$

where sup is taken over the set of all measurable functions $\varphi = \varphi(\omega)$ such that $\varphi(\omega) \le 1$.

It is well known (see, for example, Jacod and Shiryaev (1987, Section V.4a), that

$$2\rho^2(P_0, P_1) \le ||P_0 - P_1|| \le 2\sqrt{2}\rho(P_0, P_1). \tag{1.80}$$

Therefore the variation distance and the Kakutani–Hellinger distance produce the same topology on the set of probability measures:

$$\|P_n - P\| \to 0 \iff \rho^2(P_n, P) \to 0, \quad n \to \infty. \tag{1.81}$$

From (1.78) and (1.80) it follows also that

$$2\,[1 - H(P_0, P_1)] \le \|P_0 - P_1\| \le \sqrt{8[1 - H(P_0, P_1)]}. \tag{1.82}$$

Let us mention some of the most important and easily verified properties of the *Hellinger transformation* $H(a;\mathcal{E})$:

1) $0 \le H(a;\mathcal{E}) \le 1$;
2) $\log H(a;\mathcal{E})$ is a convex function, i.e. for $0 \le \lambda \le 1$

$$\log H(\lambda a + (1-\lambda)b;\mathcal{E}) \le \lambda \log H(a;\mathcal{E}) + (1-\lambda)\log H(b;\mathcal{E});$$

3) if $\mathcal{E} \times \widetilde{\mathcal{E}} = (\Omega \times \widetilde{\Omega}, \mathcal{F} \times \widetilde{\mathcal{F}}; P_i \times \widetilde{P}_i \;\; i = 0,1,\ldots,k)$ is the direct product of experiments $\mathcal{E} = (\Omega, \mathcal{F}; P_i, i = 0,1,\ldots,k)$ and $\widetilde{\mathcal{E}} = (\widetilde{\Omega}, \widetilde{\mathcal{F}};\, \widetilde{P}_i, i = 0,1,\ldots,k)$ then

$$H(a;\mathcal{E} \times \widetilde{\mathcal{E}}) = H(a;\mathcal{E})\, H(a;\widetilde{\mathcal{E}})$$

(the multiplicative property).

Now we turn to the *Mellin transformation* (Strasser (1985, Section 5)) $M(a;\mathcal{E})$ of an experiment $\mathcal{E} = (\Omega, \mathcal{F};\, P_0, P_1, \ldots, P_k)$. If we define

$$Z_i = \frac{dP_i}{dP_0}, \quad i = 1,\ldots,k,$$

and denote by $\mu = \mathcal{L}(Z_1,\ldots,Z_k \mid P_0)$ the distribution of the *likelihood* vector $Z = (Z_1,\ldots,Z_k)$ with respect to the measure P_0 then we see that $\mu \in \mathcal{M}_k$ and, hence, for $b \in \Delta_k$, the Mellin transformation $M(b;\mu)$ is well defined. We denote it by $M(b;\mathcal{E})$ and call *the Mellin transformation of the experiment* $\mathcal{E}$ (with respect to the measure P_0). Thus for $b = (b_1,\ldots,b_k) \in \Delta_k$

$$M(b;\mathcal{E}) = \int_{R_+^k} x_1^{b_1}\ldots x_k^{b_k}\,\mu(dx) = \int_\Omega Z_1^{b_1}\ldots Z_k^{b_k}\,P_0(d\omega) \quad \left(= E_0 Z_1^{b_1}\ldots Z_k^{b_k}\right).$$

In order to prove relations between the Hellinger transformation

$$H(a;\mathcal{E}) = \int_\Omega (P_0(d\omega))^{a_0}\ldots(P_k(d\omega))^{a_k} \tag{1.83}$$

with $a = (a_0, \dots, a_k) \in \Sigma_{k+1}$, and the Mellin transformation $M(b;\mathcal{E})$, $b = (b_1, \dots, b_k) \in \Delta_k$ we notice that if $a_0 > 0$, then

$$\begin{aligned} H(a;\mathcal{E}) &= \int_\Omega (P_0(d\omega))^{1-(a_1+\dots+a_k)} (P_1(d\omega))^{a_1} \dots (P_k(d\omega))^{a_k} \\ &= \int_\Omega \left(\frac{dP_1}{dP_0}\right)^{a_1} \dots \left(\frac{dP_k}{dP_0}\right)^{a_k} P_0(d\omega) \\ &= E_0\, Z_1^{a_1} \dots Z_k^{a_k} \\ &= M(b;\mathcal{E}) \end{aligned}$$

for $b = (b_1, \dots, b_k) = (a_1, \dots, a_k)$.

It is also useful to notice that if

$$L_i = \log Z_i\,, \quad i = 1, \dots, k,$$

then

$$E_0\, Z_1^{a_1} \dots Z_k^{a_k} = E_0 \exp\left\{\sum_{i=1}^{k} a_i\, L_i\right\}.$$

In other words, the Hellinger transformation $H(a;\mathcal{E})$ coincides with the Laplace transformation of the distribution $\mathcal{L}(Z_1, \dots, Z_k \mid P_0)$.

For the case of binary experiments $\mathcal{E} = (\Omega, \mathcal{F}; P_0, P_1)$, it holds for every $0 \le \alpha \le 1$

$$H(\alpha) = E_Q\, \zeta_0^\alpha \zeta_1^{1-\alpha} = E_0\, Z_1^{1-\alpha} \quad (= M(1-\alpha;\mathcal{E}))\,.$$

3. Here we present some examples where explicit expressions for the Hellinger and Mellin transformations are available.

Example 1.24 Let

$$P = P_1 \times P_2 \times \dots\,, \qquad \widetilde{P} = \widetilde{P}_1 \times \widetilde{P}_2 \times \dots$$

with Gaussian measures P_k and $\widetilde{P}_k$ on $(R, \mathcal{B}(R))$ be such that $P_k \sim \mathcal{N}(a_k, 1)$, $\widetilde{P}_k \sim \mathcal{N}(\widetilde{a}_k, 1)$. It is clear that

$$H(\alpha; P \times \widetilde{P}) = \prod_{k=1}^{\infty} H(\alpha; P_k \times \widetilde{P}_k)$$

where we find by direct calculation for $0 \le \alpha \le 1$

$$H(\alpha; P_k \times \widetilde{P}_k) = \exp\left\{-\frac{\alpha(1-\alpha)}{2}(a_k - \widetilde{a}_k)^2\right\}.$$

Hence

$$H(\alpha; P \times \widetilde{P}) = \exp\left\{-\frac{\alpha(1-\alpha)}{2}\sum_{k=1}^{\infty}(a_k - \widetilde{a}_k)^2\right\}. \tag{1.84}$$

Example 1.25 Set again $P = P_1 \times P_2 \times \dots, \widetilde{P} = \widetilde{P}_1 \times \widetilde{P}_2 \times \dots$ but let now P_k and $\widetilde{P}_k$ be the Poisson distributions with the parameters λ_k and $\widetilde{\lambda}_k$ respectively. By simple calculation one has

$$H\left(\frac{1}{2}; P, \widetilde{P}\right) = \exp\left\{-\frac{1}{2}\sum_{k=1}^{\infty}\left(\sqrt{\lambda_k} - \sqrt{\widetilde{\lambda}_k}\right)^2\right\}.$$

Example 1.26 Let $(\Omega, \mathcal{F}, P_0)$ be a probability space and $X = (X_1, \dots, X_k)$ be a Gaussian vector such that $E_0 X_i = 0$ and $E_0 X_i X_j = K(i,j)$, $i,j = 1, \dots, k$. Define the measures $P_1, \dots, P_k$ by expressions

$$dP_i = \exp\left\{X_i - \frac{1}{2}E_0 X_i^2\right\} dP_0.$$

According to the definition from Section 1.1 the experiment $\mathcal{E} = (\Omega, \mathcal{F}; P_0, P_1, \dots, P_k)$ is a Gaussian shift.

Let $Z_i = dP_i/dP_0$ and $\mu = \mathcal{L}(Z_1, \dots, Z_k \mid P_0)$. Using that X is a Gaussian vector X under the measure P_0 we easily derive the expression for the Mellin transformation $M(a; \mathcal{E}) = M(a; \mu)$, $a \in \Delta_k$:

$$\begin{aligned}
M(a; \mathcal{E}) &= \int_\Omega \prod_{i=1}^k Z_i^{a_i}\, dP_0 \\
&= E_0 \exp\left\{\sum_{i=1}^k a_i \left(X_i - \frac{1}{2}E_0 X_i^2\right)\right\} \\
&= \exp\left\{-\frac{1}{2}\sum_{i=1}^k a_i E_0 X_i^2 + \frac{1}{2}\sum_{i,j=1}^k a_i a_j K(i,j)\right\} \\
&= \exp\left\{-\frac{1}{2}\sum_{i,j=1}^k a_i a_j K(i,j) - \frac{1}{2}\sum_{i=1}^k a_i K(i,i)\right\}.
\end{aligned}$$

1.12 Absolutely Continuous and Contiguous Probability Measures

1. Let $\mathcal{E} = (\Omega, \mathcal{F}; P, \widetilde{P})$ be a binary experiment. For various statistical problems related to such kind of experiments it is important to find simple criteria of *absolute continuity*, *equivalence* (mutual absolute continuity), *singularity* of the the measures P and $\widetilde{P}$. The notion of the Hellinger integral $H(\alpha) = H(\alpha; P, \widetilde{P})$ introduced in Section 1.11 gives us a convenient analytic tool to investigate these questions.

Recall that the measure $\widetilde{P}$ is *absolutely continuous* with respect to the measure P ($\widetilde{P} \ll P$) iff $P(A) = 0 \iff \widetilde{P}(A) = 0$ for $A \in \mathcal{F}$; the measures P and $\widetilde{P}$ are *equivalent* iff $P \ll \widetilde{P}$ and $\widetilde{P} \ll P$; the measures P and $\widetilde{P}$ are *singular* or *orthogonal* ($P \perp \widetilde{P}$) if there exists such a set $A \in \mathcal{F}$ that $P(A) = 1$ and $\widetilde{P}(\Omega \backslash A) = 1$.

Let P and $\widetilde{P}$ be two probability measure on $(\Omega, \mathcal{F})$ and let $\zeta = dP/dQ$, $\widetilde{\zeta} = d\widetilde{P}/dQ$ with $Q = (P + \widetilde{P})/2$.

Theorem 1.19 *The following conditions are equivalent:*

(a) $\widetilde{P} \ll P$;

(b) $\widetilde{P}(\zeta > 0) = 1$;

(c) $H(\alpha; P, \widetilde{P}) \to 1, \quad \alpha \downarrow 0$.

Theorem 1.20 *The following conditions are equivalent:*

(a) $\widetilde{P} \perp P$;

(b) $\widetilde{P}(\zeta > 0) = 0$;

(c) $H(\alpha; P, \widetilde{P}) \to 0, \quad \alpha \downarrow 0$;

(d) $H(\alpha; P, \widetilde{P}) = 0$ *for all* $\alpha \in (0,1)$;

(e) $H(\alpha; P, \widetilde{P}) = 0$ *for some* $\alpha \in (0,1)$.

Proof. See, for example, in Jacod and Shiryaev (1988, Section V.1a). □

Example 1.27 Let P and $\widetilde{P}$ be two Gaussian measures on $(R^\infty, \mathcal{B}(R^\infty))$ defined in Example 1.24, Section 1.11. Using the equivalence of the assertions (a) and (c) in the theorems above we obtain from (1.84) the following result on equivalence and singularity of the measures P and $\widetilde{P}$:

$$P \sim \widetilde{P} \iff \sum_{k=1}^{\infty} (a_k - \widetilde{a}_k)^2 < \infty,$$

$$P \perp \widetilde{P} \iff \sum_{k=1}^{\infty} (a_k - \widetilde{a}_k)^2 = \infty.$$

2. Let $\mathcal{E}^n = (\Omega^n, \mathcal{F}^n; P^n, \widetilde{P}^n)$, $n \geq 1$, be a sequence of binary statistical experiments.

Definition 1.23 A sequence of measures $(\widetilde{P}^n)_{n\geq 1}$ is *contiguous* to a sequence $(P^n)_{n\geq 1}$ (denoted by $(\widetilde{P}^n) \lhd (P^n)$) if for any sequence of sets $A^n \in \mathcal{F}^n$, $n \geq 1$, with $P^n(A^n) \to 0$, one has $\widetilde{P}^n(A^n) \to 0$. If $(\widetilde{P}^n) \lhd (P^n)$ and $(P^n) \lhd (\widetilde{P}^n)$, then the sequences $(\widetilde{P}^n)$ and (P^n) are *mutually contiguous* (denoted by $(\widetilde{P}^n) \lhd\rhd (P^n)$).

Definition 1.24 Two sequences $(\widetilde{P}^n)$ and (P^n) are *entirely separated* if there exist a subsequence (n') and sets $A^{n'} \in \mathcal{F}^{n'}$ such that $P^{n'}(A^{n'}) \to 1$ and $\widetilde{P}^{n'}(A^{n'}) \to 0$ (denoted by $(\widetilde{P}^n)\Delta(P^n)$).

It is clear that the notions of "contiguity", "mutual contiguity" and "entire separation" are asymptotic analogs of the notions of "absolute continuity", "equivalence" and "singularity".

Denote $Q^n = (P^n + \widetilde{P}^n)/2$, $\zeta^n = dP^n/dQ^n$, $\widetilde{\zeta^n} = d\widetilde{P}^n/dQ^n$, $Z^n = \widetilde{\zeta^n}/\zeta^n$.

Theorem 1.21 *The following conditions are equivalent:*

(*a*) $(\widetilde{P}^n) \lhd (P^n)$;

(*b*) *the family of distributions* $\left\{\mathcal{L}(\zeta^{-n} \mid \widetilde{P}^n)\right\}_{n\geq 1}$ *is tight, i.e.*

$$\lim_{N\to\infty} \limsup_n \widetilde{P}^n\left(\frac{1}{\zeta^n} > N\right) = 0;$$

(*b'*) *the family of distributions* $\left\{\mathcal{L}(Z^n \mid \widetilde{P}^n)\right\}_{n\geq 1}$ *is tight, i.e.*

$$\lim_{N\to\infty} \limsup_n \widetilde{P}^n\left(Z^n > N\right) = 0;$$

(*c*) $\lim_{\alpha\downarrow 0} \liminf_n H(\alpha; P^n, \widetilde{P}^n) = 1$.

Theorem 1.22 *The following assertions are equivalent:*

(*a*) $(\widetilde{P}^n)\,\Delta\,(P^n)$;

(*b*) $\liminf_n \widetilde{P}^n\{\zeta^n \geq \varepsilon\} = 0, \quad \forall\varepsilon > 0$;

(*b'*) $\limsup_n \widetilde{P}^n\{Z^n \leq N\} = 0, \quad \forall N > 0$;

(*c*) $\lim_{\alpha\downarrow 0} \liminf_n H(\alpha; P^n, \widetilde{P}^n) = 0$;

(*d*) *for all* $\alpha \in (0,1)$ $\quad \liminf_n H(\alpha; P^n, \widetilde{P}^n) = 1$;

(*e*) *for some* $\alpha \in (0,1)$ $\quad \liminf_n H(\alpha; P^n, \widetilde{P}^n) = 1$.

Proof. See, for example, in Jacod and Shiryaev (1988, Section V.1a). □

3. The following assertions stated by Le Cam (1960) play an important role in many questions of asymptotic statistics.

Let $\mathcal{E}^n = (\Omega^n, \mathcal{F}^n; P^n, \widetilde{P}^n)$, $n \geq 1$ be a sequence of binary statistical experiments, let $Q^n = \frac{1}{2}(P^n + \widetilde{P}^n)$, $\zeta^n = dP^n/dQ^n$, $\widetilde{\zeta^n} = d\widetilde{P}^n/dQ^n$, $Z^n = \widetilde{\zeta^n}/\zeta^n$.

Lemma 1.5 *(The "first Le Cam Lemma") Let the distributions* $\mu^n = \mathcal{L}(Z^n \mid P^n$, $n \geq 1$, *weakly converge to a probability distribution* μ *on* $(R, \mathcal{B}(R))$. *Then* $(\widetilde{P}^n) \lhd (P^n)$ *iff* $\int_R x\,\mu(dx) = 1$.

Lemma 1.6 *(The "third Le Cam Lemma")* *(A) The conditions* (i) *and* (ii) *below are equivalent:*

(i) *The distributions* $\mu^n = \mathcal{L}(Z^n \mid P^n)$, $n \geq 1$, *weakly converge to a probability measure* μ *on* $(R, \mathcal{B}(R))$ *and* $(\widetilde{P}^n) \lhd (P^n)$.

(ii) *The distributions* $\widetilde{\mu}^n \doteq \mathcal{L}(Z^n \mid \widetilde{P}^n)$, $n \geq 1$, *weakly converge to a probability measure* $\widetilde{\mu}$ *on* $(R, \mathcal{B}(R))$.

(B) Let any of the conditions (i) *or* (ii) *hold. Then* $\widetilde{\mu}(dz) = z\,\mu(dz)$. *Moreover, if* X^n *is a random variable on* $(\Omega^n, \mathcal{F}^n)$, $n \geq 1$, *with values in some metric space* E *and* $\mathcal{L}((Z^n, X^n) \mid P^n)$ *weakly converges to some probability measure* $\nu = \nu(dz, dx)$ *on* $R_+ \times E$ *then* $\mathcal{L}((Z^n, X^n) \mid \widetilde{P}^n)$ *weakly converges to* $\widetilde{\nu} = \widetilde{\nu}(dz, dx)$ *with* $\widetilde{\nu}(dz, dx) = z\,\nu(dz, dx)$.

Proof. See, e.g. the monographs Hajek and Sidak (1969) or Strasser (1985). □

Remark 1.8 The so called "second Le Cam Lemma" describes an asymptotic expansion of the likelihood ratios for differentiable (in L_2) experiments; for more details see Section 2.10.

Chapter 2

Convergence of Statistical Experiments

2.1 Strong ($\Delta-$) Convergence

1. The present chapter is devoted to the systematic study of various kinds of convergence of statistical experiments $\mathcal{E}^n$, $n \to \infty$, to a "limit" experiment $\mathcal{E}$ and their consequences related to the question of "extension" of some optimal properties of the limit experiments $\mathcal{E}$ to the experiments $\mathcal{E}^n$ for "small", "mean" and "large" values of n.

We begin with the *strong* convergence of experiments, then pass to the so-called *weak* convergence and, finally, introduce the (new) notion of the λ-convergence which particular case is the well known condition of *Local Asymptotic Normality* (LAN). It is natural that a great attention is paid to the question of relations between these kinds of convergence, to their probabilistic and statistical sense and to their applications.

Definition 2.1 A sequence of statistical experiments

$$\mathcal{E}^n = (\Omega^n, \mathcal{F}^n;\ P_\theta^n, \theta \in \Theta),\ n \geq 1,$$

converges strongly to an experiment $\mathcal{E} = (\Omega, \mathcal{F};\ P_\theta, \theta \in \Theta)$ if

$$\Delta(\mathcal{E}^n, \mathcal{E}) \to 0, \quad n \to \infty. \tag{2.1}$$

The convergence (2.1) was introduced in Le Cam (1972, 1979) and it is also called Δ-*convergence* and is denoted as

$$\mathcal{E}^n \xrightarrow{\Delta} \mathcal{E}, \quad n \to \infty.$$

For many statistical problems the condition of Δ-convergence turns out to be rather strong very often it is simply not satisfied. Nevertheless, one can use the following less restrictive definition of strong convergence on some classes $\mathcal{K} = \{K : K \subseteq \Theta\}$ of subsets K in the parameter set Θ. This condition is a natural generalization of Definition 2.1 but it is often satisfied and usually sufficient for many statistical applications.

Definition 2.2 Let $K \subseteq \Theta$ and $\mathcal{E}^n_K = (\Omega^n, \mathcal{F}^n; P^n_\theta, \theta \in K)$, $n \geq 1$, $\mathcal{E}_K = (\Omega, \mathcal{F}; P_\theta, \theta \in K)$. If $\Delta(\mathcal{E}^n_K, \mathcal{E}_K) \to 0$, then we say that the experiments $\mathcal{E}^n$ converge to $\mathcal{E}$ on the set K. This convergence is denoted as

$$\Delta(\mathcal{E}^n, \mathcal{E}; K) \to 0 \quad \text{or} \quad \mathcal{E}^n \overset{\Delta(K)}{\longrightarrow} \mathcal{E}, \ n \to \infty.$$

If $\Delta(\mathcal{E}^n_K, \mathcal{E}_K) \to 0$ for *all* sets K from some class (collection) of sets $\mathcal{K}$ then we say that the experiments $\mathcal{E}^n$ *converge strongly* to $\mathcal{E}$ on the class $\mathcal{K}$. This convergence is denoted as

$$\Delta(\mathcal{E}^n, \mathcal{E}; \mathcal{K}) \to 0 \quad \text{or} \quad \mathcal{E}^n \overset{\Delta(\mathcal{K})}{\longrightarrow} \mathcal{E}, \ n \to \infty.$$

Let now $\mathbf{A}$ be a topological *decision space* with a Baire σ-field $\mathcal{A} = \mathcal{B}_0(\mathbf{A})$. Define

$\mathcal{W}^b_c$, the class of *continuous* (in a) *bounded* loss functions $W = W_\theta(a)$;
$\mathcal{W}^b_{lc}$, the class of *lower semicontinuous* (in a) *bounded* loss functions $W = W_\theta(a)$;
$\mathcal{W}_{lc}$, the class of *lower semicontinuous* (in a) loss functions $W = W_\theta(a)$.

(Recall that due to Section 1.4 all the considered loss functions $W = W_\theta(a)$ are supposed to be bounded from below: $\inf_a W_\theta(a) > -\infty$, $\theta \in \Theta$.)

If $\beta(W_\theta, P_\theta)$ is a bilinear functional, $W \in \mathcal{W}^b_c$, then (according to Section 1.4) the value $R_\theta(\mathcal{E}, W, \beta) = \beta(W_\theta, P_\theta)$ is called the *risk of the decision* β in the experiment $\mathcal{E}$ at a point θ for the loss function W; if $W \in \mathcal{W}_{lc}$ then by definition $R_\theta(\mathcal{E}, W, \beta) = \sup\{R_\theta(\mathcal{E}, V, \beta) : V \leq W, V \in \mathcal{W}^b_c\}$. (If $\mathbf{A}$ is a separable metric space then for $W \in \mathcal{W}_{lc}$ there always exists a sequence $V_n \in \mathcal{W}^b_c$ such that $V_n \uparrow W$.)

Recall also notation used in Chapter 1:

$$\mathcal{R}(\mathcal{E}, W) = \inf_\beta \sup_{\theta \in \Theta} R_\theta(\mathcal{E}, W, \beta) \tag{2.2}$$

for the *minimax* risk;

$$\mathcal{R}_\pi(\mathcal{E}, W) = \inf_\beta \int R_\theta(\mathcal{E}, W, \beta)\, \pi(d\theta), \quad \pi \in \Pi^0, \tag{2.3}$$

for π-*Bayes* (or, simply, for *Bayes*) risk.

If $K \subseteq \Theta$ then we define

$$\mathcal{R}(\mathcal{E}, W; K) = \mathcal{R}(\mathcal{E}_K, W) \quad \left(= \inf_\beta \sup_{\theta \in K} R_\theta(\mathcal{E}, W, \beta)\right) \tag{2.4}$$

and

$$\mathcal{R}(\mathcal{E}, W; \mathcal{K}) = \sup_{K \in \mathcal{K}} \mathcal{R}(\mathcal{E}_K, W) \quad \left(= \sup_{K \in \mathcal{K}} \inf_\beta \sup_{\theta \in K} R_\theta(\mathcal{E}, W, \beta)\right). \tag{2.5}$$

Similar notations $\mathcal{R}_\pi(\mathcal{E}, W; K)$ and $\mathcal{R}_\pi(\mathcal{E}, W; \mathcal{K})$ are used for the π-Bayes risks.

Let also $\Pi = \Pi(\Theta)$ be the set of *all* probability measures on $(\Theta, \mathcal{B}(\Theta))$ and let $\Pi^0 = \Pi^0(\Theta)$ be the *subset* in Π of measures on Θ with finite supports.

Theorem 1.10 in Section 1.7 yields the following assertions.

Theorem 2.1 *1. Let $K \subseteq \Theta$ and $\Delta(\mathcal{E}^n, \mathcal{E}; K) \to 0$.*

(i) *If $W \in \mathcal{W}_c^b$ then*

$$\lim_{n\to\infty} \mathcal{R}(\mathcal{E}^n, W; K) = \mathcal{R}(\mathcal{E}, W; K) \tag{2.6}$$

and

$$\lim_{n\to\infty} \mathcal{R}_\pi(\mathcal{E}^n, W; K) = \mathcal{R}_\pi(\mathcal{E}, W; K), \quad \pi \in \Pi^0. \tag{2.7}$$

(ii) *If $W \in \mathcal{W}_{lc}$ then*

$$\lim_{n\to\infty} \mathcal{R}(\mathcal{E}^n, W; K) \geq \mathcal{R}(\mathcal{E}, W; K) \tag{2.8}$$

and

$$\lim_{n\to\infty} \mathcal{R}_\pi(\mathcal{E}^n, W; K) \geq \mathcal{R}_\pi(\mathcal{E}, W; K), \quad \pi \in \Pi^0. \tag{2.9}$$

2. If $\mathcal{K}$ is a class of subsets in Θ and $\Delta(\mathcal{E}^n, \mathcal{E}; K) \to 0$ for all $K \in \mathcal{K}$ then the assertions (2.6) - (2.9) remain valid after substituting K by $\mathcal{K}$.

Proof. The assertions (2.6) and (2.7) are the direct consequence of the implications $(D) \Longleftrightarrow (B)$ and $(D) \Longleftrightarrow (C)$ of Theorem 1.10, Section 1.7. The assertion (2.8) (and similarly (2.9)) follows from the fact that if $V \in \mathcal{W}_c^b$ and $V \leq W$ then

$$\begin{aligned}
\mathcal{R}(\mathcal{E}, V; K) &= \inf_\beta \sup_{\theta\in K} R_\theta(\mathcal{E}, V, \beta) \\
&= \lim_{n\to\infty} \inf_\beta \sup_{\theta\in K} R_\theta(\mathcal{E}^n, V, \beta) \\
&\leq \liminf_{n\to\infty} \inf_\beta \sup_{\theta\in K} R_\theta(\mathcal{E}^n, W, \beta) \\
&= \liminf_{n\to\infty} \mathcal{R}(\mathcal{E}^n, W; K),
\end{aligned}$$

and, therefore, for $\mathcal{R}(\mathcal{E}, W; K) = \sup\{\mathcal{R}(\mathcal{E}, V; K) : V \leq W,\ V \in \mathcal{W}_c^b\}$ we obtain the required inequality (2.8). □

2. In statistical applications one often consider "conditional" settings to define optimal or asymptotic optimal decisions when there are some constraints ("conditions") on the class of considered decisions. One of the possible examples, related to the hypothesis testing, is considered at the end of this section.

Let $\mathcal{E} = (\Omega, \mathcal{F}; P_\theta, \theta \in \Theta)$ be a statistical experiment, let $\mathbf{A}$ be a decision space, and let $\mathfrak{B}(\mathcal{E}, \mathbf{A})$ be the corresponding set of decisions. Let then a collection of functions $U = \{U_\theta(a), \theta \in \Theta\}$ on $\mathbf{A}$ be given. Suppose, first, that $U \in \mathcal{W}_c^b$. Define in the set $\mathfrak{B}(\mathcal{E}, \mathbf{A})$ the subset $\mathfrak{B}_U(\mathcal{E}, \mathbf{A})$ of such decisions β that $\beta(U_\theta, P_\theta) \leq 0$, $\theta \in \Theta$.

Similarly, if $U \in \mathcal{W}_{lc}$, then $\mathfrak{B}_U(\mathcal{E}, \mathbf{A})$ is treated as the set $\bigcap \mathfrak{B}_V(\mathcal{E}, \mathbf{A})$ with the intersection over all the functions V with $V \le U$ and $V \in \mathcal{W}_c^b$. Obviously the set $\mathfrak{B}_U(\mathcal{E}, \mathbf{A})$ is compact and convex (the same as for $\mathfrak{B}(\mathcal{E}, \mathbf{A})$).

Introduce now for an experiment $\mathcal{E}$, for a loss function $W \in \mathcal{W}_{lc}$ and for a "constraint" function $U \in \mathcal{W}_{lc}$ the "conditional" minimax and the "conditional" Bayes risks

$$\mathcal{R}_U(\mathcal{E}, W) = \inf_{\beta \in \mathfrak{B}_U(\mathcal{E},\mathbf{A})} \sup_{\theta \in \Theta} R_\theta(\mathcal{E}, W, \beta),$$

$$\mathcal{R}_{\pi,U}(\mathcal{E}, W) = \inf_{\beta \in \mathfrak{B}_U(\mathcal{E},\mathbf{A})} \left[\int R_\theta(\mathcal{E}, W, \beta)\, \pi(d\theta) \right], \quad \pi \in \Pi.$$

Let a sequence of experiments $\mathcal{E}^n$ be such that $\Delta(\mathcal{E}^n, \mathcal{E}) \to 0$. Then all the assertions of Theorem 2.1 remain valid if one substitutes the risks $\mathcal{R}(\mathcal{E}, W)$ and $\mathcal{R}_\pi(\mathcal{E}, W)$ by the "conditional" risks $\mathcal{R}_U(\mathcal{E}, W)$ and $\mathcal{R}_{\pi,U}(\mathcal{E}, W)$ and similarly for the experiments $\mathcal{E}^n$. Strictly speaking, for proving this assertion we have to formulate the corresponding "conditional" version of Theorem 1.10 in Section 1.7. But the inspection of its proof shows that we really use only the properties of compactness and convexity of the sets $\mathfrak{B}(\mathcal{E}, \mathbf{A})$ and $\mathfrak{B}(\mathcal{E}^*, \mathbf{A})$ and these properties remain true if we pass to the smaller sets $\mathfrak{B}_U(\mathcal{E}, \mathbf{A})$ and $\mathfrak{B}_U(\mathcal{E}^*, \mathbf{A})$.

3. It follows from (2.8) and (2.9) that when we extend the results (2.6) and (2.7) to the case of an *unbounded* loss function W the corresponding asymptotic *equalities* for the risks become, in general, *inequalities*. However, it turns out that for many unbounded loss functions $W = W_\theta(a)$ (like $W_\theta(a) = \|a - \theta\|^p$, $p > 0$ typically used in estimation problems) the assertions like (2.6), (2.7) remain valid as well under some additional conditions.

Let $K \subseteq \Theta$. The assumption $\theta \in K$ means that the statistician has the a priori information that "the unknown value of the parameter θ belongs to the set K". We say that a decision a from a decision space $\mathbf{A}$ is *K-inadmissible* if there exists another decision $b \in \mathbf{A}$ such that

$$\inf_{\theta \in K} W_\theta(a) > \sup_{\theta \in K} W_\theta(b). \tag{2.10}$$

A point a is *K-admissible* if it is not K-inadmissible. The set of all K-admissible points in $\mathbf{A}$ for a loss function W is denoted by $D(W, K)$. It is easy to see that

$$D(W, K) = \left\{ a \in \mathbf{A} : \inf_{\theta \in K} W_\theta(a) \le \inf_{b \in \mathbf{A}} \sup_{\theta \in K} W_\theta(b) \right\}.$$

This definition is of clear meaning: if the statistician has a priori information that $\theta \in K$ it is worthwhile to restrict oneself to the set of the decisions with values in the set $D(W, K)$.

Let $\mathcal{K}$ be a class of subsets in Θ, let W be a loss function. Introduce the following condition:

($W\mathcal{K}$) For any $K \in \mathcal{K}$

$$\sup_{b \in D(W,K)} \sup_{\theta \in K} W_\theta(b) < \infty.$$

Remark 2.1 The condition ($W\mathcal{K}$) is obviously satisfied for a bounded loss function W ($\|W\| < \infty$). But a loss function W can be also unbounded on the whole decision space $\mathbf{A}$; however, under this condition it is bounded if we restrict the parameter set to $K \in \mathcal{K}$ and the decision space to $D(W, K)$.

Example 2.1 Let Θ be a linear Banach space, let $\mathbf{A} = \Theta$,

$$W_\theta(a) = g(\|a - \theta\|),$$

where $g = g(x)$ is a monotonous function on R_+ and let a class of sets $\mathcal{K}$ be such that for any $K \in \mathcal{K}$

$$\sup_{a,b \in K} \|a - b\| < \infty.$$

We want to show that under these assumptions the condition ($W\mathcal{K}$) is satisfied. In fact, let $K \in \mathcal{K}$, let a be an arbitrary point from K and let $A_K(a) = \sup_{b \in K} \|a - b\|$.

Let a point $c \in \mathbf{A}$ be such that $\|c - a\| > 2\,A_K(a)$. Then by virtue of the properties of the function $g = g(x)$

$$\begin{aligned}
\inf_{\theta \in K} W_\theta(c) &= \inf_{\theta \in K} g(\|c - \theta\|) \\
&\geq \inf_{\theta \in K} g(\|c - a\| - \|a - \theta\|) \\
&> g(A_K(a)) = g(\sup_{\theta \in K} \|a - \theta\|) \\
&\geq \sup_{\theta \in K} g(\|a - \theta\|) \\
&= \sup_{\theta \in K} W_\theta(a).
\end{aligned}$$

Hence, each point c with $\|c - a\| > 2\,A_K(a)$ for some fixed point a from K is K-inadmissible and, therefore, such points are not included in the set $D(W, K)$. On the other hand, if c is such that $\|c - a\| \leq 2\,A_K(a)$ then

$$\sup_{\theta \in K} W_\theta(c) = \sup_{\theta \in K} g(\|c - \theta\|) \leq \sup_{\theta \in K} g(\|c - a\| + \|a - \theta\|) \leq g(3\,A_K(a)),$$

and therefore

$$\begin{aligned}
\sup_{c \in D(W,K)} \sup_{\theta \in K} W_\theta(c) &\leq \sup_{a \in K} \sup_{\{c:\, \|c-a\| \leq 2A_K(a)\}} W_\theta(c) \\
&\leq \sup_{a \in K} g(3\,A_K(a)) \\
&\leq g(3 \sup_{a,b \in K} \|a - b\|) < \infty.
\end{aligned}$$

Hence, the condition $(\mathcal{WK})$ does hold.

Theorem 2.2 *Let $\mathcal{K}$ be a class of subsets from Θ, let $\Delta(\mathcal{E}^n, \mathcal{E}; \mathcal{K}) \to 0$ and let a loss function W be continuous and satisfy the condition $(W\mathcal{K})$. Then*

$$\sup_{K \in \mathcal{K}} \lim_{n \to \infty} \mathcal{R}(\mathcal{E}^n, W; K) = \sup_{K \in \mathcal{K}} \mathcal{R}(\mathcal{E}, W; K) \tag{2.11}$$

and

$$\sup_{K \in \mathcal{K}} \lim_{n \to \infty} \mathcal{R}_\pi(\mathcal{E}^n, W; K) = \sup_{K \in \mathcal{K}} \mathcal{R}_\pi(\mathcal{E}, W; K), \quad \pi \in \Pi. \tag{2.12}$$

Proof. Let $K \in \mathcal{K}$ and let $D(W, K)$ be the set of all K-admissible points in $\mathbf{A}$. For each K-inadmissible point $a \in \mathbf{A}$ there exists a point $b \in \mathbf{A}$ such that (2.10) holds. Denote it by $b = T_K(a)$. For all K-admissible points a put $T_K(a) = a$. It is easy to see that the mapping T_K can be chosen to be continuous. From the construction we have immediately

$$W_\theta(T_K(a)) \le W_\theta(a), \quad \theta \in K, a \in \mathbf{A}.$$

This implies for any decision $\beta \in \mathfrak{B}(\mathcal{E}, W)$

$$\begin{aligned} \sup_{\theta \in K} R_\theta(\mathcal{E}, W, \beta) &= \sup_{\theta \in K} \beta(W_\theta, P_\theta) \\ &\ge \sup_{\theta \in K} \beta(W_\theta \circ T_K, P_\theta) \\ &= \sup_{\theta \in K} R_\theta(\mathcal{E}, W \circ T_K, \beta). \end{aligned}$$

Therefore, one may consider only such minimax decisions that have their values in the set $D(W, K)$. But on the other hand the loss function W is bounded on this set and Theorem 2.1 applies; this proves (2.11). In a similar way the assertion (2.12) can be proved. □

4. Now we consider some applications of the above results to the problem of *testing two* (simple or composite) *hypotheses.*

Let (Θ_0, Θ_1) be a partition of a parameter set Θ, i.e.

$$\Theta = \Theta_0 + \Theta_1$$

with some non-empty sets Θ_0 and Θ_1, $\Theta_0 \cap \Theta_1 = \emptyset$, and let $\mathbf{A} = \{0, 1\}$ be the decision space consisting of two points representing the acceptance of the *hypothesis* $H : \theta \in \Theta_0$ or of the *alternative* $K : \theta \in \Theta_1$, respectively.

Let $\rho = \rho(\omega, a)$ be a Markov kernel, $\omega \in \Omega$, $a \in \mathbf{A}$. Since $\rho(\omega, 1) + \rho(\omega, 0) = 1$,, this kernel can be characterized by the function $\rho(\omega) = \rho(\omega, 1)$ called *a test* (for the *testing problem* with two hypotheses H and K). It is clear that the risk $\beta_\rho(W_\theta, P_\theta)$ corresponding to the test ρ fulfills

$$\beta_\rho(W_\theta, P_\theta) = W_\theta(1)\, E_\theta\, \rho(\omega) + W_\theta(0)\, \big(1 - E_\theta\, \rho(\omega)\big).$$

Hence, if

$$W_\theta(0) = \begin{cases} 0, & \theta \in \Theta_0\,, \\ 1, & \theta \in \Theta_1\,, \end{cases} \qquad W_\theta(1) = \begin{cases} 1, & \theta \in \Theta_0\,, \\ 0, & \theta \in \Theta_1\,, \end{cases}$$

then

$$\beta_\rho(W_\theta, P_\theta) = 1 - E_\theta\, \rho(\omega), \quad \theta \in \Theta_1,$$

and

$$\beta_\rho(W_\theta, P_\theta) = E_\theta\, \rho(\omega), \quad \theta \in \Theta_0.$$

Remark 2.2 The function $\Psi_\theta(\rho) = E_\theta \rho(\omega)$, $\theta \in \Theta$, is called *the power function* of the test $\rho = \rho(\omega)$. The test $\rho^* = \rho^*(\omega)$ with the power function $\Psi_\theta(\rho^*) = 1$ for $\theta \in \Theta_1$ and $\Psi_\theta(\rho^*) = 0$ for $\theta \in \Theta_0$ would be "ideal". But usually such "ideal" tests do not exist.

If we have a sequence of experiments $\mathcal{E}^n \xrightarrow{\Delta} \mathcal{E}$ then it follows from Theorem 2.1 that

$$\lim_{n\to\infty} \inf_{\rho^n \in \mathfrak{B}(\mathcal{E}^n, \mathrm{A})} \max \left[\sup_{\theta\in\Theta_0} E_\theta\, \rho^n(\omega),\ \sup_{\theta\in\Theta_1} (1 - E_\theta\, \rho^n(\omega)) \right]$$
$$= \inf_{\rho\in\mathfrak{B}(\mathcal{E}, \mathrm{A})} \max \left[\sup_{\theta\in\Theta_0} E_\theta\, \rho(\omega),\ \sup_{\theta\in\Theta_1} (1 - E_\theta\, \rho(\omega)) \right]. \qquad (2.13)$$

Let

$$\begin{aligned} \Phi(\alpha; \mathcal{E}^n) &= \{\rho^n : E_\theta\, \rho^n \le \alpha,\ \theta \in \Theta_0\}\,, \quad n \ge 1, \\ \Phi(\alpha; \mathcal{E}) &= \{\rho : E_\theta\, \rho \le \alpha,\ \theta \in \Theta_0\} \end{aligned}$$

be the classes of tests of the level α, $\alpha \in [0, 1]$.

Then the following *conditional version* of (2.13) follows from the previous considerations: *for any* $\alpha \in [0, 1]$

$$\lim_{n\to\infty} \inf_{\rho^n\in\Phi(\alpha, \mathcal{E}^n)} \sup_{\theta\in\Theta_1} (1 - E_\theta\, \rho^n) = \inf_{\rho\in\Phi(\alpha; \mathcal{E})} \sup_{\theta\in\Theta_1} (1 - E_\theta\, \rho) \qquad (2.14)$$

or, equivalently,

$$\lim_{n\to\infty} \sup_{\rho^n\in\Phi(\alpha, \mathcal{E}^n)} \inf_{\theta\in\Theta_1} E_\theta\, \rho^n = \sup_{\rho\in\Phi(\alpha; \mathcal{E})} \inf_{\theta\in\Theta_1} E_\theta\, \rho. \qquad (2.15)$$

In particular, if the hypotheses considered are simple, ($\Theta_0 = \theta_0$ and $\Theta_1 = \theta_1$), *then*

$$\lim_{n\to\infty} \sup_{\{\rho^n :\, E^n_{\theta_0} \rho^n \le \alpha\}} E_{\theta_1}\, \rho^n = \sup_{\{\rho :\, E_{\theta_0} \rho \le \alpha\}} E_{\theta_1}\, \rho.$$

2.2 Weak ($w-$) Convergence

1. Let $\mathcal{E}^n = (\Omega^n, \mathcal{F}^n; P_\theta^n, \theta \in \sigma)$, $n \geq 1$, and $\mathcal{E} = (\Omega, \mathcal{F}; P_\theta, \theta \in \sigma)$ be experiments with a *finite* parameter set $\sigma = \{\theta_1, \dots, \theta_k\}$. According to Section 1.3, denote by μ_σ^n and μ_σ the standard measures of the experiments $\mathcal{E}^n$ and $\mathcal{E}$.

Definition 2.3 (Blackwell (1951, 1953); Le Cam (1964, 1986)) The sequence of experiments $\mathcal{E}^n$ *weakly converges* to the experiment $\mathcal{E}$ ($\mathcal{E}^n \xrightarrow{w} \mathcal{E}$) if the measures μ_σ^n weakly converge to the measure μ_σ ($\mu_\sigma^n \xrightarrow{w} \mu_\sigma$).

For the case of a finite parameter set, it follows from Theorem 1.9, Section 1.7 that the weak and strong convergence are *equivalent*:

$$\mathcal{E}^n \xrightarrow{\Delta} \mathcal{E} \iff \mathcal{E}^n \xrightarrow{w} \mathcal{E}. \tag{2.16}$$

It is clear that for a finite set σ there exist dominating probability measures Q^n ($P_\theta^n \ll Q^n$, $\theta \in \sigma$) and Q ($P_\theta \ll Q$, $\theta \in \sigma$). Let

$$Z_\theta^n = \frac{dP_\theta^n}{dQ^n}, \quad Z_\theta = \frac{dP_\theta}{dQ} \tag{2.17}$$

be the corresponding derivatives.

If the measures Q^n and Q are chosen of the form $Q^n = \frac{1}{k}(P_{\theta_1}^n + \dots + P_{\theta_k}^n)$ and $Q = \frac{1}{k}(P_{\theta_1} + \dots + P_{\theta_k})$ then the previously given definition of the weak convergence $\mathcal{E}^n \xrightarrow{w} \mathcal{E}$ really means the convergence of the laws $\mathcal{L}(Z_{\theta_1}^n, \dots, Z_{\theta_k}^n \mid Q^n)$ of the vectors $Z_\sigma^n \equiv (Z_{\theta_1}^n, \dots, Z_{\theta_k}^n)$ with respect to the measures Q^n to the distribution $\mathcal{L}(Z_{\theta_1}, \dots, Z_{\theta_k} \mid Q)$ of the vector $Z_\sigma \equiv (Z_{\theta_1}, \dots, Z_{\theta_k})$ with respect to the measure Q:

$$\mathcal{L}(Z_{\theta_1}^n, \dots, Z_{\theta_k}^n \mid Q^n) \xrightarrow{w} \mathcal{L}(Z_{\theta_1}, \dots, Z_{\theta_k} \mid Q) \iff \mathcal{E}^n \xrightarrow{w} \mathcal{E}. \tag{2.18}$$

It is not difficult to show that this assertion remains valid also for arbitrary dominating measures Q^n and Q.

Now we present the definition of weak convergence of statistical experiments for the case of an arbitrary parameter set Θ.

Definition 2.4 A sequence of experiments $\mathcal{E}^n = (\Omega^n, \mathcal{F}^n; P_\theta^n, \theta \in \Theta)$, $n \geq 1$, *weakly converges* to an experiment $\mathcal{E} = (\Omega, \mathcal{F}; P_\theta, \theta \in \Theta)$ ($\mathcal{E}^n \xrightarrow{w} \mathcal{E}$) if for any finite subset $\sigma \subseteq \Theta$ the weak convergence $\mathcal{E}_\sigma^n \xrightarrow{w} \mathcal{E}_\sigma$ takes place with $\mathcal{E}_\sigma^n = (\Omega^n, \mathcal{F}^n; P_\theta^n, \theta \in \sigma)$, $\mathcal{E}_\sigma = (\Omega, \mathcal{F}; P_\theta, \theta \in \sigma)$.

If for the experiments $\mathcal{E}^n$ and $\mathcal{E}$ are dominated and if Q^n and Q are the corresponding dominating measures (i.e. $P_\theta^n \ll Q^n$ and $P_\theta \ll Q$, $\theta \in \Theta$), then the weak convergence $\mathcal{E}^n \xrightarrow{w} \mathcal{E}$ follows from the finite-dimensional convergence in distributions $Z^n \xrightarrow{d_f} Z$ of the random fields $Z^n = (Z_\theta^n)_{\theta \in \Theta}$ to $Z = (Z_\theta)_{\theta \in \Theta}$, i.e. that for any $k \geq 1$ and each subset $\{\theta_1, \dots, \theta_k\}$ from Θ

$$\mathcal{L}(Z_{\theta_1}^n, \dots, Z_{\theta_k}^n \mid Q^n) \xrightarrow{w} \mathcal{L}(Z_{\theta_1}, \dots, Z_{\theta_k} \mid Q). \tag{2.19}$$

2. It follows from (2.16) and from the previous definitions that (strong) convergence $\mathcal{E}^n \xrightarrow{\Delta} \mathcal{E}$ is a stricter condition than (weak) convergence $\mathcal{E}^n \xrightarrow{w} \mathcal{E}$:

$$\mathcal{E}^n \xrightarrow{\Delta} \mathcal{E} \quad \Rightarrow \quad \mathcal{E}^n \xrightarrow{w} \mathcal{E}. \tag{2.20}$$

This leads to the natural question about possibility to extend the results of Theorem 2.1 from the case of strong convergence to the case of weak one.

Theorem 2.3 *Let $K \subseteq \Theta$ and $\mathcal{E}^n_K \xrightarrow{w} \mathcal{E}_K$. If $W \in \mathcal{W}^b_{lc}$ or $W \in \mathcal{W}_{lc}$ then*

$$\liminf_{n\to\infty} \mathcal{R}(\mathcal{E}^n, W; K) \;\ge\; \mathcal{R}(\mathcal{E}, W; K) \tag{2.21}$$

$$\liminf_{n\to\infty} \mathcal{R}_\pi(\mathcal{E}^n, W; K) \;\ge\; \mathcal{R}_\pi(\mathcal{E}, W; K), \quad \pi \in \Pi^0. \tag{2.22}$$

Proof. We present two variants of the proof. The *first* one is based on the property (2.20) and it utilizes the result of Theorem 2.1. The *second* proof is direct without addressing to the corresponding results under the assumption of strong convergence.

Suppose first that $W \in \mathcal{W}^b_c$ (i.e. W is bounded and continuous). If σ is a finite subset of a set K then by Theorem 2.1

$$\lim_n \inf_{\beta\in\mathfrak{B}(\mathcal{E}^n,\mathbf{A})} \sup_{\theta\in\sigma} R_\theta(\mathcal{E}^n, W, \beta) = \inf_{\beta\in\mathfrak{B}(\mathcal{E},\mathbf{A})} R_\theta(\mathcal{E}, W, \beta). \tag{2.23}$$

This results implies ($\Pi^0(\sigma)$ is the set of probability measures on σ)

$$\begin{aligned}
\liminf_n \inf_{\beta\in\mathfrak{B}(\mathcal{E}^n,\mathbf{A})} \sup_{\theta\in K} R_\theta(\mathcal{E}^n, W, \beta) &\ge \lim_n \inf_{\beta\in\mathfrak{B}(\mathcal{E}^n,\mathbf{A})} \sup_{\theta\in\sigma} R_\theta(\mathcal{E}^n, W, \beta)\\
&= \inf_{\beta\in\mathfrak{B}(\mathcal{E},\mathbf{A})} \sup_{\theta\in\sigma} R_\theta(\mathcal{E}, W, \beta)\\
&\ge \inf_{\beta\in\mathfrak{B}(\mathcal{E},\mathbf{A})} \sup_{\pi\in\Pi^0(\sigma)} R_\pi(\mathcal{E}, W, \beta)\\
&= \sup_{\pi\in\Pi^0(\sigma)} \inf_{\beta\in\mathfrak{B}(\mathcal{E},\mathbf{A})} R_\pi(\mathcal{E}, W, \beta)
\end{aligned} \tag{2.24}$$

where the last equality is a direct consequence of the following version of the classical minimax theorem (see, e.g. Parthasarathy and Raghavan (1971, Chapter 5) or Strasser (1985, Theorem 45.2)).

Theorem 2.4 *Let X, Y be convex subsets of two vector spaces, and let also X be compact. Let then $f = f(x, y)$ be a function on $X \times Y$ with values in $[0, \infty]$ such that*

(a) *$f(\cdot, y)$ is convex for each $y \in Y$;*
(b) *$f(x, \cdot)$ is concave for each $x \in X$;*
(c) *$f(\cdot, y)$ is lower semicontinuous for each $y \in Y$.*

Then

$$\inf_{x\in X} \sup_{y\in Y} f(x, y) = \sup_{y\in Y} \inf_{x\in X} f(x, y). \tag{2.25}$$

Since a finite set $\sigma \subseteq K$ in (2.24) can be chosen arbitrarily, this assertion and again Theorem 2.4 yield

$$\begin{aligned}
\liminf_n \inf_{\beta\in\mathfrak{B}(\mathcal{E}^n,\mathbf{A})} \sup_{\theta\in K} R_\theta(\mathcal{E}^n,W,\beta) &\geq \sup_{\pi\in\Pi^0(\sigma)} \inf_{\beta\in\mathfrak{B}(\mathcal{E},\mathbf{A})} R_\pi(\mathcal{E},W,\beta)\\
&= \inf_{\beta\in\mathfrak{B}(\mathcal{E},\mathbf{A})} \sup_{\pi\in\Pi^0(\sigma)} R_\pi(\mathcal{E},W,\beta)\\
&= \inf_{\beta\in\mathfrak{B}(\mathcal{E},\mathbf{A})} \sup_{\theta\in K} R_\theta(\mathcal{E},W,\beta) \qquad (2.26)
\end{aligned}$$

which proves (2.21). The assertion (2.22) can be proved similarly.

Now we give the second proof. We begin with the assertion (2.22). Suppose first that $W \in \mathcal{W}_c^b$ and let the support of a measure $\pi \in \Pi^0(K)$ be concentrated at points $\theta_1,\dots,\theta_k$. Denote (compare with Section 1.3)

$$\nu = \sum_{i=1}^k P_{\theta_i}, \quad f_i = \frac{dP_{\theta_i}}{d\nu}.$$

According to Theorem 1.2, Section 1.4, it is sufficient for calculation of the Bayes risks

$$\mathcal{R}_\pi(\mathcal{E},W;K) = \inf_{\beta\in\mathfrak{B}(\mathcal{E},\mathbf{A})} \int R_\theta(\mathcal{E},W,\beta)\pi(d\theta), \quad \pi\in\Pi^0(K),$$

to take infimum not over the whole class $\mathfrak{B}(\mathcal{E},\mathbf{A})$ but only over the class $\mathcal{R}(\mathcal{E},\mathbf{A})$ of Markov kernels $\rho = \rho(\omega,da)$. Hence, for $\pi \in \Pi^0(K)$

$$\begin{aligned}
\mathcal{R}_\pi(\mathcal{E},W;K) &= \inf_\rho \sum_{i=1}^k \pi_{\theta_i} \int_\Omega \int_{\mathbf{A}} W_{\theta_i}(a)\,\rho(\omega,da)\,P_{\theta_i}(d\omega)\\
&= \inf_\rho \int_\Omega \int_{\mathbf{A}} \sum_{i=1}^k \pi_{\theta_i} W_{\theta_i}(a)\,\rho(\omega,da)\,f_i(\omega)\,\nu(d\omega)\\
&= \int_\Omega \nu(d\omega) \inf_\rho \left\{ \int_{\mathbf{A}} \left[\sum_{i=1}^k \pi_{\theta_i} W_{\theta_i}(a)\,f_i(\omega)\right] \rho(\omega,da)\right\}\\
&= \int_\Omega h(f(\omega))\,\nu(d\omega) \qquad (2.27)
\end{aligned}$$

with

$$h(f) = \inf_{a\in\mathbf{A}} \left\{\sum_{i=1}^k \pi_{\theta_i} W_{\theta_i}(a)\,f_i\right\}, \quad f\in\Sigma_k, \qquad (2.28)$$

and

$$\Sigma_k = \left\{ f = (f_1,\dots,f_k),\ f_i \geq 0,\ \sum_{i=1}^k f_i = 1\right\}.$$

Changing the variables in (2.27) gives

$$\mathcal{R}_\pi(\mathcal{E},W;K) = \int_{\Sigma_k} h(f)\,\mu(df) \tag{2.29}$$

with $\mu(B) = \nu(f^{-1}(B))$, $B \in \mathcal{B}(\Sigma_k)$ (see Section 1.3). If the function $W_\theta(a)$ is bounded and continuous in a then the function $h(f)$ defined in (2.28) is continuous and bounded on Σ_k (since it is the infimum of continuous bounded concave functions).

Since $\mathcal{E}_K^n \xrightarrow{w} \mathcal{E}_K$, the weak convergence of the corresponding standard measures (on Σ_k) μ^n to μ takes place. Thus, for the measure π concentrated at points $\theta_1, \dots, \theta_k \in K$ and for every bounded and continuous (in a) function $W_\theta(a)$ we obtain from (2.29)

$$\lim_n \mathcal{R}_\pi(\mathcal{E}^n, W; K) = \mathcal{R}_\pi(\mathcal{E}, W; K) \tag{2.30}$$

which proves the required relation (2.22). (Note that under the assumption $W \in \mathcal{W}_c^b$ one has really the *equality* in (2.22).)

We now show that (2.21) implies (2.22) (in the case $W \in \mathcal{W}_c^b$) Notice that (similarly to (2.24))

$$\mathcal{R}(\mathcal{E},W;K) = \inf_\beta \sup_{\theta\in K} R_\theta(\mathcal{E},W,\beta) = \sup_{\pi\in\Pi^0(K)} \inf_\beta R_\pi(\mathcal{E},W,\beta). \tag{2.31}$$

Let $\varepsilon > 0$ and let $\pi_\varepsilon^0 \in \Pi^0(K)$ be such that

$$\inf_\beta R_{\pi_\varepsilon^0}(\mathcal{E},W,\beta) \geq \sup_{\pi\in\Pi^0(K)} \inf_\beta R_\pi(\mathcal{E},W,\beta) - \varepsilon.$$

This and (2.31) yield, for sufficiently large n,

$$\begin{aligned}
\mathcal{R}(\mathcal{E}^n,W;K) &= \sup_{\pi\in\Pi^0(K)} \inf_\beta R_\pi(\mathcal{E}^n,W,\beta) \\
&\geq \inf_\beta R_{\pi_\varepsilon^0}(\mathcal{E}^n,W,\beta) \\
&\geq \inf_\beta R_{\pi_\varepsilon^0}(\mathcal{E},W,\beta) - \varepsilon \\
&\geq \sup_{\pi\in\Pi^0(K)} \inf_\beta R_\pi(\mathcal{E},W,\beta) - 2\varepsilon \\
&= \mathcal{R}(\mathcal{E}^n,W;K) - 2\varepsilon.
\end{aligned}$$

Since $\varepsilon > 0$ was selected arbitrary this proves the desired inequality (2.21).

Finally, the extension of the proof to the case of functions W from the classes $\mathcal{W}_{lc}^b$ and $\mathcal{W}_{lc}$ can be easily done taking into account Definition 1.9, Section 1.4. □

Remark 2.3 Both proofs have used the fact (following from the classical minimax theorem) that

$$\inf_\beta \sup_{\theta\in K} R_\theta(\mathcal{E},W,\beta) = \sup_{\pi\in\Pi^0(K)} \inf_\beta R_\pi(\mathcal{E},W,\beta).$$

It is worth mentioning that this result remains valid if $\inf_\beta$ is taken not over all $\beta \in \mathfrak{B}(\mathcal{E}, \mathbf{A})$ but only over some subset $\mathfrak{B}_0(\mathcal{E}, \mathbf{A})$ which is convex and weakly compact. (For the case of a metric space $\mathbf{A}$ it is sufficient for $\mathfrak{B}_0(\mathcal{E}, \mathbf{A})$ to be a closed subset of $\mathfrak{B}(\mathcal{E}, \mathbf{A})$).

2.3 Reasons for Introducing the λ-Convergence (Sufficient Statistics, Localization, Reparametrization, LAN Property)

1. In the next section we will introduce the notion of *λ-convergence* of statistical experiments ($\mathcal{E}^n \xrightarrow{\lambda} \mathcal{E}$). This notion expresses the idea that for the experiments $\mathcal{E}^n$ there exist asymptotically sufficient statistics exist allowing for "reduction of data without loss of information". This notion (having an intermediate position between the weak and strong convergence and admitting for many cases an efficient verification) gives us the possibility to obtain precise lower bounds for asymptotic minimax decision problems. In this section we explain the ideas and reasons for introducing the λ-convergence.

2. Localization. Suppose for simplicity that Θ is an open subset in R^d and let

$$\mathbb{E}^n = (\Omega^n, \mathcal{F}^n; \mathbf{P}_\theta^n, \theta \in \Theta), \quad n \geq 1,$$

be a sequence of statistical experiments.

Considering the problem of estimating the true parameter (true "theory") θ, one may sometimes assume that some a priori information is available about where the true value of the parameter θ "is". Such information, for example, can be of the form of "the true value is in a neighborhood of a point θ_0" which plays the role of an initial approximation. For many situations this value θ_0 can be taken as a result of a preliminary "crude" estimation of θ.

Assume therefore that the true value θ is in a neighborhood of the point θ_0. It is natural to redefine the n-th experiment $\mathbb{E}^n$ utilizing this a priori information about "rough" initial approximation θ_0. Represent θ in the form

$$\theta = \theta_0 + \varphi(n, \theta_0)\alpha, \tag{2.32}$$

where $\varphi(n, \theta_0)$ is chosen by dimension reasons ("the more observations in the experiment $\mathbb{E}^n$ is available, the more precise the initial approximation θ_0 has to be") in such a way that $\varphi(n, \theta_0) \downarrow 0$, $n \to \infty$, and α can be considered as a *new parameter* which has to be estimated. We call $(\varphi(n, \theta_0))_{n \geq 1}$ the *localizing sequence.* Below, for simplicity of exposition we assume this sequence to be numeric an extension to the case when, for example, $\varphi(n, \theta_0)$ is a matrix sequence with norms converging to zero, is straightforward.

Such a *reparametrization* (transition from "θ-model" to "α-model") determines for each experiment $\mathbb{E}^n$ a new (*localized*) experiment

$$\mathcal{E}^n = (\Omega^n, \mathcal{F}^n; P_\alpha^n, \alpha \in \mathcal{A}^n), \quad n \geq 1,$$

with $\mathcal{A}^n = \{\alpha \in R^d : \theta_0 + \varphi(n,\theta_0)\alpha \in \Theta\}$ and $P_\alpha^n = \mathbf{P}^n_{\theta_0+\varphi(n,\theta_0)\alpha}$. Since $\varphi(n,\theta_0) \downarrow 0$, it clearly follows $\mathcal{A}^n \uparrow \mathcal{A} = R^d$. Sometimes, in order to emphasize which localization point θ_0 is chosen we write $\mathcal{E}^n(\theta_0)$ instead of $\mathcal{E}^n$.

Let

$$Z_\alpha^n(\omega) = \frac{dP_\alpha^n}{dP_0^n}(\omega) \quad \left(= \frac{d\mathbf{P}^n_{\theta_0+\varphi(n,\theta_0)\alpha}}{d\mathbf{P}^n_{\theta_0}} \right) \tag{2.33}$$

be the *likelihood*, i.e. Radon-Nikodym derivative of the measure P_α^n with respect to the measure P_0^n.

It is worth noting that if $\widehat{\theta}^n$ is the *maximum likelihood estimator* for "θ-model" and $\widehat{\alpha}^n$ is the similar estimator for "α-model" then

$$\widehat{\theta}^n = \theta_0 + \varphi(n,\theta_0)\widehat{\alpha}^n \tag{2.34}$$

and hence, by (2.32),

$$\widehat{\theta}^n - \theta = \varphi(n,\theta_0)(\widehat{\alpha}^n - \alpha).$$

Therefore, if a loss function $W = W_\theta(a)$ is of the form $W_\theta(a) = W(\alpha - \theta)$ then (with $\theta = \theta_0 + \varphi(n,\theta_0)\alpha$)

$$\mathbf{E}_\theta^n W(\varphi^{-1}(n,\theta_0)(\widehat{\theta}^n - \theta)) = E_\alpha^n W(\widehat{\alpha}^n - \alpha), \tag{2.35}$$

where $\mathbf{E}_\theta^n$ (respectively E_α^n) mean the integration with respect to the measures $\mathbf{P}_\theta^n$ (respectively P_α^n), and hence, for any $b > 0$

$$\begin{aligned} &\sup_{\{\theta:\, |\varphi^{-1}(n,\theta_0)(\theta-\theta_0)| \leq b\}} \mathbf{E}_\theta^n W(\varphi^{-1}(n,\theta_0)(\widehat{\theta}^n - \theta)) \\ &= \sup_{\{\alpha:\, |\alpha| \leq b\}} E_\alpha^n W(\widehat{\alpha}^n - \alpha). \end{aligned} \tag{2.36}$$

This formulae should be taken into account when "transferring" the asymptotic results from "α-models" to "θ-models".

The method of transition to a local α-scale depends on the localizing sequence $\varphi(n,\theta_0)$. A choice of this localizing sequence $\varphi(n,\theta_0)$ is determined by the aim to get the sequence of localized experiments $\mathcal{E}^n$ converging (in some sense) to a non-trivial limit experiment, say $\mathcal{E}$. If this limit experiment $\mathcal{E}$ is of a simple structure and the problem of finding the true "theory" α can be solved for it, then one may hope to obtain "close" results also for the experiments $\mathcal{E}^n$, at least for a large number n.

In fact, the requirement of "non-triviality" for the limit experiment $\mathcal{E}$ prompts to apply the idea of contiguity ("drawing together") of two families of measures $(\mathbf{P}^n_{\theta_0+\varphi(n,\theta_0)\alpha})_{n\geq 1}$ and $(\mathbf{P}^n_{\theta_0})_{n\geq 1}$, $\left((\mathbf{P}^n_{\theta_0+\varphi(n,\theta_0)\alpha}) \triangleleft (\mathbf{P}^n_{\theta_0})\right)$ when we look for a "right" choice of the sequence $\varphi(n, \theta_0)$. (The questions related to the contiguity problem were discussed in more details in Section 1.12.)

3. Let again $\mathcal{E}^n = (\Omega^n, \mathcal{F}^n; P_\alpha^n, \alpha \in \mathcal{A}^n)$ be the reparametrized (localized) experiments, $n \geq 1$. The well-known condition of asymptotic normality of the experiments $\mathbb{E}^n$ can be formulated in the following way.

Let $\mathcal{E} = (R^d, \mathcal{B}(R^d); P_\alpha, \alpha \in R^d)$ be a Gaussian shift experiment such that $P_\alpha \sim \mathcal{N}(\alpha, I)$ with the unit matrix $I = I(d \times d)$. We shall suppose that each measure P_α is the distribution of an "observed" random variable $\lambda = \lambda(x)$ given in "coordinate" manner, i.e. $\lambda(x) = x$. Then it holds for $Z_\alpha(\lambda) = dP_\alpha/dP_0(\lambda)$

$$Z_\alpha(\lambda) = \exp\left\{(\alpha, \lambda) - \frac{1}{2}|\alpha|^2\right\}. \tag{2.37}$$

Definition 2.5 (Le Cam (1960)) A sequence of experiments $(\mathbb{E}^n)_{n\geq 1}$ is *locally* (at a point θ_0) *asymptotically normal* (LAN) and the corresponding (localized) experiments $(\mathcal{E}^n(\theta_0))_{n\geq 1}$ are *asymptotically normal* (AN) if there is a sequence of random variables $\lambda^n = \lambda^n(\omega) : \Omega^n \to R^d$ such that

$$\mathcal{L}(\lambda^n \mid P_0^n) \xrightarrow{w} \mathcal{L}(\lambda \mid P_0) \quad (= P_0), \tag{2.38}$$

$$Z_\alpha^n(\omega) - Z_\alpha(\lambda^n(\omega)) \xrightarrow{P_0^n} 0, \quad \alpha \in R^d, \tag{2.39}$$

with $Z_\alpha^n(\omega)$ defined in (2.33) and $Z_\alpha(\lambda)$ from (2.37).

Remark 2.4 In (2.39) the convergence $\xrightarrow{P_0^n}$ means that

$$P_0^n \{|Z_\alpha^n(\omega) - Z_\alpha(\lambda^n(\omega))| > \varepsilon\} \to 0, \quad n \to \infty, \tag{2.40}$$

for every $\varepsilon > 0$ and any $\alpha \in R^d$. (If α is fixed then the condition $\mathcal{A}^n \uparrow R^d$ ensures the existence of a number $n^0(\alpha)$ such that $\alpha \in \mathcal{A}^n$ for all $n \geq n^0(\alpha)$. Therefore the left-hand sides of (2.38) and (2.40) are well defined for every $\alpha \in R^d$ and all sufficiently large n.)

In the next section we introduce the so-called λ-*convergence*, which is a straightforward generalization of the previous definition to the case of a limit experiment which is *not* necessarily a Gaussian shift.

4. Two experiments $\mathcal{E}$ and $\mathcal{E}^*$ are *strongly* $(\Delta-)$ *equivalent* ("$\mathcal{E} \overset{\Delta}{\sim} \mathcal{E}^*$") if $\Delta(\mathcal{E}, \mathcal{E}^*) = 0$. The experiments $\mathcal{E}$ and $\mathcal{E}^*$ are *weakly* $(w-)$ *equivalent* ("$\mathcal{E} \overset{w}{\sim} \mathcal{E}^*$") if for any finite subset $\sigma \subseteq \Theta$ the corresponding standard measures μ_σ and μ_σ^* coincide. Because $\Delta(\mathcal{E}, \mathcal{E}^*) = \sup_\sigma \Delta(\mathcal{E}_\sigma, \mathcal{E}_\sigma^*)$ when σ runs over all finite subsets in Θ and $\Delta(\mathcal{E}_\sigma, \mathcal{E}_\sigma^*) = 0 \iff \mathcal{E}_\sigma \overset{w}{\sim} \mathcal{E}_\sigma^*$ we conclude

$$\mathcal{E} \overset{\Delta}{\sim} \mathcal{E}^* \iff \mathcal{E} \overset{w}{\sim} \mathcal{E}^*.$$

Now we describe also one more kind of equivalence of experiments related to the idea of sufficiency. Let $\mathcal{E} = (\Omega, \mathcal{F};\ P_\theta\,, \theta \in \Theta)$ be an experiment where the dimension of the sample space Ω can be fairly arbitrary. Hence, it is natural to ask whether one can reduce data described by this experiment in such a way that the dimension of the sample space will be "smaller" but containing the same information as for the original experiment.

Useful models of "small" dimension are provided by canonical experiments $\mathcal{E} = (X, \mathfrak{X};\ P_\theta\,, \theta \in \Theta)$ with a complete separable metric space $(X, \mathfrak{X})$ (Polish space). The typical examples of such spaces are (see Section 1.1) the spaces R^d, R^∞, **C**, **D**.

Let $\mathcal{E} = (\Omega, \mathcal{F};\ P_\theta\,, \theta \in \Theta)$ be a statistical experiment and let $\lambda = \lambda(\omega)$ be a random element defined on a measurable space $(\Omega, \mathcal{F})$ with values in a Polish space $(X, \mathfrak{X})$,. Define the new experiment

$$\mathcal{E}^\lambda = (X, \mathfrak{X};\ P_\theta^\lambda\,, \theta \in \Theta)$$

with

$$P_\theta^\lambda(B) = P_\theta(\omega : \lambda(\omega) \in B) \quad \left(= P_\theta \circ \lambda^{-1}(B)\right),\ B \in \mathfrak{X}.$$

Since the experiment $\mathcal{E}^\lambda$ has been constructed from $\mathcal{E}$ by the mapping λ, it holds $\delta(\mathcal{E}, \mathcal{E}^\lambda) = 0$. If $\mathcal{E}$ is a dominated experiment ($P_\theta \ll Q$, $\theta \in \Theta$) then the experiment $\mathcal{E}^\lambda$ is also dominated with $(P_\theta^\lambda \ll Q^\lambda,$, $\theta \in \Theta$ where $Q^\lambda = Q \circ \lambda^{-1}$. Denote

$$Z_\theta(\omega) = \frac{dP_\theta}{dQ}(\omega), \quad Z_\theta^\lambda(x) = \frac{dP_\theta^\lambda}{dQ^\lambda}(x).$$

Define also $\mathcal{F}^\lambda = \sigma\{\omega : \lambda(\omega)\}$ and suppose that for every $\theta \in \Theta$

$$Z_\theta(\omega) = Z_\theta^\lambda(\lambda(\omega)). \tag{2.41}$$

We claim that the assumption (2.41) implies the equality $\delta(\mathcal{E}^\lambda, \mathcal{E}) = 0$. Indeed, let

$$\mathcal{E} \mid \mathcal{F}^\lambda = (\Omega, \mathcal{F}^\lambda;\ P_\theta \mid \mathcal{F}^\lambda,\ \theta \in \Theta).$$

It is clear that $\delta(\mathcal{E}, \mathcal{E}|\mathcal{F}^\lambda) = 0$. From (2.41) for each $A \in \mathcal{F}$ (Q-a.s.)

$$\begin{aligned} E_\theta\left[I(A) \mid \mathcal{F}^\lambda\right] &= E_Q\left[Z_\theta \mid \mathcal{F}^\lambda\right]^\oplus E_Q\left[I(A)\, Z_\theta \mid \mathcal{F}^\lambda\right] \\ &= Z_\theta(\lambda)^\oplus Z_\theta(\lambda)\, E_Q\left[I(A) \mid \mathcal{F}^\lambda\right] \\ &= E_Q\left[I(A) \mid \mathcal{F}^\lambda\right]. \end{aligned}$$

In other words, the σ-field $\mathcal{F}^\lambda$ is sufficient for the family $\mathcal{P} = \{P_\theta\,, \theta \in \Theta\}$. As in Section 1.2, under the assumption of existence of regular conditional distributions $P_\theta(\cdot \mid \mathcal{F}^\lambda)$, $\theta \in \Theta$, the experiment $\mathcal{E} \mid \mathcal{F}^\lambda$ is *exhaustive* for $\mathcal{E}$, i.e. $\delta(\mathcal{E}|\mathcal{F}^\lambda, \mathcal{E}) = 0$. Therefore $\Delta(\mathcal{E}|\mathcal{F}^\lambda, \mathcal{E}) = 0$ and under (2.41) it holds $\mathcal{E}|\mathcal{F}^\lambda \overset{\Delta}{\sim} \mathcal{E}$. Since obviously $\mathcal{E}|\mathcal{F}^\lambda \overset{\Delta}{\sim} \mathcal{E}^\lambda$, it holds $\mathcal{E} \overset{\Delta}{\sim} \mathcal{E}^\lambda$. Therefore, the following assertion is proved.

Lemma 2.1 *Let (2.41) hold and let there exist a family of regular conditional distributions* $P_\theta(\cdot \mid \mathcal{F}^\lambda)$, $\theta \in \Theta$. *Then* $\mathcal{E} \overset{\Delta}{\sim} \mathcal{E}^\lambda$.

It is useful to represent this assertion in the following form:

$$\left.\begin{array}{l} \mathcal{L}(\lambda \mid Q) = Q^\lambda \\ Z_\theta(\omega) - Z_\theta^\lambda(\lambda(\omega)) = 0 \quad (Q\text{–a.s.}) \end{array}\right\} \quad \Rightarrow \quad \mathcal{E} \overset{\Delta}{\sim} \mathcal{E}^\lambda. \tag{2.42}$$

Note that the assumption of existence of regular conditional distributions $P_\theta(\cdot \mid \mathcal{F}^\lambda)$, $\theta \in \Theta$, is always satisfied if the experiment $\mathcal{E}^\lambda$ is *canonical* (i.e. the space $(X, \mathfrak{X})$ is Polish) and the experiment $\mathcal{E}$ is dominated. The assertion (2.42) implies also that in the case of "small" dimension of X "the statistic" $\lambda = \lambda(\omega)$ really provides a reduction of data without loss of information (in the sense that $\Delta(\mathcal{E}, \mathcal{E}^\lambda) = 0$).

2.4 λ-Convergence (Definition, Examples)

1. Let $\mathcal{E} = (X, \mathfrak{X}; P_\alpha, \alpha \in \mathcal{A})$ be a canonical experiment with a complete separable space $(X, \mathfrak{X})$. We will suppose that the family $\mathcal{P} = \{P_\alpha, \alpha \in \mathcal{A}\}$ is dominated. Due to Section 1.1 a dominating measure, say P, can be chosen of the form

$$P = \sum_{i=1}^{\infty} c_i P_{\alpha_i}, \tag{2.43}$$

with $c_i \geq 0$, $\sum_{i=1}^\infty c_i = 1$, and α_i being some points from $\mathcal{A}$, $i \geq 1$. Let

$$Z_\alpha(x) = \frac{dP_\alpha}{dP}(x).$$

Suppose now that statistical experiments

$$\mathcal{E}^n = (\Omega^n, \mathcal{F}^n; P_\alpha^n, \alpha \in \mathcal{A}^n), \quad n \geq 1,$$

are given with $\mathcal{A}^n \uparrow \mathcal{A}$. Denote

$$P^n = \sum_{\{i:\, \alpha_i \in \mathcal{A}^n\}} c_i P_{\alpha_i}^n \Big/ \sum_{\{i:\, \alpha_i \in \mathcal{A}^n\}} c_i, \tag{2.44}$$

with the point $\alpha_i \in \mathcal{A}$ from (2.43), $i \geq 1$.

Definition 2.6 We say that experiments $(\mathcal{E}^n)_{n \geq 1}$ *λ-converge* to the dominated canonical experiment $\mathcal{E}$ if there exist $\mathcal{F}^n/\mathfrak{X}$-measurable random elements $\lambda^n = \lambda^n(\omega)$, $\omega \in \Omega^n$ such that

$$\begin{array}{rcll} \mathcal{L}(\lambda^n \mid P^n) & \xrightarrow{w} & \mathcal{L}(\lambda \mid P) & (= P) \\ Z_\alpha^n(\omega) - Z_\alpha(\lambda^n(\omega)) & \xrightarrow{P^n} & 0, & \alpha \in \mathcal{A}, \end{array} \tag{2.45}$$

with the canonical random element $\lambda = \lambda(x)$, i.e. $\lambda(x) \equiv x$. We denote this convergence by

$$\mathcal{E}^n \xrightarrow{\lambda} \mathcal{E}.$$

Remark 2.5 If for all n beginning from some n_0 the strengthening of the property (2.45) holds in the sense that

$$\mathcal{L}(\lambda^n \mid P^n) = \mathcal{L}(\lambda \mid P) \quad (= P)$$
$$Z_\alpha^n(\omega) - Z_\alpha(\lambda^n(\omega)) = 0, \quad \alpha \in \mathcal{A}, \ (P^n\text{-a.s.})$$

then we say that a *λ-coincidence* of the experiments $\mathcal{E}^n$ and $\mathcal{E}$ takes place for all $n \geq n_0$.

Example 2.2 If $\mathcal{E} = (R^d, \mathcal{B}(R^d); P_\alpha, \alpha \in R^d)$ is a Gaussian shift experiment, $P_\alpha \sim \mathcal{N}(\alpha, I)$, $I = I(d \times d)$, then $Z_\alpha(\lambda)$ is defined by the formula (2.37) with $\lambda(x) = x$, and hence (2.45) becomes just the condition of asymptotic normality with $P = P_0$ and $P^n = P_0^n$.

Example 2.3 Let $\mathcal{M}^d$ be the set of all symmetric non-negative $d \times d$-matrices. Consider an experiment

$$\mathcal{E} = (R^d \times \mathcal{M}^d, \mathcal{B}(R^d) \otimes \mathcal{B}(\mathcal{M}^d); P_\alpha, \alpha \in R^d).$$

We will denote $\lambda = (\gamma, \Gamma)$ with $\gamma \in R^d$, $\Gamma \in \mathcal{M}^d$, and let

$$Z_\alpha(\lambda) = \exp\{(\Gamma\gamma, \alpha) - \frac{1}{2}|\Gamma\alpha|^2\}. \tag{2.46}$$

Let also γ and Γ be *independent* under the measure P_0 and $\mathcal{L}(\gamma \mid P_0) = \mathcal{N}(0,1)$. Such an experiment, being a particular case of the (γ, Γ)-models studied in detail in Chapter 3, describes *mixed Gaussian shift* experiments. The requirement of λ-convergence of experiments $\mathcal{E}^n = (\Omega^n, \mathcal{F}^n; P_\alpha^n, \alpha \in R^d)$, $n \geq 1$, to $\mathcal{E}$ means here existence of random elements $\gamma^n = \gamma^n(\omega) \in R^d$ and $\Gamma^n = \Gamma^n(\omega) \in \mathcal{M}^d$ such that

$$\mathcal{L}(\gamma^n, \Gamma^n \mid P_0^n) \xrightarrow{w} \mathcal{L}(\lambda, \Gamma \mid P_0)$$

and

$$Z_\alpha^n(\omega) - \exp\{(\Gamma^n\gamma^n, \alpha) - \frac{1}{2}|\Gamma^n\alpha|^2\} \xrightarrow{P_0^n} 0, \quad \alpha \in R^d.$$

Example 2.4 Let $\mathbb{E}^T = (\mathbf{C}^T, \mathcal{C}^T; \mathbf{P}_\theta^T, \theta \in R)$, $T > 0$, be an experiment, corresponding to the random process $X^T = (X_t)_{t \leq T}$ which is a solution of the stochastic differential equation

$$dX_t = -\theta X_t \, dt + dW_t, \quad X_0 = 0, \quad 0 \leq t \leq T. \tag{2.47}$$

Hence, $\mathbf{P}_\theta^T$ is the distribution of this process for the given value of θ. If $\theta = 0$ then $\mathbf{P}_0^T$ is exactly the Wiener measure on the space of continuous functions $\mathbf{C}^T$.

It is well known (see Liptser and Shiryaev (1978, Section 17.2)) that $\mathbf{P}_\theta^T \ll \mathbf{P}_0^T$ and the density $Z_\theta^T(X) = d\mathbf{P}_\theta^T/d\mathbf{P}_0^T(X)$ admits representation

$$Z_\theta^T(X) = \exp\left\{-\theta\int_0^T X_s\,dX_s - \frac{\theta^2}{2}\int_0^T X_s^2\,ds\right\}. \tag{2.48}$$

If $T \to \infty$ then the measures $\mathbf{P}_\theta^T$ and $\mathbf{P}_0^T$ are asymptotically singular (i.e. there exists $A^T \in \mathcal{C}^T$ such that $\mathbf{P}_\theta^T(A^T) \to 1$ and $\mathbf{P}_0^T(A^T) \to 0$). Thus, to be able to speak about existence of a "limit" (as $T \to \infty$) experiment the original experiments should be reparametrized.

It has been already noted that the transition from "θ-model" to "α-model" depends on an initial approximation θ_0. Since the probability properties of the process X are distinguished for three different cases, when $\theta_0 < 0$, $\theta_0 > 0$, and $\theta_0 = 0$, we will consider these cases separately.

A. $\theta_0 > 0$. Let

$$Z^T(\theta_0, \theta; X) = \frac{d\mathbf{P}_\theta^T}{d\mathbf{P}_{\theta_0}^T}(X).$$

Then from (2.48)

$$Z^T(\theta_0, \theta; X) = \exp\left\{(\theta - \theta_0)\int_0^T X_s\,dX_s + \frac{1}{2}(\theta^2 - \theta_0^2)\int_0^T X_s^2\,ds\right\}. \tag{2.49}$$

Define

$$\theta = \theta_0 + \alpha\sqrt{2\,\theta_0/T}. \tag{2.50}$$

For such a reparametrization it holds

$$Z^T\left(\theta_0, \theta_0 + \alpha\sqrt{2\,\theta_0/T}; X\right)$$
$$= \exp\left\{\alpha\sqrt{2\,\theta_0/T}\int_0^T X_s\,dX_s - \left[\theta_0\alpha\sqrt{\frac{2\,\theta_0}{T}} + \frac{\theta_0\alpha^2}{T}\right]\int_0^T X_s^2\,ds\right\}.$$

It is not difficult to prove (see, for example, Anderson (1959)) that for $T \to \infty$ and $\theta_0 > 0$

$$\frac{1}{T}\int_0^T X_s^2\,ds \quad \overset{\mathbf{P}_{\theta_0}^T}{\longrightarrow} \quad \frac{1}{2\,\theta_0}, \tag{2.51}$$

$$\frac{1}{\sqrt{T}}\int_0^T X_s\,dW_s \quad \overset{d(\mathbf{P}_{\theta_0}^T)}{\longrightarrow} \quad \mathcal{N}\left(\frac{1}{2\,\theta_0}, 1\right), \tag{2.52}$$

where the symbol $d(\mathbf{P}_{\theta_0}^T)$ means the convergence in *distribution* with respect to the measures $\mathbf{P}_{\theta_0}^T$.

If X is considered under the measure $\mathbf{P}^T_{\theta_0}$ then

$$Z^T\left(\theta_0, \theta_0 + \alpha\sqrt{2\theta_0/T}; X\right)$$
$$= \exp\left\{\alpha\sqrt{2\theta_0/T}\int_0^T X_s\,dW_s - \frac{2\theta_0\alpha^2}{T}\int_0^T X_s^2\,ds\right\}$$
$$= \exp\left\{\alpha\sqrt{2\theta_0/T}\int_0^T X_s\,dW_s - \frac{\alpha^2}{2} + \frac{\alpha^2}{2}\left[1 - \frac{2\theta_0}{T}\right]\int_0^T X_s^2\,ds\right\}.$$

If we define

$$\lambda^T = \sqrt{2\theta_0/T}\int_0^T X_s\,dW_s$$

and (similarly to (2.37))

$$Z_\alpha(\lambda) = \exp\{\alpha\lambda - \frac{1}{2}\alpha^2\},$$

then

$$Z^T\left(\theta_0, \theta_0 + \alpha\sqrt{2\theta_0/T}; X\right) - Z_\alpha\left(\sqrt{2\theta_0/T}\int_0^T X_s\,dW_s\right)$$
$$= Z_\alpha(\lambda^T)\left[\exp\left\{\frac{\alpha^2}{2}\left(1 - \frac{2\theta_0}{T}\right)\int_0^T X_s^2\,ds\right\} - 1\right].$$

By virtue of (2.51) the expression in brackets tends to zero in probability (with respect to $\mathbf{P}^T_{\theta_0}$). Hence

$$Z^T\left(\theta_0, \theta_0 + \alpha\sqrt{2\theta_0/T}; X\right) - Z_\alpha(\lambda^T) \xrightarrow{\mathbf{P}^T_{\theta_0}} 0.$$

Moreover, from (2.52)

$$\mathcal{L}(\lambda^T \mid \mathbf{P}^T_{\theta_0}) \to \mathcal{N}(0,1).$$

This means that for $\theta_0 > 0$ we have *λ-convergence* of the experiments $\mathcal{E}^T = (\mathbf{C}^T, \mathcal{C}^T;\ P^T_\alpha, \alpha \in R)$ with $P^T_\alpha = \mathbf{P}^T_{\theta_0 + \alpha\sqrt{2\theta_0/T}}$, to the *Gaussian shift experiment* $\mathcal{E}$ (see Section 2.3).

B. $\theta_0 < 0$. It is known (see, for example, Anderson(1959)) that

$$\mathcal{L}\left\{\frac{e^{\theta_0 T}}{2|\theta_0|}\int_0^T X_s\,dW_s,\ \frac{e^{2\theta_0 T}}{(2\theta_0)^2}\int_0^T X_s^2\,ds \mid \mathbf{P}^T_{\theta_0}\right\} \xrightarrow{w} \mathcal{L}(\xi\eta, \xi^2), \tag{2.53}$$

with *independent* standard Gaussian random variable ξ, η (with the parameters 0 and 1). Thus, if we define

$$\theta = \theta_0 + \alpha \frac{e^{\theta_0 T}}{2|\theta_0|},$$

then from (2.49)

$$Z^T(\theta_0, \theta; X) = \exp\left\{-\alpha \frac{e^{\theta_0 T}}{2|\theta_0|} \int_0^T X_s\, dW_s - \frac{\alpha^2}{2} \frac{e^{2\theta_0 T}}{(2\,\theta_0)^2} \int_0^T X_s^2\, ds\right\}.$$

Let $\lambda^T = (\gamma^T, \Gamma^T)$ with

$$(\Gamma^T)^2 = \frac{e^{2\,\theta_0 T}}{(2\,\theta_0)^2} \int_0^T X_s^2\, ds, \quad \gamma^T = (\Gamma^T)^{-1} \frac{e^{\theta_0 T}}{2|\theta_0|} \int_0^T X_s\, dW_s\,.$$

Next, let $\lambda = (\gamma, \Gamma)$ where γ and Γ are defined by the relation $\gamma = -\alpha\Gamma + \varepsilon$ with $\varepsilon \sim \mathcal{N}(0,1)$, $\Gamma \sim \mathcal{N}(0,1)$, and Γ, ε are independent. If P_α is the distribution of the pair $\lambda = (\gamma, \Gamma)$ corresponding to the parameter $\alpha \in R$ then

$$Z_\alpha(\lambda) = \frac{dP_\alpha}{dP_0} = \exp(\alpha\,\Gamma\gamma - \frac{1}{2}\Gamma^2\alpha^2).$$

The statement (2.53) and the equality $Z^T(\theta_0, \theta_0 + \alpha \frac{e^{\theta_0 T}}{2|\theta_0|}; X) = Z_\alpha(\lambda^T)$ imply that the experiments $\mathcal{E}^T(\theta_0)$ λ-converge as $T \to \infty$ to the experiment $\mathcal{E} = (R^2, \mathcal{B}(R^2); P_\alpha, \alpha \in R)$.

C. $\theta_0 = 0$ This is in some sense a "frontier" case since it separates the above considered cases of $\theta_0 > 0$ and $\theta_0 < 0$. Under the assumption $\theta_0 = 0$ one has

$$Z^T(0, \theta; X) = \frac{d\mathbf{P}_\theta^T}{d\mathbf{P}_0^T}(X) = \exp\left\{-\theta \int_0^T X_s\, dX_s - \frac{\theta^2}{2} \int_0^T X_s^2\, ds\right\}.$$

Put

$$\theta = \frac{\alpha}{T}, \quad P_\alpha^T = \mathbf{P}_{\alpha/T}^T,$$

and

$$\lambda^T = \left(\frac{1}{T} \int_0^T X_s\, dX_s,\, \frac{1}{T^2} \int_0^T X_s^2\, ds\right).$$

Then

$$Z^T(0, \frac{\alpha}{T}; X) = \exp\left\{-\frac{\alpha}{T} \int_0^T X_s\, dX_s - \frac{\alpha^2}{2T^2} \int_0^T X_s^2\, ds\right\}.$$

Consider the properties of λ^T with respect to the measure $P_0^T = \mathbf{P}_0^T$. It is clear that

$$\mathcal{L}(\lambda^T \mid \mathbf{P}_0^T) = \mathcal{L}\left\{\frac{1}{T}\int_0^T W_s\, dW_s,\ \frac{1}{T^2}\int_0^T W_s^2\, ds\right\}.$$

By virtue of the self-similarity property of the standard Wiener process W (which means that $(c^{-1/2}W_{cs})_{s\geq 0}$ is again a standard Wiener process for every $c > 0$)

$$\begin{aligned}\mathcal{L}\left\{\frac{1}{T}\int_0^T W_s\, dW_s,\ \frac{1}{T^2}\int_0^T W_s^2\, ds\right\} &= \mathcal{L}\left\{\int_0^1 W_s\, dW_s,\ \int_0^1 W_s^2\, ds\right\}\\ &= \mathcal{L}\left\{\frac{1}{2}(W_1^2-1),\ \int_0^1 W_s^2\, ds\right\}.\end{aligned}$$

Hence, under the measure $\mathbf{P}_0^T$ for any $T > 0$

$$Z^T(0, \frac{\alpha}{T}; X) = Z^1(0, \alpha; X), \tag{2.54}$$

and one can say that here we have not only λ-convergence of the experiments $\mathcal{E}^T(0)$ (to $\mathcal{E}^1(0)$) but simply their λ-coincidence. Note that the property (2.54) can be considered as some kind of self-similarity of the likelihood for the model (2.47) near the point $\theta_0 = 0$.

Example 2.5 This example illustrates situations when the experiment from Example 2.4 (generated by a diffusion process) appears as a limit experiment.

Let

$$\mathbb{E}^n = (R^n, \mathcal{B}(R^n); \mathbf{P}_\theta^n, |\theta| \leq 1), \quad n \geq 1,$$

be the experiment corresponding to a sequence $(x_1, \ldots, x_n)$ generated by the first order autoregression model

$$x_k = \theta x_{k-1} + \varepsilon_k, \quad x_0 = 0, \tag{2.55}$$

with independent normally distributed random errors $(\varepsilon_k)_{k\geq 1}$, $\varepsilon_k \sim \mathcal{N}(0,1)$.

Rewrite the model (2.55) in the form

$$\frac{x_k - x_{k-1}}{\sqrt{n}} = -\frac{1-\theta}{\Delta}\frac{x_{k-1}}{\sqrt{n}}\Delta + \frac{\varepsilon_k}{\sqrt{n}} \tag{2.56}$$

with $\Delta = 1/n$.

Choose as an initial point θ_0 for the parameter $\theta \in [0,1]$ the point $\theta_0 = 1$ and introduce a new parameter α putting

$$\theta = 1 - \frac{\alpha}{n},$$

where $\alpha \in [0, 2n]$.

It is convenient to introduce new variables

$$X^{(n)}\Big(\frac{k}{n}\Big) = \frac{x_k}{\sqrt{n}}, \quad W^{(n)}\Big(\frac{k}{n}\Big) = \frac{1}{\sqrt{n}}\sum_{i=1}^{k} \varepsilon_i, \quad 0 \le k \le n,$$

and to denote for $t \in [0,1]$

$$X^{(n)}(t) = X^{(n)}\Big(\frac{k}{n}\Big), \quad W^{(n)}(t) = W^{(n)}\Big(\frac{k}{n}\Big), \quad \frac{k}{n} \le t < \frac{k+1}{n}.$$

Under the notation introduced the model (2.56) can be rewritten in the "differential-difference" form

$$dX^{(n)}(t) = -\alpha\, X^{(n)}(t)\, dA^{(n)}(t) + dW^{(n)}(t) \tag{2.57}$$

where $A^{(n)}(t) = \frac{k}{n}$ for $\frac{k}{n} \le t < \frac{k+1}{n}$.

It is clear that the original experiment $\mathbb{E}^n$ is "equivalent" to the experiment

$$\widetilde{\mathcal{E}}^n = (\mathbf{D}, \mathcal{D};\ P^n_\alpha\,,\, 0 \le \alpha \le 2n),$$

where $\mathbf{D}$ is the space of functions on [0,1] continuous from the right and having left limits, and P^n_α is the measure on $(\mathbf{D}, \mathcal{D})$ induced from $\mathbf{P}^n_{1-\alpha/n}$ by the mapping $(x_1, \dots, x_n) \to (X^{(n)}(t),\, t \le 1)$ introduced above.

It follows "almost evidently" from equation (2.57) that the limit experiment $\mathcal{E} = (\mathbf{D}, \mathcal{D};\ P_\alpha\,, \alpha \ge 0)$ (for $\widetilde{\mathcal{E}}^n$ as $n \to \infty$) is described by the process $X = (X(t),\, t \le 1)$ obeying the stochastic equation

$$dX(t) = -\alpha\, X(t)\, dt + dW(t), \tag{2.58}$$

with a standard Wiener process W, i.e. the measure P_α is the distribution on $(\mathbf{D}, \mathcal{D})$ of the process X from (2.58).

(Formally, λ-convergence $\widetilde{\mathcal{E}}^n \xrightarrow{\lambda} \mathcal{E}$ can be easily established here because this model allows an explicit expression of likelihood for the experiment $\mathcal{E}$ as well as for the experiments $\widetilde{\mathcal{E}}^n$, see Anderson (1959).)

Example 2.6 This example is a complication of the previous one. Instead of (2.55) the following model is considered:

$$x_k = \theta_k\, x_{k-1} + \varepsilon_k, \quad x_0 = 0, \tag{2.59}$$

where $(\theta_k)_{k \ge 1}$ is a sequence of *independent identically distributed random variables* with $E\,\theta_k = 0$, $D\,\theta_k = \sigma^2$. We assume σ^2 to be *known* and the parameter θ to be *unknown* but $\theta^2 + \sigma^2 \le 1$. If we define

$$\eta_k = \frac{\theta_k - \theta}{\sigma},$$

then (2.59) takes the form

$$x_k = \theta\, x_{k-1} + (\varepsilon_k + \sigma\, \eta_k\, x_{k-1}).$$

Denoting $\alpha = (1-\theta)n$, $\Sigma = \sigma\sqrt{n}$ and using the notation of the previous example we obtain that the model (2.59) is equivalent to the following one:

$$dX^{(n)}(t) = -\alpha X^{(n)}(t)\,dA^{(n)}(t) + \sqrt{1+\Sigma^2\left(X^{(n)}(t)\right)}\,dW^{(n)}(t), \qquad (2.60)$$

where $0 \le t \le 1$, and α, Σ (for nth series) belong to the set

$$\left\{(\alpha,\Sigma) : \alpha = (1-\theta)\,n, \quad \Sigma = \sigma\sqrt{n} \text{ and } \theta^2+\sigma^2 \le 1\right\}.$$

Similarly to Example 2.5 one can show that the limit experiment corresponds to the model

$$dX(t) = -\alpha X(t)\,dt + \sqrt{1+\Sigma^2 X^2(t)}\,dW(t), \quad 0 \le t \le 1,$$

with $\alpha \ge 0$, $\Sigma \ge 0$, Σ being known and α being estimated.

2.5 λ-Convergence and Accompanying Experiments

1. One of the main questions under consideration in this and further sections is the relations between the previously introduced *λ-convergence*, *weak* and *strong convergence* of statistical experiments.

Definition 2.7 A dominated canonical experiment $\mathcal{E} = (X, \mathfrak{X}; P_\alpha, \alpha \in \mathcal{A}; P)$ is *regular* (see Section 1.9, c) if for every $\alpha \in \mathcal{A}$, the P-measure of the set of discontinuous points $x \in X$ of the density $Z_\alpha(x) = \dfrac{dP_\alpha}{dP}(x)$ is equal to zero (i.e. $Z_\alpha(x)$ are P-continuous functions in $x \in X$).

In what follows we consider a sequence of experiments $(\mathcal{E}^n)_{n\ge1}$, λ-converging (in the sense of Definition 2.6) to some fixed regular experiment $\mathcal{E}$. Theorem 2.5 below (and the same for the similar Theorem 2.8 in Section 2.6) plays a central role for the investigation of various questions of convergence of statistical experiments and of properties of the corresponding decisions.

Theorem 2.5 *Let $\mathcal{E}^n \xrightarrow{\lambda} \mathcal{E}$ with a regular statistical experiment $\mathcal{E}$. Then there exist extensions $(\widehat{\Omega}^n, \widehat{\mathcal{F}}^n, \widehat{P}^n)$ of the original probability spaces $(\Omega^n, \mathcal{F}^n, P^n)$, $n \ge 1$, and $\widehat{\mathcal{F}}^n/\mathfrak{X}$-measurable random elements $\widehat{\lambda}^n : \widehat{\Omega}^n \to X$ such that*

(i) $\rho(\widehat{\lambda}^n, \lambda^n) \xrightarrow{\widehat{P}^n} 0,$

(ii) $\mathcal{L}(\widehat{\lambda}^n \mid \widehat{P}^n) = \mathcal{L}(\lambda \mid P) \quad (= P),$

(iii) $Z^n_\alpha - Z_\alpha(\widehat{\lambda}^n) \xrightarrow{\widehat{P}^n} 0, \quad \alpha \in \mathcal{A},$

(iv) $\widehat{E}^n \left|Z^n_\alpha - Z_\alpha(\widehat{\lambda}^n)\right| \to 0, \quad \alpha \in \mathcal{A}.$

Here ρ is the metric in X, and $\widehat{E}^n$ means integration with respect to $\widehat{P}^n$.

Remark 2.6 An explicit construction of the extended spaces $(\widehat{\Omega}^n, \widehat{\mathcal{F}}^n, \widehat{P}^n)$ of the original probability spaces $(\Omega^n, \mathcal{F}^n, P^n)$, $n \geq 1$, is described in Section 2.10 below in the proof of the *Lemma about "Reconstruction"*. As follows from this construction these extensions can be chosen to be semi-direct products (see Section 1.4) with the help of Markov kernels K^n from $(\Omega^n, \mathcal{F}^n)$ to $(X, \mathcal{X})$, $n \geq 1$, (i.e. $(\widehat{\Omega}^n, \widehat{\mathcal{F}}^n) = (\Omega^n \times X, \mathcal{F}^n \otimes \mathcal{X})$), and the measures $\widehat{P}^n$ are semi-direct products of the measures P^n and the Markov kernels K^n. The original spaces $(\Omega^n, \mathcal{F}^n, P^n)$ are naturally embedded in these extensions $(\widehat{\Omega}^n, \widehat{\mathcal{F}}^n, \widehat{P}^n)$, and the random variables given on $(\Omega^n, \mathcal{F}^n, P^n)$ allow for a natural continuation to the random variables on the extended spaces.

Remark 2.7 According to the definition of λ-convergence (see (2.45))

$$\mathcal{L}(\lambda^n \mid P^n) \xrightarrow{w} \mathcal{L}(\lambda \mid P) \qquad (= P)$$

$$Z_\alpha^n(\omega) - Z_\alpha(\lambda^n(\omega)) \xrightarrow{P^n} 0, \qquad \alpha \in \mathcal{A}.$$

Therefore the assertion (i) of the theorem means that with the help of randomizations the random elements λ^n can be "transformed" into new random elements $\widehat{\lambda}^n$ that are asymptotically close to λ^n (in the sense that $\rho(\widehat{\lambda}^n, \lambda^n) \xrightarrow{P^n} 0$) but their distributions $\mathcal{L}(\widehat{\lambda}^n \mid P^n)$ *exactly coincide* with the distribution $\mathcal{L}(\lambda \mid P)$. The assertion (iii) means that in the definition of λ-convergence the random elements λ^n can be exchanged by $\widehat{\lambda}^n$ with the important feature that $\mathcal{L}(\widehat{\lambda}^n \mid P^n) = \mathcal{L}(\lambda \mid P)$ for all $n \geq 1$.

Now we give some consequences of Theorem 2.5 which are important for investigating the properties of statistical decisions.

Theorem 2.6 *Let $\mathcal{E}$ be a regular experiment and $\mathcal{E}^n \xrightarrow{\lambda} \mathcal{E}$. Then*

(i) $P^n_{\alpha,s}(\Omega^n) \to 0, \qquad n \to \infty, \alpha \in \mathcal{A},$
where $P^n_{\alpha,s}(\Omega^n)$ is the singular component of the measure P^n_α with respect to P^n.

(ii) $(P^n_\alpha) \triangleleft (P^n), \quad \alpha \in \mathcal{A},$
i.e. the family of measures $(P^n_\alpha)_{n\geq 1}$ is contiguous to the family $(P^n)_{n\geq 1}$.

(iii) For any uniformly bounded sequence of $\mathcal{F}^n$-measurable random variables φ_n (i.e. $|\varphi_n| \leq c$)

$$E^n_\alpha \varphi_n - \widehat{E}^n Z_\alpha(\widehat{\lambda}^n) \varphi_n \to 0, \quad n \to \infty, \alpha \in \mathcal{A}. \tag{2.61}$$

(iv) The weak convergence of statistical experiments holds:

$$\mathcal{E}^n \xrightarrow{w} \mathcal{E}. \tag{2.62}$$

2. Using the elements $\widehat{\lambda}^n$ defined on the extensions $(\widehat{\Omega}, \widehat{\mathcal{F}}, \widehat{P})$ one can construct new statistical experiments

$$\widehat{\mathcal{E}}^n = (\widehat{\Omega}^n, \widehat{\mathcal{F}}^n, \widehat{P}^n_\alpha, \alpha \in \mathcal{A}), \tag{2.63}$$

with

$$d\widehat{P}_\alpha^n = Z_\alpha(\widehat{\lambda}^n)\, d\widehat{P}^n. \tag{2.64}$$

These experiments $\widehat{\mathcal{E}}^n$ will be called *accompanying experiments* for $\mathcal{E}^n$.

Theorem 2.7 *Let $\mathcal{E}$ be a regular experiment and $\mathcal{E}^n \xrightarrow{\lambda} \mathcal{E}$.*

(i) For any $\alpha \in \mathcal{A}$

$$\|P_\alpha^n - \widehat{P}_\alpha^n\| \to 0, \quad n \to \infty. \tag{2.65}$$

(ii) For each $n \geq 1$ the experiments $\widehat{\mathcal{E}}^n$ and $\mathcal{E}$ are weakly and strongly equivalent.

The first assertion of the theorem justifies the name of *accompanying* for the experiments $\widehat{\mathcal{E}}^n$. The second one claims that the experiments $\widehat{\mathcal{E}}^n$ and $\mathcal{E}$ are equivalent for every $n \geq 1$ from the point of view of the corresponding distributions.

3. Proof of Theorem 2.5 The possibility to construct elements $\widehat{\lambda}^n$ with the desired properties is based on the following general assertion which is proved below in Section 2.10.

Lemma 2.2 ***(About "Reconstruction")*** *Let a random element ξ (respectively ξ^n, $n \geq 1$) with values in a complete metric separable space $(X, \mathfrak{X}, \rho)$ be given on a probability space $(\Omega, \mathcal{F}, P)$ (respectively $(\Omega^n, \mathcal{F}^n, P^n)$, $n \geq 1$) and let $\xi^n \xrightarrow{d} \xi$, i.e. $\mathcal{L}(\xi^n \mid P^n) \xrightarrow{w} \mathcal{L}(\xi \mid P)$, $n \to \infty$. Then there exist (Markov) extensions $(\widehat{\Omega}^n, \widehat{\mathcal{F}}^n, \widehat{P}^n)$ of the spaces $(\Omega^n, \mathcal{F}^n, P^n)$ and random elements $\widehat{\xi}^n$ given on them with values in $(X, \mathfrak{X}, \rho)$ such that*

$$\begin{aligned} \rho(\widehat{\xi}^n, \xi^n) &\xrightarrow{\widehat{P}^n} 0, \\ \mathcal{L}(\widehat{\xi}^n \mid \widehat{P}^n) &= \mathcal{L}(\xi \mid P). \end{aligned}$$

The assertions (i) and (ii) of the theorem follow directly from this lemma. We prove now the third assertion of the theorem. Define for $\delta > 0$

$$D^\delta Z_\alpha(x) = \sup_{\{y: \rho(x,y) \leq \delta\}} |Z_\alpha(x) - Z_\alpha(y)|. \tag{2.66}$$

The assumption of regularity of the experiment $\mathcal{E}$ implies that for any $\varepsilon > 0$ a $\delta > 0$ can be found such that

$$P(D^\delta Z_\alpha(\lambda) > \varepsilon) \leq \varepsilon.$$

Then for sufficiently large n

$$\widehat{P}^n(\rho(\widehat{\lambda}^n, \lambda^n) \geq \delta) \leq \varepsilon,$$

and for $\rho(\widehat{\lambda}^n, \lambda^n) \leq \delta$

$$|Z_\alpha(\widehat{\lambda}^n) - Z_\alpha(\lambda^n)| \leq D^\delta Z_\alpha(\widehat{\lambda}^n).$$

Thus, taking into account the equality $\mathcal{L}(\widehat{\lambda}^n \mid \widehat{P}^n) = \mathcal{L}(\lambda \mid P)$ we have

$$\begin{aligned}\widehat{P}^n(|Z_\alpha(\widehat{\lambda}^n) - Z_\alpha(\lambda^n)| > \varepsilon) &\leq \varepsilon + \widehat{P}^n(D^\delta Z_\alpha(\widehat{\lambda}^n) > \varepsilon) \\ &= \varepsilon + P(D^\delta Z_\alpha(\lambda) > \varepsilon) \leq 2\varepsilon,\end{aligned}$$

which together with the property $Z_\alpha^n - Z_\alpha(\widehat{\lambda}^n) \xrightarrow{\widehat{P}^n} 0$ proves assertion (iii).

To prove the last assertion of the theorem it suffices to verify the condition of uniform integrability of the families of random variables $\{Z_\alpha^n, n \geq 1\}$ and $\{Z_\alpha(\widehat{\lambda}^n), n \geq 1\}$. Because $\mathcal{L}(Z_\alpha(\widehat{\lambda}^n) \mid \widehat{P}^n) = \mathcal{L}(Z_\alpha(\lambda) \mid P)$ for any $n \geq 1$ the family $\{Z_\alpha(\widehat{\lambda}^n)\}_{n\geq 1}$ is uniform integrable.

Now we consider the sequence $\{Z_\alpha^n, n \geq 1\}$ and show that it is also uniformly integrable, i.e. that

$$\lim_{n\to\infty} E^n Z_\alpha^n I_{\{Z_\alpha^n > b\}} \to 0, b \to 0.$$

Choose for each $b > 0$ a continuous function $f_b(x)$ on R_+ such that $f_b(x) = 1$ for $x \leq b$ and $f_b(x) = 0$ for $x > b + 1$. Since $E^n Z_\alpha^n \leq 1$ it suffices to verify the condition

$$\lim_{n\to\infty} E^n f_b(Z_\alpha^n) \to 1, \quad b \to \infty. \tag{2.67}$$

It follows from the convergence $Z_\alpha^n - Z_\alpha(\widehat{\lambda}^n) \xrightarrow{\widehat{P}^n} 0$ and also from continuity and boundedness of the function $f_b(x)$ that for each b

$$\lim_{n\to\infty} \left[E^n f_b(Z_\alpha^n) - \widehat{E}^n f_b(Z_\alpha(\widehat{\lambda}^n))\right] = 0. \tag{2.68}$$

Now the condition of uniform integrability of the family $\{Z_\alpha(\widehat{\lambda}^n)\}_{n\geq 1}$ and the equalities $\widehat{E}^n Z_\alpha(\widehat{\lambda}^n) = E Z_\alpha(\lambda) = 1$ imply

$$\lim_{n\to\infty} \widehat{E}^n f_b(Z_\alpha(\widehat{\lambda}^n)) = E f_b(Z_\alpha(\lambda)) \to 1, \quad b \to \infty. \tag{2.69}$$

Combining (2.68) and (2.69) leads to (2.67) and the theorem follows.

4. Proof of Theorem 2.6 It follows from the Lebesgue decomposition of the measures P_α^n with respect to P^n (see Jacod and Shiryaev (1987, Section V.1a)) that the assertion $P_{\alpha,s}^n(\Omega^n) \to 0$ is equivalent to the property $E^n Z_\alpha^n \to 1$. But the latter follows directly from the assertions (iv) and (ii) of Theorem 2.5 and the fact that for the limit experiment $\mathcal{E}$ it holds $E Z_\alpha(\lambda) = 1$. So, the property (i) is proved.

Let sets $A^n \in \mathcal{F}^n$ be such that $P^n(A^n) \to 0$, $n \to \infty$. Then for any number $b > 0$

$$P_\alpha^n(A^n) \leq P_{\alpha,s}^n(\Omega^n) + E^n Z_\alpha^n I_{A^n} \leq P_{\alpha,s}^n(\Omega^n) + b\,P^n(A^n) + E^n Z_\alpha^n I_{\{Z_\alpha^n > b\}}.$$

Making first n and then b tend to infinity we obtain from the first assertion of the theorem and from the previously proved property of uniform integrability of the sequence $\{Z_\alpha^n\}$ that $P_\alpha^n(A^n) \to 0,\ n \to \infty$ which exactly means the contiguity $(P_\alpha^n) \triangleleft (P^n)$, i.e. the property (ii).

We now prove the third assertion of the theorem. Since

$$E_\alpha^n \varphi_n = \int \varphi_n \, dP_{\alpha,s}^n + E^n Z_\alpha^n \varphi_n,$$

and $|\int \varphi_n \, dP_{\alpha,s}^n| \leq c P_{\alpha,s}^n(\Omega^n) \to 0,\ n \to \infty$, it is sufficient to show that

$$E^n Z_\alpha^n \varphi_n - \widehat{E}^n Z_\alpha(\widehat{\lambda}^n) \varphi_n \to 0, \quad n \to \infty.$$

Since also

$$\begin{aligned} |E^n Z_\alpha^n \varphi_n - \widehat{E}^n Z_\alpha(\widehat{\lambda}^n) \varphi_n| &= |\widehat{E}^n \varphi_n (Z_\alpha^n - Z_\alpha(\widehat{\lambda}^n))| \\ &\leq c \widehat{E}^n |Z_\alpha^n - Z_\alpha(\widehat{\lambda}^n)| \to 0, \quad n \to \infty \end{aligned}$$

the statement (2.61) is proved.

Finally, to check the property (iv) that for the case of a regular experiment $\mathcal{E}$

$$\mathcal{E}^n \xrightarrow{\lambda} \mathcal{E} \quad \Rightarrow \quad \mathcal{E}^n \xrightarrow{w} \mathcal{E}$$

it suffices to verify that

$$\mathcal{L}\left(\sum_{i=1}^m c_i Z_{\alpha_i}^n \mid P^n \right) \to \mathcal{L}\left(\sum_{i=1}^m c_i Z_{\alpha_i} \mid P \right) \tag{2.70}$$

for any points α_i from $\mathcal{A}$ and any number c_i, $i = 1, \dots, m$. By Theorem 2.5

$$\sum_{i=1}^m c_i Z_{\alpha_i}^n - \sum_{i=1}^m c_i Z_{\alpha_i}(\widehat{\lambda}^n) \xrightarrow{\widehat{P}^n} 0,$$

and (2.70) follows from the condition $\mathcal{L}(Z_\alpha(\widehat{\lambda}^n) \mid \widehat{P}^n) = \mathcal{L}(Z_\alpha(\lambda) \mid P)$, $\alpha \in \mathcal{A}$. Theorem 2.6 is proved.

5. Proof of Theorem 2.7. The convergence $\|P_\alpha^n - \widehat{P}_\alpha^n\| \to 0$ follows from assertion (iii) of Theorem 2.6 because $P_\alpha^n(A) - \widehat{P}_\alpha^n(A) = \left[E_\alpha^n I(A) - \widehat{E}^n \widehat{Z}_\alpha^n I(A)\right]$, $A \in \widehat{\mathcal{F}}^n$.

In order to prove the weak equivalence $\widehat{\mathcal{E}}^n \sim \mathcal{E}$ it suffices to notice that due to the property (ii) of Theorem 2.5 for each finite set $\{\alpha_1, \dots, \alpha_m\}$

$$\mathcal{L}\left(Z_{\alpha_i}(\widehat{\lambda}_n),\ i = 1, \dots, m \mid \widehat{P}^n\right) = \mathcal{L}(Z_{\alpha_i}(\lambda),\ i = 1, \dots, m \mid P).$$

Concerning the strong equivalence $\widehat{\mathcal{E}}^n \overset{\Delta}{\approx} \mathcal{E}$ that it is equivalent to the weak one (see Section 2.3).

2.6 Uniform Versions of λ-Convergence

1. The results of the previous section only allow to conclude that λ-convergence implies weak convergence ($\mathcal{E}^n \xrightarrow{\lambda} \mathcal{E} \Rightarrow \mathcal{E}^n \xrightarrow{w} \mathcal{E}$) and, hence (from the point of view of its use in problems of statistical decision theory) it is only a useful way to verify the condition of the weak convergence in concrete problems. In this section we demonstrate that some strengthening of the condition of λ-convergence implies already the strong convergence of statistical experiments (on compacts) which in turn gives us the possibility to obtain stronger conclusions about the experiments $\mathcal{E}^n$ if we know the properties of the "limit" experiment $\mathcal{E}$.

Let $K \subseteq \mathcal{A}$ be a subset of a parameter set $\mathcal{A}$. By $\mathcal{K} = \{K : K \subseteq \mathcal{A}\}$ we shall denote one or another (depending on the problem under consideration) collection of subsets of $\mathcal{A}$.

Definition 2.8 A sequence of experiments $\mathcal{E}^n = (\Omega^n, \mathcal{F}^n; P^n_\alpha, \alpha \in \mathcal{A}^n)$, $\mathcal{A}^n \uparrow \mathcal{A}$, $n \to \infty$, *K-uniformly λ-converges* to a dominated canonical experiment $\mathcal{E} = (X, \mathfrak{X}; P_\alpha, \alpha \in \mathcal{A})$ (notation $\mathcal{E}^n \xrightarrow{\lambda(K)} \mathcal{E}$) if there exist $\mathcal{F}^n/\mathfrak{X}$-measurable elements $\lambda^n = \lambda^n(\omega) : \Omega^n \to X$ such that

$$\mathcal{L}(\lambda^n \mid P^n) \xrightarrow{w} \mathcal{L}(\lambda \mid P) \qquad (= P)$$
$$\sup_{\alpha \in K} P^n(|Z^n_\alpha - Z_\alpha(\lambda^n)| > \varepsilon) \to 0, \qquad n \to \infty, \tag{2.71}$$

for any $\varepsilon > 0$ with the measures P^n defined in (2.44).

Definition 2.9 Let $\mathcal{K} = \{K : K \subseteq \mathcal{A}\}$ be a collection of subsets in $\mathcal{A}$. We say that *$\mathcal{K}$-uniform λ-convergence* or *$\lambda(\mathcal{K})$-convergence* ($\mathcal{E}^n \xrightarrow{\lambda(\mathcal{K})} \mathcal{E}$) holds if there exist elements λ^n such that the properties (2.71) are satisfied for every $K \in \mathcal{K}$.

Definition 2.10 A canonical dominated experiment $\mathcal{E} = (X, \mathfrak{X}; P_\alpha, \alpha \in \mathcal{A}; P)$ is *$\mathcal{K}$-regular* if the density functions (Radon-Nikodym derivatives) $Z_\alpha(x) = dP_\alpha/dP(x)$, $\alpha \in \mathcal{A}$, $x \in X$, satisfy for each $K \in \mathcal{K}$ the following conditions:

(i) The family $\{Z_\alpha(x), \alpha \in K\}$ is *uniformly integrable* under the measure P, i.e.

$$\sup_{\alpha \in K} E\, Z_\alpha(x)\, I_{\{Z_\alpha(x) > b\}} \to 0, \quad b \to \infty;$$

(ii) The family $\{Z_\alpha(x), \alpha \in K\}$ is *uniformly continuous* with respect to the measure P, i.e.

$$\sup_{\alpha \in K} P(D^\delta Z_\alpha(x) > \varepsilon) \to 0, \quad \delta \to 0,$$

with

$$D^\delta Z_\alpha(x) = \sup_{\{y : \rho(x,y) \le \delta\}} |Z_\alpha(x) - Z_\alpha(y)|.$$

Remark 2.8 If the class $\mathcal{K}$ consists of one-point sets $K = \{\alpha\}$, $\alpha \in \mathcal{A}$, then the definition of $\mathcal{K}$-regularity becomes the definition of a regular experiment from Section 2.5. It is easy to see that in the case that the class $\mathcal{K}$ consists of all finite subsets of the set $\mathcal{A}$, $K = \{\alpha_1, \ldots, \alpha_m\}$, $\alpha_i \in \mathcal{A}$, the condition of $\mathcal{K}$-regularity is again equivalent to the condition of regularity of the experiment $\mathcal{E}$.

A simple sufficient condition for $\mathcal{K}$-regularity is given by the following assertion.

Lemma 2.3 *Let a parameter set* $\mathcal{A}$ *be a metric space with a metric* $\rho_{\mathcal{A}}(\alpha, \beta)$ *and let* $\mathcal{K}$ *be the class of compacts in* $\mathcal{A}$. *Let then* $\mathcal{E} = (X, \mathfrak{X}; P_\alpha, \alpha \in \mathcal{A})$ *be a dominated canonical experiment and let the density functions* $Z_\alpha(x) = dP_\alpha/dP(x)$ *be continuous in* (x, α) *on* $X \times \mathcal{A}$. *Then the experiment* $\mathcal{E}$ *is* $\mathcal{K}$*-regular.*

Proof. In order to prove the uniform continuity of the family $\{Z_\alpha(x),\ \alpha \in K\}$ under the measure P for a compact $K \in \mathcal{K}$ let us choose $\varepsilon > 0$ and let C be a compact in X such that $P(C) > 1 - \varepsilon$. Denote $F(x, \alpha) = Z_\alpha(x)$,

$$D^\delta F(x, \alpha) = \sup_{\{y, \beta:\ \rho(x,y) + \rho_{\mathcal{A}}(\alpha, \beta) \le \delta\}} |F(x, \alpha) - F(y, \beta)|.$$

Since the function $F(x, \alpha)$ is continuous on $X \times \mathcal{A}$, it holds for any $\varepsilon > 0$ and for sufficiently small $\delta > 0$

$$\sup_{x \in C,\, \alpha \in K} D^\delta F(x, \alpha) < \varepsilon.$$

From the definition of $D^\delta Z_\alpha(x)$ one has

$$D^\delta Z_\alpha(x) \le D^\delta F(x, \alpha), \qquad x \in C,\ \alpha \in K.$$

Thus, for sufficiently small δ

$$\sup_{\alpha \in K} P(D^\delta Z_\alpha(x) > \varepsilon) \le \sup_{\alpha \in K} P(D^\delta Z_\alpha(x) > \varepsilon, C) + P(\overline{C}) \le 2\varepsilon,$$

where $\overline{C}$ is the complement of C in X. This implies the required uniform continuity.

In order to prove the uniform integrability under the measure P of the family $\{Z_\alpha(x),\ \alpha \in K\}$, assume the contrary. Then there is a sequence (α_m) of points in K for which the family $\{Z_{\alpha_m}(x),\ m \ge 1\}$ is not uniformly integrable. Since K is compact, there are a subsequence $(\alpha_{m'})$ and a point $\alpha \in K$ such that $\alpha_{m'} \to \alpha$ and the family $\{Z_{\alpha_{m'}}(x)\}$ is not uniformly integrable. But from the above proved property of uniform continuity of the family $\{Z_\alpha(x),\ \alpha \in K\}$ under P we obtain that

$$Z_{\alpha_{m'}} \xrightarrow{P} Z_\alpha.$$

Then $Z_{\alpha_{m'}} \xrightarrow{d} Z_\alpha$. But $E Z_{\alpha_{m'}} = 1$, $E Z_\alpha = 1$. Thus (see Billingsley (1968, Chapter 2, Section 6)) the family $\{Z_{\alpha_{m'}}\}$ is uniformly integrable which contradicts the assumption we made and the lemma follows. □

Theorem 2.8 *Let $\mathcal{K} = \{K : K \subseteq A\}$ be some class of subsets $K \subseteq A$ and let $\mathcal{E}^n \xrightarrow{\lambda(\mathcal{K})} \mathcal{E}$ where $\mathcal{E}$ is a $\mathcal{K}$-regular experiment. Then there exist extensions $(\widehat{\Omega}^n, \widehat{\mathcal{F}}^n, \widehat{P}^n)$ of the original probability spaces $(\Omega^n, \mathcal{F}^n, P^n)$, $n \geq 1$, and $\widehat{\mathcal{F}}^n/\mathfrak{X}$-measurable elements $\widehat{\lambda}^n : \widehat{\Omega}^n \to X$ such that, as $n \to \infty$*

(i) $\rho(\widehat{\lambda}^n, \lambda^n) \xrightarrow{\widehat{P}^n} 0;$

(ii) $\mathcal{L}(\widehat{\lambda}^n \mid \widehat{P}^n) = \mathcal{L}(\lambda \mid P) \quad (= P);$

(iii) $\sup_{\alpha \in K} \widehat{P}^n(|Z_\alpha^n - Z_\alpha(\widehat{\lambda}^n)| > \varepsilon) \to 0, \quad K \in \mathcal{K};$

(iv) $\sup_{\alpha \in K} \widehat{E}^n |Z_\alpha^n - Z_\alpha(\widehat{\lambda}^n)| \to 0, \quad K \in \mathcal{K}.$

Proof. The existence of elements $\widehat{\lambda}^n$, $n \geq 1$, satisfying the properties (i) and (ii) follows from the lemma about "Reconstruction" (see Section 2.5). The assertion (iii) and (iv) are proved (using the property of $\mathcal{K}$-regularity of the experiment $\mathcal{E}$ and the second property in (2.71)) in the same way as in Theorem 2.5. As illustration we give the proof of the most complicated part claiming uniform integrability of the family $\{Z_\alpha^n, \alpha \in K, n \geq 1\}$ (for every $K \in \mathcal{K}$), i.e.

$$\lim_{n\to\infty} \sup_{\alpha \in K} E^n Z_\alpha^n I_{\{Z_\alpha^n > b\}} \to 0, \quad b \to 0. \tag{2.72}$$

Let $b > 0$ and let $f_b(x)$ be the function on R_+ of the form

$$f_b(x) = \begin{cases} 1 & x \leq b \\ 0 & x > b+1 \\ x - b & 0 < x \leq b+1 \end{cases}.$$

Then $|f_b(x) - f_b(y)| \leq |x - y|$ and from the third assertion of the theorem we get

$$\lim_{n\to\infty} \sup_{\alpha \in K} \left[E^n f_b(Z_\alpha^n) - \widehat{E}^n f_b(Z_\alpha(\widehat{\lambda}^n))\right] = 0. \tag{2.73}$$

Since $E^n Z_\alpha^n \leq 1$, to prove (2.72) it suffices to show that

$$\lim_{n\to\infty} \sup_{\alpha \in K} E^n f_b(Z_\alpha^n) \to 1, \quad b \to \infty. \tag{2.74}$$

Then the family $\{Z_\alpha(\widehat{\lambda}^n), \alpha \in K, n \geq 1\}$ is uniform integrable because of the $\mathcal{K}$-regularity of the experiment $\mathcal{E}$ and the equality $\mathcal{L}(\widehat{\lambda}^n \mid \widehat{P}^n) = \mathcal{L}(\lambda \mid P)$. Since also $\widehat{E}^n Z_\alpha(\widehat{\lambda}^n) = E\, Z_\alpha(\lambda) = 1$, this yields

$$\lim_{n\to\infty} \sup_{\alpha \in K} \widehat{E}^n f_b(Z_\alpha(\widehat{\lambda}^n)) = \sup_{\alpha \in K} E\, f_b(Z_\alpha(\lambda)) \to 1, \quad b \to \infty. \tag{2.75}$$

It remains to note that (2.75) and (2.73) evidently imply (2.74). □

2. The assertions of the next theorem are uniform versions of the analogous assertions of Theorems 2.6 and 2.7.

Theorem 2.9 *Let $\mathcal{E}$ be a $\mathcal{K}$-regular experiment and $\mathcal{E}^n \xrightarrow{\lambda(\mathcal{K})} \mathcal{E}$. Let also $\widehat{\mathcal{E}}^n$ be the accompanying experiments defined by (2.63) and (2.64). Then for any $K \in \mathcal{K}$*

(i) $\sup_{\alpha\in K} P^n_{\alpha,s}(\Omega^n) \to 0, \quad n \to \infty, \alpha \in \mathcal{A}$,
where $P^n_{\alpha,s}(\Omega^n)$ is the singular component of the measure P^n_α with respect to P^n.

(ii) The family of the measures $(P^n_\alpha)_{n\ge1}$ is contiguous to the family $(P^n)_{n\ge1}$ uniformly in $\alpha \in K$, i.e. for any sets $A^n_\alpha \in \mathcal{F}^n$, $\alpha \in K$,

$$P^n(A^n_\alpha) \to 0 \quad \Longrightarrow \quad \sup_{\alpha\in K} P^n_\alpha(A^n_\alpha) \to 0, \qquad n \to \infty.$$

(iii) For any uniformly bounded sequence of $\mathcal{F}^n$-measurable random variables φ_n ($|\varphi_n| \le c$)

$$\sup_{\alpha\in K} |E^n_\alpha \varphi_n - \widehat{E}^n Z_\alpha(\widehat{\lambda}_n)\varphi_n| \to 0, \quad n \to \infty.$$

(iv) $\sup_{\alpha\in K} \|P^n_\alpha - \widehat{P}^n_\alpha\| \to 0, \quad n \to \infty.$

(v) For each $n \ge 1$ the experiments $\widehat{\mathcal{E}}^n$ and $\mathcal{E}$ are weakly and strongly equivalent.

(vi) The uniform weak convergence $\mathcal{E}^n \xrightarrow{w(K)} \mathcal{E}$, $K \in \mathcal{K}$, holds.

(vii) The uniform strong convergence $\mathcal{E}^n \xrightarrow{\Delta(\mathcal{K})} \mathcal{E}$ holds, i.e. for each $K \in \mathcal{K}$

$$\Delta(\mathcal{E}^n, \mathcal{E}; K) \to 0, \quad n \to \infty. \tag{2.76}$$

Proof. The first five assertions of the theorem are obtained in a similar way to Theorems 2.6 and 2.7, taking into account the properties of $\mathcal{K}$-regularity of the experiment $\mathcal{E}$ and the λ-convergence, uniform in $K \in \mathcal{K}$, of the experiments $\mathcal{E}^n$ to $\mathcal{E}$ (see the proof of Theorem 2.8).

For proving the property (2.76) it is sufficient to remark that the assertions (iv) and (v) of the theorem imply

$$\Delta(\mathcal{E}^n, \mathcal{E}; K) \le \Delta(\mathcal{E}^n, \widehat{\mathcal{E}}^n; K) + \Delta(\widehat{\mathcal{E}}^n, \mathcal{E}; K) \le \sup_{\alpha\in K} \|P^n_\alpha - \widehat{P}^n_\alpha\| \to 0, \quad n \to \infty.$$

Before proving (vi) we explain how the uniform weak convergence $\mathcal{E}^n \xrightarrow{w(K)} \mathcal{E}, K \in \mathcal{K}$, is treated. Let $\mathcal{E}^n_K$ and $\mathcal{E}_K$ be the restriction of the experiments $\mathcal{E}^n$ and $\mathcal{E}$ on K. Denote by $S_m(K)$ the class of all finite subsets $\sigma = \{\alpha_1, \dots, \alpha_m\}$ consisting of m points $\alpha_i \in K$. For the set $\sigma = \{\alpha_1, \dots, \alpha_m\} \in S_m(K)$ put

$$Z_\sigma = (Z_{\alpha_1}, \dots, Z_{\alpha_m}), \quad Z^n_\sigma = (Z^n_{\alpha_1}, \dots, Z^n_{\alpha_m}), \quad \widehat{Z}^n_\sigma = (Z_{\alpha_1}(\widehat{\lambda}^n), \dots, Z_{\alpha_m}(\widehat{\lambda}^n)).$$

The convergence $\mathcal{E}^n \xrightarrow{w(K)} \mathcal{E}$ means that the following property holds for each $m \geq 1$:

$$\sup_{\sigma \in S_m(K)} |E^n f(Z_\sigma^n) - E f(Z_\sigma)| \to 0, \quad n \to \infty, \tag{2.77}$$

with any real-valued continuous bounded function $f = f(z)$ on R^m. Due to the property (ii) of Theorem 2.8 the assertion (2.77) is equivalent to the convergence

$$\sup_{\sigma \in S_m(K)} \widehat{E}^n \left[f(Z_\sigma^n) - f(\widehat{Z}_\sigma^n) \right] \to 0, \quad n \to \infty. \tag{2.78}$$

Suppose first that the function $f = f(z)$, $z \in R^m$, is uniformly continuous on the whole space R^m, i.e. for any $\varepsilon > 0$ and for sufficiently small $\delta > 0$ the condition $\|z - z'\| < \delta$ implies $|f(z) - f(z')| \leq \varepsilon$. Then

$$\begin{aligned} \sup_{\sigma \in S_m(K)} \widehat{E}^n \left[f(Z_\sigma^n) - f(\widehat{Z}_\sigma^n) \right] &\leq \varepsilon + \sup_{\sigma \in S_m(K)} \widehat{P}^n \left(\|Z_\sigma^n - \widehat{Z}_\sigma^n\| > \delta \right) \\ &\leq \varepsilon + \sup_{\alpha \in K} \widehat{P}^n \left(|Z_\alpha^n - \widehat{Z}_\alpha^n| > \frac{\delta}{m} \right) \end{aligned}$$

and (2.78) directly follows from the assertion (iv) of Theorem 2.8.

In the general case a continuous function $f(z)$ is uniformly continuous on each ball $B_b = \{z \in R^m : \|z\| \leq b\}$. If $w_\delta(f, b) = \sup\{|f(z) - f(z')| : z, z' \in B_b, |z - z'| \leq \delta\}$ is the module of continuity of the function f on the ball B_b then $w_\delta(f, b) \to 0$ as $\delta \to 0$ and similarly to the above

$$\begin{aligned} &\sup_{\sigma \in S_m(K)} \widehat{E}^n \left[f(Z_\sigma^n) - f(\widehat{Z}_\sigma^n) \right] \\ &\quad \leq w_\delta(f, b) + \sup_{\alpha \in K} \widehat{P}^n \left(|Z_\alpha^n - \widehat{Z}_\alpha^n| > \frac{\delta}{m} \right) \\ &\quad + \sup_{\alpha \in K} \left[P^n(Z_\alpha^n > b) + \widehat{P}^n(\widehat{Z}_\alpha^n > b) \right]. \end{aligned} \tag{2.79}$$

Because of the uniform integrability of the families $\{Z_\alpha^n\}$, $\{\widehat{Z}_\alpha^n\}$ in $\alpha \in K$ and $n \to \infty$ the last term of (2.79) tends to zero as first n and then b tend to infinity. This yields (2.78), similarly to the case considered above. □

2.7 $\lambda(\mathcal{K})$-Convergence and Asymptotic Minimax Theorems

1. Let K be a subset of a parameter set $\mathcal{A}$ and let $\mathcal{E}$ be a K-regular experiment. If $\mathcal{E}^n \xrightarrow{\lambda(K)} \mathcal{E}$, then due to Theorem 2.9 the convergence $\mathcal{E}^n \xrightarrow{\Delta(K)} \mathcal{E}$ holds, i.e.

$$\mathcal{E}^n \xrightarrow{\lambda(K)} \mathcal{E} \quad \Rightarrow \quad \mathcal{E}^n \xrightarrow{\Delta(K)} \mathcal{E}, \tag{2.80}$$

and Theorem 2.1 implies for any function $W \in \mathcal{W}_c^b$ (continuous and bounded)

$$\mathcal{E}^n \xrightarrow{\lambda(K)} \mathcal{E} \quad \Rightarrow \quad \lim_{n\to\infty} \mathcal{R}(\mathcal{E}^n, W; K) = \mathcal{R}(\mathcal{E}, W; K). \tag{2.81}$$

At the same time λ-convergence implies weak convergence

$$\mathcal{E}^n \xrightarrow{\lambda} \mathcal{E} \quad \Rightarrow \quad \mathcal{E}^n \xrightarrow{w} \mathcal{E}, \tag{2.82}$$

and, thus, by Theorem 2.3 for $W \in \mathcal{W}_{lc}$ (bounded from below)

$$\mathcal{E}^n \xrightarrow{\lambda} \mathcal{E} \quad \Rightarrow \quad \lim_{n\to\infty} \mathcal{R}(\mathcal{E}^n, W; K) \geq \mathcal{R}(\mathcal{E}, W; K). \tag{2.83}$$

The assertion (2.80) means that $\lambda(K)$-convergence is a stronger condition than $\Delta(K)$-convergence. That is why results like (2.81) are not surprising. But, and it is important to stress it now, the property of $\lambda(K)$-convergence is not only a useful way to verify $\Delta(K)$-convergence. As follows from the exposition below, $\lambda(K)$-convergence and its versions, formulated in fairly natural terms, allow us firstly to obtain asymptotic minimax theorems for a *wider* class of loss functions W and secondly, these kinds of convergence permit us to construct explicitly, from "optimal" decisions for limit experiments, "asymptotic optimal" decisions for experiments $\mathcal{E}^n$ using elements $\widehat{\lambda}^n$ and the accompanying experiments $\widehat{\mathcal{E}}^n$. (For more details see also Chapter 4.)

2. Suppose that $\mathcal{E}^n = (\Omega^n, \mathcal{F}^n; P_\alpha^n, \alpha \in \mathcal{A}^n)$, $n \geq 1$, are statistical experiments where the parameter sets $\mathcal{A}^n$, $n \geq 1$, are such that $\mathcal{A}^n \uparrow \mathcal{A}$. By $\mathcal{E} = (X, \mathfrak{X}; P_\alpha, \alpha \in \mathcal{A})$ we denote a canonical dominated (by a measure P) experiment with the parameter set $\mathcal{A}$. Let $K \subseteq \mathcal{A}$ be some fixed parameter subset. We will assume $\mathcal{E}$ to be a K-regular experiment. Let then $(\mathbf{A}, \mathcal{A})$ be a measurable space of possible decisions with respect to the parameter α. By $W_\alpha(a)$ we denote a loss function which characterizes the "quality" of a decision a if the true value of the parameter is α. Also we suppose that $0 \leq W_\alpha(a) \leq C < \infty$, and that for each $\alpha \in \mathcal{A}$ the function $W_\alpha(a)$, $a \in \mathbf{A}$, is $\mathcal{A}$-measurable.

Since we do not suppose for $\mathbf{A}$ any topological structure and it is not assumed, say, $W \in \mathcal{W}_c^b$, for definitions of risks we will restrict ourselves to Markov decisions described by Markov kernels $\varkappa = \varkappa(\omega, A)$ (see Section 1.4). Define

$$W_\alpha(\varkappa) = \int_{\mathbf{A}} W_\alpha(a)\, \varkappa(\omega, da),$$

and let

$$R(\mathcal{E}, W, \varkappa; K) = \sup_{\alpha \in K} E_\alpha W_\alpha(\varkappa) \quad \left(= \sup_{\alpha \in K} \int_\Omega \int_{\mathbf{A}} W_\alpha(a)\, \varkappa(\omega, da)\, P_\alpha(d\omega) \right) \tag{2.84}$$

be the *risk* corresponding to the decision $\varkappa$ for the experiment $\mathcal{E} = (\Omega, \mathcal{F}; P_\alpha, \alpha \in$

$\mathcal{A}$) on the set K, and let

$$\mathcal{R}(\mathcal{E}, W; K) = \inf_{\varkappa} \sup_{\alpha \in K} E_\alpha W_\alpha(\varkappa)$$

be the minimax risk for the experiment $\mathcal{E} = (\Omega, \mathcal{F}; P_\alpha, \alpha \in \mathcal{A})$ on the set K.

The expressions $R(\mathcal{E}, W, \varkappa)$ and $\mathcal{R}(\mathcal{E}, W)$ we use for the risk of a decision $\varkappa$ and the minimax risk on the *whole* space $\mathcal{A}$.

Theorem 2.10 *Let $\mathcal{E}^n \xrightarrow{\lambda(K)} \mathcal{E}$ for a canonical dominated K-regular experiment $\mathcal{E}$ and $\|W\| < \infty$. Then*

$$\lim_{n \to \infty} \inf_{\varkappa^n} \sup_{\alpha \in K} E_\alpha^n W_\alpha(\varkappa^n) = \inf_{\varkappa} \sup_{\alpha \in K} E_\alpha W_\alpha(\varkappa), \tag{2.85}$$

i.e.

$$\lim_{n \to \infty} \mathcal{R}(E^n, W; K) = \mathcal{R}(\mathcal{E}, W; K).$$

Proof. First of all, it is worth mentioning that the assertion (2.85) is not a consequence of the minimax theorems under the condition of $\Delta(K)$-convergence because in the considered context no assumptions of topological character are made on the decision space and on the properties of the loss function (of the type of $W \in \mathcal{W}_c^b$).

The main point of the proof of (2.85) is based on the use of accompanying experiments $\widehat{\mathcal{E}}^n$ and Theorems 2.8 and 2.9. Let $\varepsilon > 0$. Then for sufficiently large n such that (see (2.73))

$$\sup_{\alpha \in K} \|P_\alpha^n - \widehat{P}_\alpha^n\| \le \varepsilon$$

we have

$$\begin{aligned}
&\sup_{\alpha \in K} E_\alpha^n W_\alpha(\varkappa^n) \\
&\quad = \sup_{\alpha \in K} \int_{\Omega^n} \int_{\mathbf{A}} W_\alpha(a)\, \varkappa^n(\omega, da)\, P_\alpha^n(d\omega) \\
&\quad \ge \sup_{\alpha \in K} \int_{\widehat{\Omega}^n} \int_{\mathbf{A}} W_\alpha(a)\, \varkappa^n(\omega, da)\, \widehat{P}_\alpha^n(d\omega) - \varepsilon \|W\| \\
&\quad = \sup_{\alpha \in K} \int_{\widehat{\Omega}^n} \int_{\mathbf{A}} W_\alpha(a)\, \varkappa^n(\omega, da)\, Z_\alpha(\widehat{\lambda}^n)\, \widehat{P}^n(d\omega) - \varepsilon \|W\|.
\end{aligned} \tag{2.86}$$

Since $\mathbf{A}$ is a Polish space, for the measures $\widehat{P}^n$ there exist regular conditional probabilities $\widehat{P}^n(\cdot \mid \widehat{\lambda}^n)$ with respect to the elements $\widehat{\lambda}^n : \widehat{\Omega}^n \to \mathbf{A}$. Define, for $A \in \mathcal{A}$,

$$\widehat{\varkappa}^n(\widehat{\lambda}^n, A) = \widehat{E}^n \left[\varkappa^n(\omega, A) \mid \widehat{\lambda}^n \right] = \int \varkappa^n(\omega, A)\, \widehat{P}^n(d\omega \mid \widehat{\lambda}^n).$$

It is clear that $\widehat{\varkappa}^n(\widehat{\lambda}^n, A)$ is again a Markov kernel. Thus, from (2.86) and from the property (ii) of Theorem 2.8

$$\begin{aligned}
&\sup_{\alpha\in K} E_\alpha^n W_\alpha(\varkappa^n) \\
&\quad \geq \sup_{\alpha\in K} \int_{\widehat{\Omega}^n}\int_{\mathbf{A}} W_\alpha(a)\, \widehat{\varkappa}^n(\widehat{\lambda}^n, da)\, Z_\alpha(\widehat{\lambda}^n)\, \widehat{P}^n(d\omega) - \varepsilon\|W\| \\
&\quad = \sup_{\alpha\in K} \int_X\int_{\mathbf{A}} W_\alpha(a)\, \widehat{\varkappa}^n(\lambda, da)\, Z_\alpha(\lambda)\, P(dx) - \varepsilon\|W\|, \qquad (2.87)
\end{aligned}$$

i.e.

$$\sup_{\alpha\in K} E_\alpha^n W_\alpha(\varkappa^n) \geq \sup_{\alpha\in K} E_\alpha^n W_\alpha(\widehat{\varkappa}^n) - \varepsilon\|W\| \geq \inf_{\varkappa} \sup_{\alpha\in K} E_\alpha W_\alpha(\varkappa) - \varepsilon\|W\|.$$

Since $\varepsilon > 0$ was chosen arbitrarily this implies the inequality

$$\liminf_{n\to\infty} \inf_{\varkappa^n} \sup_{\alpha\in K} E_\alpha^n W_\alpha(\varkappa^n) \geq \inf_{\varkappa} \sup_{\alpha\in K} E_\alpha W_\alpha(\varkappa). \qquad (2.88)$$

To complete the proof let us consider a sequence of Markov decisions $\varkappa^{(n)}$, $n \geq 1$, such that

$$\lim_{n\to\infty} \sup_{\alpha\in K} E_\alpha W_\alpha(\varkappa^{(n)}) = \inf_{\varkappa} \sup_{\alpha\in K} R_\alpha(\mathcal{E}, W, \varkappa) \quad (= \mathcal{R}(\mathcal{E}, W; K)).$$

Using the decisions $\varkappa^{(n)} = \varkappa^{(n)}(x, A)$ and the elements $\widehat{\lambda}^n$ from Theorem 2.8 we can construct for the accompanying experiments $\widehat{\mathcal{E}}^n$ new decisions

$$\widehat{\varkappa}^n = \varkappa^{(n)}(\widehat{\lambda}^n, A).$$

Then from the boundedness of the function $W_\alpha(a)$ and the property $\sup_{\alpha\in K} \|P_\alpha^n - \widehat{P}_\alpha^n\| \to 0$, $n \to \infty$ we obtain

$$\begin{aligned}
\lim_{n\to\infty} \sup_{\alpha\in K} E_\alpha^n W_\alpha(\widehat{\varkappa}^n) &= \lim_{n\to\infty} \sup_{\alpha\in K} \widehat{E}_\alpha^n W_\alpha(\widehat{\varkappa}^n) \\
&= \lim_{n\to\infty} \sup_{\alpha\in K} E_\alpha W_\alpha(\varkappa^{(n)}), \qquad (2.89)
\end{aligned}$$

which along with (2.88) yields the desired equality (2.85). □

Remark 2.9 If for the limit experiment $\mathcal{E}$ there exists a minimax decision $\varkappa = \varkappa(x, A)$, then it follows from the consideration above that $\widehat{\varkappa}^n = \varkappa(\widehat{\lambda}^n, A)$ are asymptotically minimax for the experiments $\mathcal{E}^n$.

3. The following result is an immediate consequence of Theorem 2.10.

Theorem 2.11 *Let $\mathcal{K} = \{K : K \subseteq \mathcal{A}\}$ be a collection of subsets of $\mathcal{A}$, let $\mathcal{E}^n \xrightarrow{\lambda(\mathcal{K})} \mathcal{E}$ with some $\mathcal{K}$-regular experiment $\mathcal{E}$, and let $\|W\| < \infty$. Then*

$$\sup_{K\in\mathcal{K}} \liminf_{n} \inf_{\varkappa^n} \sup_{\alpha\in K} E_\alpha^n W_\alpha(\varkappa^n) = \sup_{K\in\mathcal{K}} \inf_{\varkappa} \sup_{\alpha\in K} E_\alpha W_\alpha(\varkappa), \qquad (2.90)$$

i.e.

$$\sup_{K\in\mathcal{K}} \lim_{n} \mathcal{R}(\mathcal{E}^n, W; K) = \sup_{K\in\mathcal{K}} \mathcal{R}(\mathcal{E}, W; K).$$

Note that if the class of subsets $\mathcal{K}$ is rich enough, i.e. for any finite subset $\sigma = \{\alpha_1, \dots, \alpha_k)$ in $\mathcal{A}$ there exists an element K from $\mathcal{K}$ containing it and if the loss function W is lower semicontinuous then the right-hand side of (2.90) coincides with the minimax risk on the whole space $\mathcal{A}$. Indeed, (Strasser (1985, Section 46))

$$\sup_{\sigma\in S(\mathcal{A})} \inf_{\varkappa} \sup_{\alpha\in\sigma} E_\alpha W_\alpha(\varkappa) = \inf_{\varkappa} \sup_{\alpha\in\mathcal{A}} E_\alpha W_\alpha(\varkappa) \quad (= \mathcal{R}(\mathcal{E}, W))$$

where $S(\mathcal{A})$ is the collection of all finite subset in $\mathcal{A}$. We give also a result for *Bayes decisions* similar to Theorem 2.10.

Theorem 2.12 *Let $\mathcal{E}^n \xrightarrow{\lambda(K)} \mathcal{E}$ with a canonical dominated K-regular experiment $\mathcal{E}$, let $\pi = \pi(d\alpha)$ be an a priori distribution on K, and let $\|W\| < \infty$. Then*

$$\lim_{n\to\infty} \inf_{\varkappa^n} \int E_\alpha^n W_\alpha(\varkappa^n)\,\pi(d\alpha) = \inf_{\varkappa} \int E_\alpha W_\alpha(\varkappa)\,\pi(d\alpha). \tag{2.91}$$

4. If a loss function $W = W_\alpha(a)$ is continuous in a and a minimax decision $\varkappa = \varkappa(x, A)$ is continuous in $x \in X$, then one may expect that the decisions $\tilde{\varkappa}^n = \varkappa(\lambda^n, A)$ will be also asymptotically minimax, since they were constructed from $\varkappa = \varkappa(x, A)$ by the original elements λ^n (not by the auxiliary $\widehat{\lambda}^n$) entering in the definition of λ-convergence. And the "continuity" conditions on the loss function W and a decision $\varkappa$ are similar to the condition of regularity of the densities $Z_\alpha(x)$. We give one more result in this direction for the case of bounded loss functions and non-randomized decisions.

Theorem 2.13 *Let $\mathcal{E}^n \xrightarrow{\lambda(\mathcal{K})} \mathcal{E}$ with a canonical dominated $\mathcal{K}$-regular experiment $\mathcal{E}$, and $\|W\| < \infty$. Let $\varkappa = \varkappa(x)$ be a non-randomized decision for the limit experiment $\mathcal{E}$ with values in a Polish decision space* **A**, *and let for any $K \in \mathcal{K}$ and each $\varepsilon > 0$*

$$\sup_{\alpha\in K} P\left(D^\delta W_\alpha(\varkappa(\lambda)) > \varepsilon\right) \to 0, \quad \delta \to 0, \tag{2.92}$$

with

$$D^\delta W_\alpha(\varkappa(x)) = \sup_{\{y:\,\rho(x,y)\le\delta\}} |W_\alpha(\varkappa(x)) - W_\alpha(\varkappa(y))|, \quad x \in X, \alpha \in \mathcal{A}.$$

Then

$$\sup_{\alpha\in K} |E_\alpha^n W_\alpha(\varkappa(\widehat{\lambda}^n)) - E_\alpha^n W_\alpha(\varkappa(\lambda^n))| \to 0, \quad n \to \infty, \quad K \in \mathcal{K}. \tag{2.93}$$

In particular, if the decision $\varkappa$ is minimax for the experiment $\mathcal{E}$ and the loss function W, then the decisions $\varkappa(\lambda^n)$ are asymptotically minimax for $\mathcal{E}^n$.

Proof. It is similar to the proof of Theorems 2.10 and 2.11. The most important moment here is the verification of the convergence,

$$W_\alpha(\varkappa(\widehat{\lambda}^n)) - W_\alpha(\varkappa(\lambda^n)) \to 0$$

uniformly in α on any set K from $\mathcal{K}$ under the measure $\widehat{P}^n$. But it follows from the condition (2.92) and the assertion (i) of Theorem 2.10 (about drawing together of elements λ^n and $\widehat{\lambda}^n$). The assertion (2.93) follows from this one because the loss function W is bounded and because of the uniform (in $\alpha \in K$) contiguity of the measures $\widehat{P}^n_\alpha$ to the measures $\widehat{P}^n$ (the assertion (ii) of Theorem 2.11). □

5. Let us consider now the case of an unbounded loss function W. We shall suppose only that the function W satisfies the condition $(W\mathcal{K})$ (see Section 2.2). Recall the meaning of this condition: if the parameter set $\mathcal{A}$ is restricted to some subset $K \in \mathcal{K}$ then it is natural to restrict ourselves to the given loss function W by decisions with values only in the "set of all K-admissible points $D(W,K)$" from $\mathbf{A}$, and the function W is bounded for $\alpha \in K$ and $a \in D(W,K)$.

Theorem 2.14 *Let $\mathcal{K} = \{K : K \subseteq \mathcal{A}\}$ be a collection of subsets of $\mathcal{A}$, $\mathcal{E}^n \xrightarrow{\lambda(\mathcal{K})} \mathcal{E}$ with a canonical dominated $\mathcal{K}$-regular experiment $\mathcal{E}$, and let a loss function W satisfy the condition $(W\mathcal{K})$. Then for any $K \in \mathcal{K}$*

$$\lim_{n\to\infty} \inf_{\varkappa^n} \sup_{\alpha\in K} E^n_\alpha W_\alpha(\varkappa^n) = \inf_{\varkappa} \sup_{\alpha\in K} E_\alpha W_\alpha(\varkappa)$$

and

$$\sup_{K\in\mathcal{K}} \liminf_{n} \inf_{\varkappa^n} \sup_{\alpha\in K} E^n_\alpha W_\alpha(\varkappa^n) = \sup_{K\in\mathcal{K}} \inf_{\varkappa} \sup_{\alpha\in K} E_\alpha W_\alpha(\varkappa),$$

i.e.

$$\lim_{n} \mathcal{R}(\mathcal{E}^n, W; K) = \mathcal{R}(\mathcal{E}, W; K), \quad K \in \mathcal{K},$$

and

$$\sup_{K\in\mathcal{K}} \lim_{n} \mathcal{R}(\mathcal{E}^n, W; K) = \sup_{K\in\mathcal{K}} \mathcal{R}(\mathcal{E}, W; K).$$

Proof. Let $K \in \mathcal{K}$ and $D(W,K)$ be the set of all K-admissible points in $\mathbf{A}$. Construct, similarly to the proof of Theorem 2.2 in Section 2.1, a measurable mapping $T_K : \mathbf{A} \to \mathbf{A}$ such that $T_K(a) = a$ for $a \in D(W,K)$ and

$$W_\alpha(a) \geq W_\alpha(T_K(a)), \quad \alpha \in K, a \in \mathbf{A}.$$

Denoting $W^K = (W_\alpha^K(a), \alpha \in K)$ with $W_\alpha^K(a) = W_\alpha(T_K(a))$ we get by Theorem 2.10 for any sequence of decisions $\varkappa^n$ for the experiments $\mathcal{E}^n$,

$$\begin{aligned}\lim_{n\to\infty} \inf_{\varkappa^n} \sup_{\alpha\in K} E_\alpha^n W_\alpha(\varkappa^n) &\geq \lim_{n\to\infty} \inf_{\varkappa^n} \sup_{\alpha\in K} E_\alpha^n W_\alpha^K(\varkappa^n) \\ &= \inf_{\varkappa} \sup_{\alpha\in K} E_\alpha W_\alpha^K(\varkappa) \\ &= \inf_{\varkappa} \sup_{\alpha\in K} E_\alpha W_\alpha(\varkappa).\end{aligned}$$

Let now $\varkappa$ be an arbitrary decision for the experiment $\mathcal{E}$. Denote by $T_K \circ \varkappa$ the decision of the form $T_K \circ \varkappa(x, A) = \varkappa(x, T_K(A))$. With this, $E_\alpha W_\alpha(T_K \circ \varkappa) = E_\alpha W_\alpha^K(\varkappa)$. If $\varkappa^{(n)}$ is a sequence of decisions for the limit experiment $\mathcal{E}$ such that

$$\lim_{n\to\infty} \sup_{\alpha\in K} E_\alpha W_\alpha(\varkappa^{(n)}) = \inf_{\varkappa} \sup_{\alpha\in K} E_\alpha W_\alpha(\varkappa) \quad (= \mathcal{R}(\mathcal{E}, W; K)),$$

then we also have

$$\lim_{n\to\infty} \sup_{\alpha\in K} E_\alpha W_\alpha(T_K \circ \varkappa^{(n)}) = \mathcal{R}(\mathcal{E}, W; K).$$

Using the decisions $T_K \circ \varkappa^{(n)}$ for the experiment $\mathcal{E}$ we construct the decision $\widehat{\varkappa}^n$ for the accompanying experiments $\widehat{\mathcal{E}}^n$ similarly to Theorem 2.10, i.e. with $\widehat{\varkappa}^n = T_K \circ \varkappa^{(n)}(\widehat{\lambda}^n, A)$. Since the decisions $\widehat{\varkappa}^n$ have values in $D(W, K)$, it follows $W_\alpha(\widehat{\varkappa}^n) = W_\alpha^K(\widehat{\varkappa}^n)$ and

$$\begin{aligned}\lim_{n\to\infty} \sup_{\alpha\in K} E_\alpha^n W_\alpha(\widehat{\varkappa}^n) &= \lim_{n\to\infty} \sup_{\alpha\in K} E_\alpha^n W_\alpha^K(\widehat{\varkappa}^n) \\ &= \lim_{n\to\infty} \sup_{\alpha\in K} \widehat{E}_\alpha^n W_\alpha^K(\widehat{\varkappa}^n) \\ &= \lim_{n\to\infty} \sup_{\alpha\in K} E_\alpha W_\alpha(T_K \circ \varkappa^{(n)}) \\ &= \mathcal{R}(\mathcal{E}, W; K).\end{aligned}$$

This implies the assertion of the lemma. □

Unfortunately, the assertion of Theorem 2.13 cannot be accepted as a completely satisfactory one. Although the minimax risk for the limit experiment $\mathcal{E}$ coincides with the asymptotic minimax risk for the experiments $\mathcal{E}^n$ on every set K from $\mathcal{K}$, the structure of asymptotically minimax decisions for the experiments $\mathcal{E}^n$ depends on the choice of $K \in \mathcal{K}$ (by the mapping T_K) even in the case that there exists a minimax decision for $\mathcal{E}$ on the whole parameter space $\mathcal{A}$. This problem can be bypassed if the following additional condition on the collection of sets $\mathcal{K}$ holds:

(K) *There exists a sequence of sets $K_m \in \mathcal{K}$, $m \to \infty$ increasing to $\mathcal{A}$ (i.e. $K_m \subseteq K_{m+1}, \lim_m K_m = \mathcal{A}$) such that for any $K \in \mathcal{K}$ and for some number $m(K)$ the inclusion $K \subseteq K_m$ holds if $m \geq m(K)$.*

This condition is usually satisfied if the parameter space $\mathcal{A}$ is finite dimensional and $\mathcal{K}$ is a class of compact sets. Then one can select the sets K_m as the balls with radius m and center at zero.

Theorem 2.15 *Let $\mathcal{K} = \{K : K \subseteq \mathcal{A}\}$ be a collection of subsets in $\mathcal{A}$ obeying the condition (K), and $\mathcal{E}^n \xrightarrow{\lambda(\mathcal{K})} \mathcal{E}$ for some $\mathcal{K}$-regular experiment $\mathcal{E}$, and let a loss function W satisfy the condition $(W\mathcal{K})$. Then there are decisions $\widehat{\varkappa}^n$ (for extended experiments $\widehat{\mathcal{E}}^n$) such that*

$$\sup_{K\in\mathcal{K}} \lim_n \sup_{\alpha\in K} \widehat{E}^n_\alpha W_\alpha(\widehat{\varkappa}^n) = \inf_\varkappa \sup_{\alpha\in\mathcal{A}} E_\alpha W_\alpha(\varkappa) \quad (= \mathcal{R}(\mathcal{E}, W)).$$

Proof. Note first that from the condition $K_m \uparrow \mathcal{A}$

$$\sup_{K\in\mathcal{K}} \mathcal{R}(\mathcal{E}, W; K) = \lim_{m\to\infty} \mathcal{R}(\mathcal{E}, W; K_m) = \mathcal{R}(\mathcal{E}, W).$$

Thus from the previous theorem

$$\sup_{K\in\mathcal{K}} \lim_n \mathcal{R}(\mathcal{E}^n, W; K) \geq \mathcal{R}(\mathcal{E}, W),$$

and it remains to prove the inverse inequality.

Let $(\varkappa_m)_{m\geq 1}$ be a sequence of decisions for the experiment $\mathcal{E}$ such that

$$\lim_{m\to\infty} \sup_{\alpha\in\mathcal{A}} E_\alpha W_\alpha(\varkappa_m) = \mathcal{R}(\mathcal{E}, W).$$

Let then the increasing sequence of sets K_m be defined in the condition (K) and let the mapping T_m from $\mathbf{A}$ to the set $D(W, K_m)$ of K_m-admissible points be constructed similarly to Theorem 2.13 such that

$$W_\alpha(a) \geq W_\alpha(T_m(a)), \quad \alpha \in K_m, \, a \in \mathbf{A}.$$

Then for the decisions $T_m \circ \varkappa_m$ one has

$$\sup_{\alpha\in K_m} E_\alpha W_\alpha(T_m \circ \varkappa_m) \leq \sup_{\alpha\in K_m} E_\alpha W_\alpha(\varkappa_m),$$

and from the condition $K_m \uparrow \mathcal{A}$ we obtain

$$\lim_{m\to\infty} \sup_{\alpha\in K_m} E_\alpha W_\alpha(T_m \circ \varkappa_m) = \mathcal{R}(\mathcal{E}, W).$$

Now in the same way as in Theorem 2.13 we can construct for fixed m decisions $(\widehat{\varkappa}^n_m)_{m\geq 1}$ for the experiments $\mathcal{E}^n$ such that

$$\lim_n \sup_{\alpha\in K_m} E^n_\alpha W_\alpha(\widehat{\varkappa}^n_m) = \sup_{\alpha\in K_m} E_\alpha W_\alpha(T_m \circ \varkappa_m).$$

Finally, we construct the desired sequence $\widehat{\varkappa}^n$ with the help of the diagonal procedure. Let for every $m \geq 1$ a number $n(m)$ be chosen such that the following condition is satisfied for $n \geq n(m)$:

$$\sup_{\alpha \in K_m} E_\alpha^n W_\alpha(\widehat{\varkappa}_m^n) \leq \sup_{\alpha \in K_m} E_\alpha W_\alpha(T_m \circ \varkappa_m) + \frac{1}{m}.$$

Now define

$$\widehat{\varkappa}^n = \widehat{\varkappa}_m^n \quad \text{for} \quad n(m) \leq n < n(m+1).$$

If $K \in \mathcal{K}$ then $K \in K_m$ for all m greater than some m_0. Thus, for $m \geq m_0$ and $n \geq n(m)$

$$\sup_{\alpha \in K} E_\alpha^n W_\alpha(\widehat{\varkappa}^n) \leq \sup_{\alpha \in K_m} E_\alpha W_\alpha(T_m \circ \varkappa_m) + \frac{1}{m},$$

hence

$$\limsup_n \sup_{\alpha \in K} E_\alpha^n W_\alpha(\widehat{\varkappa}^n) \leq \lim_{m \to \infty} \sup_{\alpha \in K_m} E_\alpha W_\alpha(T_m \circ \varkappa_m) = \mathcal{R}(\mathcal{E}, W),$$

and the assertion follows. □

6. Now we give one useful generalization of the results of the previous theorems if some more general form of the loss function is considered. More precisely, we assume that the loss function is of the form $W_\alpha(a, \lambda)$ with $a \in \mathbf{A}$, $\lambda \in X$. Then the risk for a decision $\varkappa^n$ in the experiment $\mathcal{E}^n$ is defined by the equality

$$E_\alpha^n W_\alpha(\varkappa^n, \lambda^n) = \int_{\Omega^n} \int_{\mathbf{A}} W_\alpha(a, \lambda^n)\, \varkappa^n(\omega, da)\, P_\alpha^n(d\omega),$$

and in the experiment $\mathcal{E}$

$$E_\alpha W_\alpha(\varkappa, \lambda) = \int_X \int_{\mathbf{A}} W_\alpha(a, x)\, \varkappa(x, da)\, P_\alpha(dx).$$

Let $\mathcal{K} = \{K : K \subseteq \mathcal{A}\}$ be a collection of subsets in $\mathcal{A}$, let $\mathcal{E}^n \xrightarrow{\lambda(\mathcal{K})} \mathcal{E}$ with a $\mathcal{K}$-regular experiment $\mathcal{E}$ and let $K \in \mathcal{K}$. As in (2.86) we find that for any $\varepsilon > 0$ for sufficiently large n

$$\begin{aligned}
&\sup_{\alpha \in K} E_\alpha^n W_\alpha(\varkappa^n) \\
&\quad = \sup_{\alpha \in K} \int_{\Omega^n} \int_{\mathbf{A}} W_\alpha(a, \lambda^n)\, \varkappa^n(\omega, da)\, P_\alpha^n(d\omega) \\
&\quad \geq \sup_{\alpha \in K} \int_{\widehat{\Omega}^n} \int_{\mathbf{A}} W_\alpha(a, \lambda^n)\, \varkappa^n(\omega, da)\, \widehat{P}_\alpha^n(d\omega) - \varepsilon \|W\| \\
&\quad = \sup_{\alpha \in K} \Big[\int_{\widehat{\Omega}^n} \int_{\mathbf{A}} W_\alpha(a, \widehat{\lambda}^n)\, \varkappa^n(\omega, da)\, \widehat{P}_\alpha^n(d\omega) - \varepsilon \|W\| \\
&\qquad + \int_{\widehat{\Omega}^n} \int_{\mathbf{A}} \Big(W_\alpha(a, \lambda^n) - W_\alpha(a, \widehat{\lambda}^n) \Big)\, \varkappa^n(\omega, da)\, \widehat{P}_\alpha^n(d\omega) \Big]. \qquad (2.94)
\end{aligned}$$

Similarly to Theorem 2.10 we obtain from the condition $\mathcal{L}(\widehat{\lambda}^n \mid \widehat{P}^n_\alpha) = \mathcal{L}(\lambda \mid P_\alpha)$

$$\begin{aligned}
&\inf_{\varkappa^n} \sup_{\alpha\in K} \int_{\widehat{\Omega}^n} \int_{\mathbf{A}} W_\alpha(a, \widehat{\lambda}^n)\, \varkappa^n(\omega, da)\, \widehat{P}^n_\alpha(d\omega) \\
&\quad = \inf_{\varkappa} \sup_{\alpha\in K} \int_X \int_{\mathbf{A}} W_\alpha(a, x)\, \varkappa(x, da)\, P_\alpha(dx) = \\
&\quad = \mathcal{R}(\mathcal{E}, W; K).
\end{aligned} \tag{2.95}$$

We now introduce an additional condition on the loss function W (compare with the condition of $\mathcal{K}$-regularity of the experiment $\mathcal{E}$). Let $\delta > 0$ and

$$D^\delta W_\alpha(a, x) = \sup_{\{y:\, \rho(x,y)\le\delta\}} |W_\alpha(a, x) - W_\alpha(a, y)|, \quad a \in \mathbf{A},\, \alpha \in \mathcal{A}.$$

Suppose that for any $K \in \mathcal{K}$ and any $\varepsilon > 0$

$$\sup_{\alpha\in K} \sup_{a\in\mathbf{A}} P_\alpha\left(D^\delta W_\alpha(a, \lambda) > \varepsilon\right) \to 0, \quad \delta \to 0. \tag{2.96}$$

It follows now from the condition $\widehat{P}^n(\rho(\widehat{\lambda}^n, \lambda^n) > \delta) < \varepsilon$ that

$$\begin{aligned}
&\sup_{\alpha\in K} \int_{\widehat{\Omega}^n} \int_{\mathbf{A}} |W_\alpha(a, \lambda^n) - W_\alpha(a, \widehat{\lambda}^n)|\, \varkappa^n(\omega, da)\, \widehat{P}^n_\alpha(d\omega) \\
&\quad \le \varepsilon \|W\| + \sup_{\alpha\in K} \int_{\widehat{\Omega}^n} \int_{\mathbf{A}} D^\delta W_\alpha(a, \widehat{\lambda}^n)\, \varkappa^n(\omega, da)\, \widehat{P}^n_\alpha(d\omega).
\end{aligned}$$

Since $D^\delta W_\alpha(a, x) \le 2\|W\|$ and $\mathcal{L}(\widehat{\lambda}^n \mid \widehat{P}^n_\alpha) = \mathcal{L}(\lambda \mid P_\alpha)$, we derive

$$\begin{aligned}
&\sup_{\alpha\in K} \int_{\widehat{\Omega}^n} \int_{\mathbf{A}} D^\delta W_\alpha(a, \widehat{\lambda}^n)\, \varkappa^n(\omega, da)\, \widehat{P}^n_\alpha(d\omega) \\
&\quad \le \sup_{\alpha\in K} \sup_{a\in\mathbf{A}} 2\|W\|\, \widehat{P}^n_\alpha\left(D^\delta W_\alpha(a, \widehat{\lambda}^n) > \varepsilon\right) + \varepsilon \\
&\quad = \sup_{\alpha\in K} \sup_{a\in\mathbf{A}} 2\|W\|\, P_\alpha\left(D^\delta W_\alpha(a, \lambda) > \varepsilon\right) + \varepsilon.
\end{aligned}$$

Therefore, by (2.96) and because $\varepsilon > 0$ was taken arbitrarily we get

$$\lim_n \sup_{\alpha\in K} \int_{\widehat{\Omega}^n} \int_{\mathbf{A}} |W_\alpha(a, \lambda^n) - W_\alpha(a, \widehat{\lambda}^n)|\, \varkappa^n(\omega, da)\, \widehat{P}^n_\alpha(d\omega) = 0.$$

This, together with (2.94) and (2.95), implies that if a (generalized) loss function $W = (W_\alpha(a, \lambda))$ obeys (2.96) then the following inequality holds:

$$\lim_n \sup_{\alpha\in K} E^n_\alpha W_\alpha(\varkappa^n, \lambda^n) \ge \inf_{\varkappa} \sup_{\alpha\in K} E_\alpha W_\alpha(\varkappa, \lambda) = \mathcal{R}(\mathcal{E}, W; K).$$

In a similar way the inverse inequality is proved. Hence, we have proved the following

Theorem 2.16 *Let $\mathcal{E}^n \xrightarrow{\lambda(\mathcal{K})} \mathcal{E}$ for some $\mathcal{K}$-regular experiment $\mathcal{E}$ and let a (generalized) loss function $W = (W_\alpha(a,\lambda))$ be bounded and satisfy the condition (2.96). Then for any $K \in \mathcal{K}$*

$$\lim_{n\to\infty} \inf_{\varkappa^n} \sup_{\alpha\in K} E_\alpha^n W_\alpha(\varkappa^n, \lambda^n) = \inf_{\varkappa} \sup_{\alpha\in K} E_\alpha W_\alpha(\varkappa, \lambda)$$

and

$$\sup_{K\in\mathcal{K}} \liminf_{n} \inf_{\varkappa^n} \sup_{\alpha\in K} E_\alpha^n W_\alpha(\varkappa^n, \lambda^n) = \sup_{K\in\mathcal{K}} \inf_{\varkappa} \sup_{\alpha\in K} E_\alpha W_\alpha(\varkappa, \lambda).$$

2.8 Comparison of Various Kinds of Convergence of Statistical Experiments

1. If $\mathcal{E} = (\Omega, \mathcal{F}; P_\alpha, \alpha \in \sigma)$ and $\mathcal{E}^n = (\Omega^n, \mathcal{F}^n; P_\alpha^n, \alpha \in \sigma)$, $n \geq 1$, are arbitrary statistical experiments with a *finite* parameter set $\sigma = \{1, \ldots, k\}$, then by Theorem 2.9 the weak ($\mathcal{E}^n \xrightarrow{w} \mathcal{E}$) and strong ($\mathcal{E}^n \xrightarrow{\Delta} \mathcal{E}$) convergence are equivalent.

Let now the weak convergence $\mathcal{E}^n \xrightarrow{w} \mathcal{E}$ hold. Consider for the experiment $\mathcal{E}$ its standard representation (see Section 1.1) $\mathcal{E}_k = (\Sigma, \mathcal{B}(\Sigma); \mu_i, i = 1, \ldots, k)$ with $\mu_i = \mathcal{L}(z \mid P_i)$ and $z = (z_1, \ldots, z_k)$, $z_i = \dfrac{dP_i}{dP}$, $P = \sum_{i=1}^k P_i$. Recall that $\Delta(\mathcal{E}, \mathcal{E}_k) = 0$ or $\mathcal{E} \sim \mathcal{E}_k$.

If $\mu = \sum_{i=1}^k \mu_i$ then for $x \in \Sigma$

$$Z_i(x) = \frac{d\mu_i}{d\mu}(x) = x_i.$$

Denote also for the experiments $\mathcal{E}^n$

$$P^n = P_1^n + \ldots + P_k^n, \qquad Z_i^n(\omega) = \frac{dP_i^n}{dP^n}(\omega), \quad i = 1, \ldots, k.$$

If now we define $\lambda^n(\omega) = Z^n(\omega)$, then the equality $Z_i^n(\omega) - Z_i(\lambda^n(\omega)) = 0$ holds P^n-a.s. and, moreover, the convergence $\mathcal{E}^n \xrightarrow{w} \mathcal{E}$ yields $\mathcal{L}(\lambda^n \mid P^n) \xrightarrow{w} \mu$. In view of (2.45) this means that the λ-convergence $\mathcal{E}^n \xrightarrow{\lambda} \mathcal{E}_k$ holds as well. This leads to the following

Theorem 2.17 *Let $\mathcal{E}$ and $\mathcal{E}^n, n \geq 1$, be statistical experiments with a finite parameter set $S = \{1, \ldots, k\}$. Then the conditions*

$$\mathcal{E}^n \xrightarrow{w} \mathcal{E}, \qquad \mathcal{E}^n \xrightarrow{\Delta} \mathcal{E}, \qquad \mathcal{E}^n \xrightarrow{\lambda} \mathcal{E}_k$$

are equivalent.

2. Now we turn to the case of an arbitrary parameter set $\mathcal{A}$. We suppose that the "limit" experiment $\mathcal{E} = (X, \mathfrak{X}; P_\alpha, \alpha \in \mathcal{A}; P)$ is *canonical* and *dominated*. The results of Section 2.6 claim that if $\mathcal{E}$ is a *regular* experiment, then $\mathcal{E}^n \xrightarrow{\lambda} \mathcal{E} \Rightarrow \mathcal{E}^n \xrightarrow{w} \mathcal{E}$. If $\mathcal{E}$ is $\mathcal{K}$-regular, then (for every $K \in \mathcal{K}$)

$$\mathcal{E}^n \xrightarrow{\lambda(K)} \mathcal{E} \quad \Rightarrow \quad \mathcal{E}^n \xrightarrow{\Delta(K)} \mathcal{E} \quad \Rightarrow \quad \mathcal{E}^n \xrightarrow{w(K)} \mathcal{E}.$$

It is of interest the question about a possibility of inverting these implications, i.e. when does the weak or the strong convergence imply one or another version of λ-convergence.

3. Consider first the question under which conditions weak convergence implies λ-convergence. We begin with some preliminary consideration. Let a dominating measure P for the experiment $\mathcal{E}$ be chosen as a finite or sigma-finite convex combination of the measures P_α and let the measures P^n for the experiments $\mathcal{E}^n$ be constructed in a similar way. Denote, as usual, $Z_\alpha = \frac{dP_\alpha}{dP}$, $Z_\alpha^n = \frac{dP_\alpha^n}{dP^n}$, $\alpha \in \mathcal{A}$.

The weak convergence $\mathcal{E}^n \xrightarrow{w} \mathcal{E}$ means that for any finite subset $\sigma = \{\alpha_1, \dots, \alpha_m\}$ from $\mathcal{A}$ the distributions of the vectors $Z_\sigma^n = (Z_{\alpha_1}^n, \dots, Z_{\alpha_m}^n)$ converge to the distribution of the vector $Z_\sigma = (Z_{\alpha_1}, \dots, Z_{\alpha_m})$. But in order to prove λ-convergence we must construct mappings λ^n from Ω^n to X with distributions converging to the measure P (the distribution of the identity mapping $\lambda(x) = x$). Therefore, the necessary condition to obtain λ-convergence from the weak one is the possibility to reconstruct the measure P on $(X, \mathfrak{X})$ from the distributions of vectors $Z_\sigma = (Z_{\alpha_1}, \dots, Z_{\alpha_m})$ when σ runs over all finite subsets in $\mathcal{A}$. Before introducing conditions of this sort let us consider the following example.

Example 2.7 Let $\mathcal{E} = (R^m, \mathcal{B}(R^m); P_\alpha, \alpha \in R^m)$ be a Gaussian shift experiment. Then $P = P_0$ and

$$Z_\alpha(x) = \frac{dP_\alpha}{dP} = \exp\{(x, \alpha) - \frac{1}{2}\|\alpha\|^2\}, \quad \alpha \in R^m, x \in R^m.$$

Choose $\sigma = \{e_1, \dots, e_m\}$ as the canonical base in R^m, $i = 1, .., m$. Hence for $x = (x^1, \dots, x^m) \in R^m$

$$Z_{e_i}(x) = \exp\{x^i - \frac{1}{2}\}, \quad i = 1, .., m.$$

Put for $z = (z^1, \dots, z^m) \in R_+^m$

$$T(z) = (\frac{1}{2} + \ln z^1, \dots, \frac{1}{2} + \ln z^m).$$

Then $T(Z_\sigma(x)) = x$ and having the distribution of the vector Z_σ (under the measure $P = P_0$) we can reconstruct the measure P itself.

Notice that in this example there exists a finite subset σ such that one can reconstruct the point $x \in X$ using $Z_\sigma(x)$. We formulate such a request in the form of a condition on the (limit) experiment $\mathcal{E}$.

(E) *An experiment* $\mathcal{E} = (X, \mathfrak{X}; P_\alpha, \alpha \in \mathcal{A}; P)$ *is canonically dominated and there exists a finite subset* $\sigma = \{\alpha_1, \dots, \alpha_m\}$ *in* $\mathcal{A}$ *and a measurable mapping* $T: R_+^m \to X$ *such that*

$$T(Z_\sigma) = \lambda, \tag{2.97}$$

where λ *is the identical mapping in* X, *i.e.* $T(Z_\sigma(x)) = \lambda(x)$, $x \in X$.

Let σ be a subset in $\mathcal{A}$ shown in the condition (E). Using the mapping Z_σ we can define the experiment $Z_\sigma\mathcal{E} = (R_+^m, \mathcal{B}(R_+^m); P_\alpha \circ Z_\sigma^{-1}, \alpha \in \mathcal{A})$ as the image of the experiment $\mathcal{E}$. Next, by condition (E) the mapping Z_σ is a sufficient statistic for $\mathcal{E}$; thus (see Section 1.4) the experiments $\mathcal{E}$ and $Z_\sigma\mathcal{E}$ are equivalent. Therefore, the condition (E) means really that the observation space is of finite dimension.

Denote by $P_\sigma = P \circ Z_\sigma^{-1}$ the image of the measure P for the mapping Z_σ. Recall that P_σ is a measure on $(R_+^m, \mathcal{B}(R_+^m))$.

Theorem 2.18 *Let for a regular experiment* $\mathcal{E}$ *the condition (E) hold and, moreover, let the corresponding mapping* T *be continuous under the measure* P_σ *on* $(R_+^m, \mathcal{B}(R_+^m))$. *Then the weak convergence* $\mathcal{E}^n \xrightarrow{w} \mathcal{E}$ *implies* λ*-convergence* $\mathcal{E}^n \xrightarrow{\lambda} \mathcal{E}$.

Proof. Let $\sigma = \{\alpha_1, \dots, \alpha_m\}$ be the set entering in the condition (E) and let $Z_\sigma^n = (Z_{\alpha_1}^n, \dots, Z_{\alpha_m}^n)$. Put

$$\lambda^n = T(Z_\sigma^n) \tag{2.98}$$

and verify the condition of λ-convergence for such a choice of λ^n. The weak convergence $\mathcal{E}^n \xrightarrow{w} \mathcal{E}$ yields the convergence in distributions

$$Z_\sigma^n \xrightarrow{d(P^n)} Z_\sigma.$$

Then we have from the continuity of the mapping T under the measure P_σ (which is the distribution of the vector Z_σ)

$$\lambda^n = T(Z_\sigma^n) \xrightarrow{d(P^n)} T(Z_\sigma) = \lambda.$$

It remains to prove the property

$$Z_\alpha^n - Z_\alpha(\lambda^n) \xrightarrow{P^n} 0, \quad \alpha \in \mathcal{A}. \tag{2.99}$$

For this purpose, in turn, it is sufficient to prove the convergence of joint distributions

$$(Z_\alpha^n, Z_\alpha(\lambda^n)) \xrightarrow{d(P^n)} (Z_\alpha, Z_\alpha). \tag{2.100}$$

The weak convergence $\mathcal{E}^n \xrightarrow{w} \mathcal{E}$ implies

$$(Z_\alpha^n, Z_\sigma^n) \xrightarrow{d(P^n)} (Z_\alpha, Z_\sigma).$$

Since T is continuous under P_σ, it also holds

$$(Z^n_\alpha, T(Z^n_\sigma)) \xrightarrow{d(P^n)} (Z_\alpha, T(Z_\sigma)),$$

i.e.

$$(Z^n_\alpha, \lambda^n) \xrightarrow{d(P^n)} (Z_\alpha, \lambda).$$

Finally, the experiment $\mathcal{E}$ is regular and therefore the function Z_α is continuous under the measure P on $(X, \mathfrak{X})$ and the latter relation together with equality $Z_\alpha(\lambda(x)) = Z_\alpha(x)$ implies (2.100). The theorem follows. □

Remark 2.10 Actually, the given proof does not utilize the condition of weak convergence $\mathcal{E}^n \xrightarrow{w} \mathcal{E}$ itself but only the convergence of finite-dimensional vectors

$$(Z^n_\alpha, Z^n_\sigma) \xrightarrow{d(P^n)} (Z_\alpha, Z_\sigma), \quad \alpha \in \mathcal{A}, \tag{2.101}$$

with σ from the condition (E).

4. We now consider the question of when does the uniform weak convergence imply uniform λ-convergence. Since this would imply also the strong convergence, in this situation all the three kinds of convergence become equivalent.

For the proof of the main assertions in this subsection we will use some facts concerning the uniform weak convergence. The accurate formulations and proofs of these facts are given in the next subsection. Let us fix a set K from the collection of sets $\mathcal{K}$.

Theorem 2.19 *Let a K-regular experiment $\mathcal{E}$ satisfy condition (E) and, moreover, let the corresponding mapping T be continuous with respect to the measure P_σ on $(R^m_+, \mathcal{B}(R^m_+))$. Then the uniform weak convergence $\mathcal{E}^n \xrightarrow{w(K)} \mathcal{E}$ implies the uniform λ-convergence $\mathcal{E}^n \xrightarrow{\lambda(K)} \mathcal{E}$.*

Proof. As in Theorem 2.17 define

$$\lambda^n = T(Z^n_\sigma),$$

where T and σ are the mapping (respectively the finite set) from the condition (E). Then again

$$\lambda^n \xrightarrow{d(P^n)} \lambda.$$

Now we have to check the condition

$$\sup_{\alpha \in K} P^n(|Z^n_\alpha - Z_\alpha(\lambda^n)| > \varepsilon) \to 0, \quad \varepsilon > 0. \tag{2.102}$$

The uniform weak convergence $\mathcal{E}^n \xrightarrow{w(K)} \mathcal{E}$ implies

$$(Z^n_\alpha, Z^n_\sigma) \xrightarrow{d(P^n;K)} (Z_\alpha, Z_\sigma). \tag{2.103}$$

From this and from the continuity of the mapping T under the measure P we have

$$(Z_\alpha^n, T(Z_\sigma^n)) \xrightarrow{d(P^n;K)} (Z_\alpha, T(Z_\sigma)), \tag{2.104}$$

which can be rewritten in the form

$$(Z_\alpha^n, \lambda^n) \xrightarrow{d(P^n;K)} (Z_\alpha, \lambda). \tag{2.105}$$

This fact and Lemma 2.2 below yield

$$(Z_\alpha^n, Z_\alpha(\lambda^n)) \xrightarrow{d(P^n;K)} (Z_\alpha, Z_\alpha),$$

and the desired property (2.102) follows. The theorem is proved. □

Remark 2.11 Note that again the proof does not use the condition $\mathcal{E}^n \xrightarrow{w(K)} \mathcal{E}$ itself but a weaker condition (2.103).

The result of Theorem 2.18 can be evidently generalized to the case of $\mathcal{K}$-uniform convergence.

Theorem 2.20 *Let the $\mathcal{K}$-regular experiment $\mathcal{E}$ satisfy the condition (E) and, moreover, let the corresponding mapping T be continuous under the measure P_σ on $(R_+^m, \mathcal{B}(R_+^m))$. Let then the experiments $\mathcal{E}^n$ be such that for any $K \in \mathcal{K}$ the weak convergence holds (uniformly in $\alpha \in K$):*

$$(Z_\alpha^n, Z_\sigma^n) \xrightarrow{d(P^n;K)} (Z_\alpha, Z_\sigma),$$

with the finite set σ from the condition (E). Then

$$\mathcal{E}^n \xrightarrow{\lambda(\mathcal{K})} \mathcal{E}.$$

5. Here we give one general result related to the uniform weak convergence of random variables. This result has been already used for the proof of Theorem 2.18.

Lemma 2.4 *Let random elements λ^n with values in a Polish space $(X, \mathcal{X}, \rho)$ weakly converge to a random element λ, $\lambda^n \xrightarrow{d(P^n)} \lambda$, and let a family of functions $(Z_\alpha\ \alpha \in \mathcal{A})$ on X be uniformly continuous and uniformly integrable under the measure P which is the distribution of λ. Then the following weak convergence holds, uniformly in $\alpha \in \mathcal{A}$:*

$$Z_\alpha(\lambda^n) \xrightarrow{d(P^n;\mathcal{A})} Z_\alpha(\lambda).$$

Proof. Weak convergence $\lambda^n \xrightarrow{d(P^n)} \lambda$ and Lemma about "Reconstruction" (which will be proved in Section 2.10) imply that there exist (maybe on extensions of the original probability spaces) random elements $\widehat{\lambda}^n$ such that

$$\mathcal{L}(\widehat{\lambda}^n \mid P^n) = \mathcal{L}(\lambda \mid P) \quad \text{and} \quad \rho(\widehat{\lambda}^n, \lambda^n) \xrightarrow{P^n} 0. \tag{2.106}$$

The assertion of the lemma will be proved if we show that

$$\sup_{\alpha\in K} P^n\big(|Z_\alpha(\widehat{\lambda}^n) - Z_\alpha(\lambda^n)| > \varepsilon\big) \to 0, \quad \varepsilon > 0. \tag{2.107}$$

Let a number $\delta > 0$ be such that

$$\sup_{\alpha\in K} P(D^\delta Z_\alpha > \varepsilon) < \varepsilon, \tag{2.108}$$

where $D^\delta Z_\alpha(x)$ is the module of continuity of the function Z_α at a point $x \in X$ (such δ exists because of the uniform continuity of the function Z_α under the measure P, see Section 2.6). Let then a number $n(\varepsilon)$ be chosen such that for $n \geq n(\varepsilon)$

$$P^n\left(\rho(\widehat{\lambda}^n, \lambda^n) > \delta\right) < \varepsilon. \tag{2.109}$$

Then due to (2.106), (2.108) and (2.109) for $n \geq n(\varepsilon)$

$$\begin{aligned}
&\sup_{\alpha\in K} P^n\big(|Z_\alpha(\widehat{\lambda}^n) - Z_\alpha(\lambda^n)| > \varepsilon\big)\\
&\quad \leq P^n\big(\rho(\widehat{\lambda}^n, \lambda^n) > \delta\big) + \sup_{\alpha\in K} P^n\big(D^\delta Z_\alpha(\widehat{\lambda}^n) > \varepsilon\big)\\
&\quad = P^n\big(\rho(\widehat{\lambda}^n, \lambda^n) > \delta\big) + \sup_{\alpha\in K} P\big(D^\delta Z_\alpha > \varepsilon\big)\\
&\quad \leq 2\varepsilon,
\end{aligned}$$

which implies (2.107) because $\varepsilon > 0$ is arbitrary. The lemma follows. □

2.9 Convergence of Statistical Decisions and Estimators

1. One or another form of convergence of statistical experiments "$\mathcal{E}^n \to \mathcal{E}$" allows to draw conclusions about minimax risks (of the type of $\liminf_n \mathcal{R}(\mathcal{E}^n) \geq \mathcal{R}(\mathcal{E})$). But at the same time the questions of *convergence of statistical procedures, optimal statistical procedures, estimators of an unknown parameters,* etc. are also of great interest.

We begin with some general notions of convergence of statistical decisions. Suppose that a statistical experiment $\mathcal{E} = (\Omega, \mathcal{F}; P_\alpha, \alpha \in \mathcal{A})$ and a sequence of experiments $\mathcal{E}^n = (\Omega^n, \mathcal{F}^n; P^n_\alpha, \alpha \in \mathcal{A}^n)$, $n \geq 1$, are given with $\mathcal{A}^n \uparrow \mathcal{A}$, $n \to \infty$. By $\mathfrak{B}(\mathcal{E}, \mathbf{A})$ (respectively $\mathfrak{B}(\mathcal{E}^n, \mathbf{A})$, $n \geq 1$) we denote the sets of generalized decisions β (respectively β^n, $n \geq 1$) (see Section 1.4), and by $\mathfrak{R}(\mathcal{E}, \mathbf{A})$ (respectively $\mathfrak{R}(\mathcal{E}^n, \mathbf{A})$, $n \geq 1$) the sets of Markov decisions $\rho = \rho(\omega, da)$ (respectively $\rho^n = \rho^n(\omega, da)$, $n \geq 1$).

Definition 2.11 Let $\mathcal{E}^n \xrightarrow{w} \mathcal{E}$. A sequence of generalized decisions β^n, $n \geq 1$, converges in distribution to β if for any continuous bounded function $f = f(a)$

and any $\alpha \in \mathcal{A}$

$$\lim_n \beta(f, P_\alpha^n) = \beta(f, P_\alpha).$$

In the Markov case ($\beta^n = \rho^n, \beta = \rho$) this definition means that

$$\iint f(a)\, \rho^n(\omega, da)\, P_\alpha^n(d\omega) \to \iint f(a)\, \rho(\omega, da)\, P_\alpha(d\omega). \tag{2.110}$$

If $\rho^n = \rho^n(\omega)$, $\rho = \rho(\omega)$ are non-randomized decisions then (2.110) becomes of the form

$$\int f(\rho^n(\omega))\, P_\alpha^n(d\omega) \to \int f(\rho(\omega))\, P_\alpha(d\omega). \tag{2.111}$$

Thus, if

$$\nu_\alpha^n(A) = P_\alpha^n\{\rho^n(\omega) \in A\}, \qquad \nu_\alpha(A) = P_\alpha\{\rho(\omega) \in A\},$$

then (2.111) means just the weak convergence of the distributions ν_α^n of the "estimator" ρ^n to the distribution ν_α of the "estimator" ρ.

Definition 2.12 Let $\mathcal{E}^n \xrightarrow{w} \mathcal{E}$ and let $(\beta^n)_{n\geq 1}$ be some sequence of generalized decisions. A decision β is an *accumulation point* of the sequence $(\beta^n)_{n\geq 1}$ if for every $\alpha \in \mathcal{A}$ and any finite collection of continuous bounded functions $f_1, \dots, f_m$ one can find a subsequence $\mathrm{N}(\alpha; f_1, \dots, f_m)$ of the sequence of natural numbers N such that for each $\alpha \in \mathcal{A}$

$$\lim_{n \in \mathrm{N}(\alpha; f_1, \dots, f_m)} \beta(f_i, P_\alpha^n) = \beta(f_i, P_\alpha), \quad i = 1, \dots, m.$$

One of the most important results of Le Cam (1964) (see also Strasser (1985, Theorem 62.3)) claims that every sequence $\{\beta^n\}$ of generalized decisions $\beta^n \in \mathfrak{B}(\mathcal{E}^n, \mathbf{A})$ has at least one accumulation point.

2. Suppose now to be given (for every experiment $\mathcal{E}^n$) a *statistical estimator* $T^n = T^n(\omega)$, i.e. Markov non-randomized decision, $n \geq 1$. The above mentioned result of L. Le Cam means the following: if $\mathcal{E}^n \xrightarrow{w} \mathcal{E}$ then for a given continuous bounded function $f = f(a)$ and for fixed α there exists a subsequence (n') with

$$\int f(a)\, \nu_\alpha^{n'}(da) = \int f(T^{n'})\, P_\alpha^{n'}(d\omega) \to \beta(f, P_\alpha).$$

One more interesting question would be: *when $\beta(f, P_\alpha)$ can be represented by a Markov kernel*, i.e. when

$$\beta(f, P_\alpha) = \iint f(a)\, \rho(\omega, da)\, P_\alpha(d\omega). \tag{2.112}$$

It is well known that if, for example, $\mathcal{E}$ is a dominated experiment and $\mathbf{A}$ is a locally compact space with countable base or a Polish space (see Section 1.4) then the representation (2.112) is possible.

The next question is whether this Markov kernel ρ can be realized by some estimator T. As follows from Lemma 1.1, Section 1.4 for the case of a Polish space $(\mathbf{A}, \mathcal{A})$ the answer to this question is positive if, additionally, extensions of experiments are admissible.

Summarizing this consideration leads to the following conclusion: if $\{T^n\}$ is a sequence of Markov non-randomized estimator, $\mathcal{E}^n \xrightarrow{w} \mathcal{E}$, the space $\mathbf{A}$ is, for example, a locally compact separable metric space and $\mathcal{E}$ is a dominated experiment then for given α and for a continuous bounded function $f = f(a)$ there are a subsequence (n') and a non-randomized estimator T (for some extension of the experiment $\mathcal{E}$) such that

$$\int f(a)\, \nu_\alpha^{n'}(da) \to \int f(a)\, \nu_\alpha^T(da). \tag{2.113}$$

Of course, this result cannot be considered as completely satisfactory, because convergence in (2.113) is not universal over all continuous bounded functions $f (f \in C_c^b)$ and all $\alpha \in \mathcal{A}$. Moreover, T depends on the choice of f and α.

In this connection the following result is of interest. In order to prove this result we assume that the property of the weak convergence is exchanged by the stronger λ-convergence and also that $\mathbf{A}$ is a complete separable metric space.

Theorem 2.21 *Let $\mathcal{E}^n \xrightarrow{\lambda} \mathcal{E}$ with a regular dominated canonical experiment $\mathcal{E}$ and let $\{T^n\}$ be a sequence of estimators such that their distributions $\nu^n = \mathcal{L}(T^n \mid P^n)$ weakly converge to some distribution ν. Then there exists, maybe for an extension $\widehat{\mathcal{E}}$ of the experiment $\mathcal{E}$, a non-randomized estimator T such that $\nu = \mathcal{L}(T \mid \widehat{P})$ and for some subsequence (n')*

$$\mathcal{L}(T^{n'} \mid P_\alpha^{n'}) \to \mathcal{L}(T \mid \widehat{P}_\alpha), \quad \alpha \in \mathcal{A}. \tag{2.114}$$

Proof. The weak convergence yields tightness of the family of distributions $\mathcal{L}(T^n \mid P^n)$, $n \geq 1$. Let now elements $\widehat{\lambda}^n$ be constructed according to Theorem 2.5 from Section 2.5. Then $\mathcal{L}(\widehat{\lambda}^n \mid P^n) = \mathcal{L}(\lambda \mid P)$ and the family $(T^n, \widehat{\lambda}^n)$ is tight (with respect to P^n) and one can extract a subsequence (n') such that $\mathcal{L}(T^{n'}, \widehat{\lambda}^{n'} \mid P^{n'})$ weakly converges to some distribution, say $G(a, \lambda)$, $a \in \mathbf{A}$, $\lambda \in X$.

With $G(a, \lambda)$ we can construct the Markov kernel $\rho(\lambda, da) = G(dt \mid \lambda)$. Using this kernel one can construct, as in Lemma 1.3 of Section 1.4, a non-randomized estimator T (for an extension $\widehat{\mathcal{E}}$ of the experiment $\mathcal{E}$) such that

$$\mathcal{L}(T^{n'}, \widehat{\lambda}^{n'} \mid P^{n'}) \to \mathcal{L}(\rho, \lambda \mid P) = \mathcal{L}(T, \lambda \mid \widehat{P}). \tag{2.115}$$

To prove now that (2.115) implies the convergence (2.114) it suffices to note that for $f \in C_c^b$ from Theorem 2.7 of Section 2.5

$$\begin{aligned}\int f(T^{n'})\, P_\alpha^{n'}(d\omega) &= \int f(T^{n'})\, Z_\alpha(\widehat{\lambda}^{n'})\, d\widehat{P}^{n'} + o(1) \\ &\to \int f(a)\, Z_\alpha(\lambda)\, G(d\lambda, da) \\ &= \int f(a)\, \rho(\lambda, da)\, Z_\alpha(\lambda)\, P(d\lambda) \\ &= \int f(a)\, \rho(\lambda, da)\, P_\alpha(d\lambda) \\ &= \int f(T)\, d\widehat{P}_\alpha.\end{aligned}$$

□

Remark 2.12 The result of this theorem is very important since it explains that *the class of possible limit distributions for a sequence of estimators $\{T^n\}$ is not wider than the class of all estimator distributions for the limit experiment.* (Symbolically this assertion can be written as "$\liminf \mathcal{L}(T^n) \subseteq \mathcal{L}(T)$"). In particular, this fact explains also the results of the type of "$\liminf \mathcal{R}(\mathcal{E}^n) \geq \mathcal{R}(\mathcal{E})$".

3. In estimation theory the minimax approach is only one of the many possibilities to define the notion of asymptotic optimality, see Ibragimov and Khasminskii (1981, Introduction) for a detailed discussion. One more approach is based on on the idea to consider only a subclass of all possible estimators excluding some "exotic" ones (in the sense that they are "too good" for some values of a parameter and "too bad" for others) and searching for optimal estimators over this subclass. This idea leads us to the so-called *regular* estimators which are considered below (see also Section 3.4).

Let $\mathcal{E} = (\Omega, \mathcal{F}; P_\alpha, \alpha \in \mathcal{A})$ be a statistical experiment with a linear vector space $\mathcal{A}$. We shall assume that the space of possible decisions $\mathbf{A}$ coincides with $\mathcal{A}$.

Definition 2.13 A non-randomized estimator $\varkappa = \varkappa(\omega)$ with values in $\mathcal{A}$ is *regular* if its distribution law $\mathcal{L}(\varkappa - \alpha \mid P_\alpha)$ does not depend on the parameter α.

Definition 2.14 Let $\mathcal{E}^n = (\Omega^n, \mathcal{F}^n; P_\alpha^n, \alpha \in \mathcal{A}^n)$, $n \geq 1$, be a sequence of experiments, $\mathcal{A}^n \uparrow \mathcal{A}$. A sequence of estimators $\varkappa^n$, $n \geq 1$, is *asymptotically regular* if there exists a distribution H independent of α and such that

$$\mathcal{L}(\varkappa^n - \alpha \mid P_\alpha^n) \xrightarrow{w} H.$$

Theorem 2.22 *Let $\mathcal{E}^n \xrightarrow{\lambda} \mathcal{E}$ with a regular canonical dominated experiment $\mathcal{E}$.*

(i) Let $\varkappa$ be a regular estimator for $\mathcal{E}$, and let $\varkappa^n$, $n \geq 1$, be some sequence of estimators such that

$$\mathcal{L}(\varkappa^n, \lambda^n \mid P^n) \xrightarrow{w} \mathcal{L}(\varkappa, \lambda \mid P). \tag{2.116}$$

Then $\varkappa^n$ is asymptotically regular.

(ii) Let $\varkappa^n$, $n \geq 1$, be a sequence of asymptotically regular estimators. Then there exists a subsequence $(n') \subseteq (n)$ and a regular estimator $\varkappa$ such that

$$\mathcal{L}(\varkappa^{n'}, \lambda^{n'} \mid P^{n'}) \xrightarrow{w} \mathcal{L}(\varkappa, \lambda \mid P). \tag{2.117}$$

Proof. (i) It suffices to check that

$$\mathcal{L}(\varkappa^n - \alpha \mid P_\alpha^n) \xrightarrow{w} \mathcal{L}(\varkappa - \alpha \mid P_\alpha). \tag{2.118}$$

For $f \in C_c^b$ from (2.116) and from Theorem 2.7 of Section 2.5

$$\begin{aligned}\int f(\varkappa^n - \alpha)\, dP_\alpha^n &= \int f(\varkappa^n - \alpha)\, Z_\alpha(\widehat{\lambda}^n)\, d\widehat{P}^n + o(1)\\ &\to \int f(\varkappa - \alpha)\, Z_\alpha(\lambda)\, dP\\ &= \int f(\varkappa - \alpha)\, dP_\alpha,\end{aligned}$$

which proves (2.118).

(ii) Let $\varkappa^n$, $n \geq 1$, be a sequence of asymptotically regular estimators. Due to Theorem 2.21 there exists a randomized estimator $\varkappa$ such that $\mathcal{L}(\varkappa^{n'} \mid P_\alpha^{n'}) \to \mathcal{L}(\varkappa \mid P_\alpha)$, $\alpha \in \mathcal{A}$, for some subsequence $(n') \subseteq (n)$. Thus, $\mathcal{L}(\varkappa^{n'} - \alpha \mid P_\alpha^{n'}) \to \mathcal{L}(\varkappa - \alpha \mid P_\alpha)$ and the asymptotic regularity of $(\varkappa^n)$ implies that $\varkappa$ is a regular estimator too.

Finally, the assertion (2.117) follows from λ-convergence and from the property $\mathcal{L}(\varkappa^{n'} \mid P^{n'}) \to \mathcal{L}(\varkappa \mid P)$ (see proof of Theorem 2.21). □

Remark 2.13 The notion of a *regular estimator* is a particular case of a more general notion of an *equivariant estimator* corresponding to the case when the parameter space (which is also the decision space) is a locally compact topological group. All the results with respect to regular and asymptotically regular estimators can be extended to the case of equivariant estimators.

2.10 Proof of Lemma about "Reconstruction"

1. Suppose that $(X, \mathfrak{X}, \rho)$ is a Polish space, $\xi = \xi(x)$ and $\xi^n = \xi^n(\omega)$, $n \geq 1$, are random elements with values in X, given on probability spaces $(\Omega, \mathcal{F}, \mathbf{P})$ and $(\Omega^n, \mathcal{F}^n, \mathbf{P}^n)$, $n \geq 1$, respectively.

Lemma 2.5 *(About "Reconstruction") Let $\xi^n \xrightarrow{d} \xi$, i.e. $\mathcal{L}(\xi^n \mid \mathbf{P}^n) \to \mathcal{L}(\xi \mid \mathbf{P})$. Then there exist extensions $(\widehat{\Omega}^n, \widehat{\mathcal{F}}^n, \widehat{\mathbf{P}}^n)$ of the original probability spaces $(\Omega^n, \mathcal{F}^n, \mathbf{P}^n)$, $n \geq 1$, and random elements $\widehat{\xi}^n = \widehat{\xi}^n(\omega)$, $n \geq 1$, on it such that*

(i) $\rho(\xi^n, \widehat{\xi}^n) \xrightarrow{\widehat{\mathbf{P}}^n} 0, \quad n \to \infty.$

(ii) $\mathcal{L}(\widehat{\xi}^n \mid \widehat{\mathbf{P}}^n) = \mathcal{L}(\xi \mid \mathbf{P})\,.$

Remark 2.14 The construction of desirable extensions $(\widehat{\Omega}^n, \widehat{\mathcal{F}}^n, \widehat{\mathbf{P}}^n)$ can be made in many ways. Due to the proof below these extensions are chosen in the form

$$(\widehat{\Omega}^n, \widehat{\mathcal{F}}^n) = (\Omega^n, \mathcal{F}^n) \otimes (X, \mathfrak{X}), \qquad \widehat{\mathbf{P}}^n = \mathbf{P}^n \times K^n,$$

where $\mathbf{P}^n \times K^n$ means the semi-direct product of the measure $\mathbf{P}^n$ and a Markov kernel K^n from Ω^n to X.

Elements $\xi^n(\omega)$ defined originally on Ω^n are naturally expanded on $\widehat{\Omega}^n$: $\xi^n(\widehat{\omega}) = \xi^n(\omega)$ for $\widehat{\omega} = (\omega, \omega')$. The measures $\mathbf{P}^n$, $n \geq 1$, are also expanded on $(\widehat{\Omega}^n, \widehat{\mathcal{F}}^n)$ in a clear manner.

Remark 2.15 The following symbolic diagram explains the relations between Lemma about "Reconstruction" and the famous Skorohod's lemma which states the possibility to construct such a probability space $(\widehat{\Omega}, \widehat{\mathcal{F}}, \widehat{\mathbf{P}})$ and elements $\widehat{\xi}, \widehat{\xi}^n$ on it such that from $\xi^n \xrightarrow{d} \xi$ we get $\widehat{\xi}^n \xrightarrow{a.s} \widehat{\xi}$ and $\xi^n \stackrel{d}{=} \widehat{\xi}^n$, $\xi \stackrel{d}{=} \widehat{\xi}$:

$$\begin{array}{ccc} \xi^n & \xrightarrow{d} & \xi \\ P^n\downarrow & & \| \\ \widehat{\xi}^n & \stackrel{d}{=} & \xi \end{array} \qquad\qquad \begin{array}{ccc} \xi^n & \xrightarrow{d} & \xi \\ d\,\| & & \|\,d \\ \widehat{\xi}^n & \xrightarrow{a.s} & \widehat{\xi} \end{array}$$

Lemma about Reconstruction *Skorohod's lemma*

For *Lemma about Reconstruction* the probability spaces $(\widehat{\Omega}^n, \widehat{\mathcal{F}}^n, \widehat{\mathbf{P}}^n)$ are extensions of the original ones; in Skorohod's construction the corresponding spaces simply coincide with the space $([0,1], \mathcal{B}([0,1], \lambda)$ with the Lebesgue measure λ that does not allow to "control" the nearness between the elements ξ^n and $\widehat{\xi}^n$. At the same time, in Lemma about "Reconstruction" the control of rapprochement $(\rho(\xi^n, \widehat{\xi}^n) \xrightarrow{\widehat{\mathbf{P}}^n} 0)$ is one of the essential assertions. However, below we show that Lemma about "Reconstruction", via a special technical trick, can be proved via Skorohod's lemma. In this connection see also Rachev, Rüschendorf and Schief (1992, Theorem 1).

First we give a direct proof of Lemma about "Reconstruction", then we show how it can be derived from Skorohod's lemma.

2. The first proof. Let us fix an arbitrary number $\varepsilon > 0$ and construct elements ξ^n_ε on some extensions $(\widehat{\Omega}^n, \widehat{\mathcal{F}}^n, \widehat{\mathbf{P}}^n_\varepsilon)$ of the original probability spaces in such a manner that for sufficiently large n $(n \geq n(\varepsilon))$

$$\mathcal{L}(\xi^n_\varepsilon \mid \widehat{\mathbf{P}}^n_\varepsilon) = \mathcal{L}(\xi \mid \mathbf{P}) \quad \text{and} \quad \widehat{\mathbf{P}}^n_\varepsilon(\rho(\xi^n, \xi^n_\varepsilon) > \varepsilon) \leq \varepsilon. \tag{2.119}$$

Let $P^n = \mathcal{L}(\xi^n \mid \mathbf{P}^n)$ and $P = \mathcal{L}(\xi \mid \mathbf{P})$ be the distributions of ξ^n and ξ in $(X, \mathfrak{X})$.

From the distributional point of view the element ξ^n considered under the measure $\mathbf{P}^n$ can be identified with the element x from X having the distribution P^n.

Thus, $\xi^n \xrightarrow{d} \xi$, can be treated as $P^n \xrightarrow{w} P$ with the distributions P^n and P on $(X, \mathfrak{X})$ which is a Polish space due to the theorem's conditions. Hence, the family $\{P^n\}$ is *tight* and for any $\varepsilon > 0$ there is a compact $K = K(\varepsilon)$ such that for all n

$$P^n(X \setminus K) = \mathbf{P}^n(\xi^n \notin K) < \varepsilon.$$

We represent the compact K as a finite sum

$$K = \sum_{i=1}^{k} S_i,$$

where sets S_i are chosen in such a way that the diameter of each S_i is not greater than ε and $P(\partial S_i) = 0$, $i = 1, \ldots, k$. (The possibility of such a representation is proved, for example, in Pollard (1984).) Denoting $S_{k+1} = X \setminus K$ we have

$$P(S_{k+1}) < \varepsilon, \quad \operatorname{diam}(S_i) \leq \varepsilon, \quad i = 1, \ldots, k. \tag{2.120}$$

The construction of the required elements ξ^n_ε obeying (2.119) will be carried out in two steps. First we construct a Markov kernel $K^n(x, A)$ from X to X such that

$$\int K^n(x, A)\, P^n(dx) = P(A) \quad \text{and} \quad K^n(x, A) \geq (1 - \varepsilon)\, I_A(x)\, P(A), \quad A \in \mathfrak{X}.$$

Then we construct elements ξ^n_ε (on the corresponding extensions of the original probability spaces) using these kernels K^n.

The weak convergence $P^n \xrightarrow{w} P$ implies $P^n(S_i) \to P(S_i), i = 1, \ldots, k$, $n \to \infty$. If $P(S_i) > 0$, then also $P^n(S_i)/P(S_i) \to 1$, and hence for sufficiently large n

$$P(S_i) \geq (1 - \varepsilon)\, P^n(S_i). \tag{2.121}$$

Suppose now that this property is true for *all* $i = 1, \ldots, k + 1$. Define kernels (discrete in x)

$$K^n(x, A) = \sum_{i=1}^{k+1} I_{S_i}(x)\, K^n(A \mid S_i) \tag{2.122}$$

with

$$\begin{aligned} K^n(A \mid S_i) &= (1 - \varepsilon)\, P(A \mid S_i) + \sum_{j=1}^{k+1} P(A \mid S_j)[P(S_j) - (1 - \varepsilon)\, P^n(S_j)] \\ &= P(A) + (1 - \varepsilon) \left[P(A \mid S_i) - \sum_{j=1}^{k+1} P(A \mid S_j)\, P^n(S_j) \right]. \end{aligned} \tag{2.123}$$

The condition (2.121) easily implies that $K^n(x,A) \geq 0$ and $K^n(x,X) = 1$. Now we verify the equalities

$$\int K^n(x,A)P^n(dx) = P(A), \quad A \in \mathfrak{X}. \tag{2.124}$$

We have

$$\begin{aligned}
&\int K^n(x,A)\, P^n(dx) \\
&= \sum_{i=1}^{k+1} K^n(A \mid S_i)\, P^n(S_i) \\
&= P(A) + (1-\varepsilon)\left[\sum_{i=1}^{k+1} P(A \mid S_i)\, P^n(S_i) - \sum_{j=1}^{k+1} P(A \mid S_j)\, P^n(S_j)\right] \\
&= P(A), \qquad A \in \mathfrak{X}.
\end{aligned}$$

Then we obtain directly from (2.122) and (2.123)

$$K^n(x,A) \geq (1-\varepsilon)\, I_A(x)\, P(A), \quad A \in \mathfrak{X}. \tag{2.125}$$

Now using the Markov kernels $K^n(x,A)$ from X to X we define new Markov kernels $\widetilde{K}^n$ from Ω^n to X of the form $\widetilde{K}^n(\omega,A) = K^n(\xi^n(\omega),A)$. The extensions $(\widehat{\Omega}^n, \widehat{\mathcal{F}}^n, \widehat{\mathbf{P}}^n_\varepsilon)$ are defined as $(\Omega^n \times X, \mathcal{F}^n \otimes \mathfrak{X}, \mathbf{P}^n \times \widetilde{K}^n)$ where the measure $\mathbf{P}^n \times \widetilde{K}^n$ is the semi-direct product of the measure $\mathbf{P}^n$ and the Markov kernel $\widetilde{K}^n$,

$$\mathbf{P}^n \times \widetilde{K}^n(A \times B) = \int_A K^n(\xi^n, B)\, \mathbf{P}^n(d\omega), \quad A \in \mathcal{F},\, B \in \mathfrak{X}.$$

Finally, put $\xi^n_\varepsilon(\omega, x) = x$ and verify the conditions (2.119). From (2.124)

$$\widehat{\mathbf{P}}^n_\varepsilon(\xi^n_\varepsilon \in B) = \mathbf{P}^n \times \widetilde{K}^n(\Omega^n \times B) = \int K^n(x,B)\, \mathbf{P}^n(d\omega) = P(B).$$

Then, (2.125) yields

$$K^n(x,S_i) \geq (1-\varepsilon)\, P(S_i), \quad x \in S_i. \tag{2.126}$$

This means that "a random mechanism" corresponding to the kernel K^n keeps $x \in S_i$ in the same set S_i with probability at least $1-\varepsilon$. Since $\operatorname{diam}(S_i) \leq \varepsilon$, $i = 1, \ldots, k$, it follows from (2.126)

$$\begin{aligned}
\widehat{\mathbf{P}}^n_\varepsilon(\rho(\xi^n_\varepsilon, \xi^n) > \varepsilon) &\leq \mathbf{P}^n(\xi^n \in S_{k+1}) + \sum_{i=1}^{k} \widehat{\mathbf{P}}^n_\varepsilon(\xi^n \in S_i,\, \xi^n_\varepsilon \notin S_i) \\
&\leq \varepsilon + \sum_{i=1}^{k} \int K^n(\xi^n, X \setminus S_i)\, I_{\{\xi^n \in S_i\}}\, \mathbf{P}^n(d\omega) \\
&\leq 2\varepsilon.
\end{aligned}$$

Substituting everywhere the value $\varepsilon/2$ instead of ε leads to (2.119).

Above we supposed that $P(S_i) > 0$ for all $i = 1, \dots, k$. If it is not the case, then we can adjoin those sets S_i for which $P(S_i) = 0$ to the set S_{k+1}. Then for sufficiently large n the condition (2.120) will be again satisfied and all the subsequent considerations remain valid.

The elements $\widehat{\xi}^n$ desired in the lemma are constructed with the help of the diagonal procedure. From the above, for any $m \geq 1$, we can construct the elements $(\xi^n_m)_{n\geq 1}$ on the extensions $(\widehat{\Omega}^n, \widehat{\mathcal{F}}^n, \widehat{\mathbf{P}}^n_m)$ such that for $n \geq n(m)$

$$\mathcal{L}(\xi^n_m \mid \widehat{\mathbf{P}}^n_m) = \mathcal{L}(\xi \mid \mathbf{P}) \quad \text{and} \quad \widehat{\mathbf{P}}^n_m(\rho(\xi^n, \xi^n_m) > \frac{1}{m}) < \frac{1}{m}.$$

Define

$$\widehat{\xi}^n = \xi^n_m, \quad \widehat{\mathbf{P}}^n = \widehat{\mathbf{P}}^n_m \qquad \text{for} \quad n(m) \leq n < n(m+1).$$

By construction the elements $\widehat{\xi}^n$ obey all the conditions of the lemma.

The second proof (D.Kramkov). From Skorohod's lemma there exists a (Polish) probability space $(Y, \mathcal{Y}, \mu)$ (it can be taken as $([0,1], \mathcal{B}([0,1], \lambda)$ with the Lebesgue measure λ) and random elements $\eta = \eta(y)$, $\eta^n = \eta^n(y)$, $n \geq 1$, with values in X such that

(A) $\quad \mathcal{L}(\eta \mid \mu) = \mathcal{L}(\xi \mid P)$;

(B) $\quad \mathcal{L}(\eta^n \mid \mu) = \mathcal{L}(\xi^n \mid P^n)$;

(C) $\quad \rho(\eta^n, \eta) \to 0, \quad n \to \infty, \ \mu\text{-a.s.}$

Define on $(X \times Y, \mathcal{X} \otimes \mathcal{Y})$ the measure $R^n(dx, dy)$ putting

$$R^n(dx, dy) = \mu(dy)\, \delta_{\eta^n(y)}(dx),$$

where, as usual, $\delta_a(dx)$ is the Dirac measure concentrated at the point a.

Denote $Q^n = \mathcal{L}(\xi^n \mid P^n)$ $(= \mathcal{L}(\eta^n \mid \mu))$ and perform the decomposition of the measure R^n (under Q^n), i.e. represent it in the form

$$R^n(dx, dy) = Q^n(dx)\, K^n(x, dy),$$

with a kernel $K^n(x, dy)$ from $(X, \mathcal{X})$ to $(Y, \mathcal{Y})$. (Such a kernel exists because in Skorohod's lemma the space $(Y, \mathcal{Y})$ is Polish.)

Define now

$$\begin{aligned}
\widehat{\Omega}^n &= \Omega^n \times X \times Y \\
\widehat{\mathcal{F}}^n &= \mathcal{F}^n \otimes \mathcal{X} \otimes \mathcal{Y} \\
\widehat{\mathbf{P}}^n(d\widehat{\omega}^n) &= \widehat{\mathbf{P}}^n(d\omega^n, dx, dy) = \mathbf{P}^n(d\omega^n)\, \delta_{\xi^n(\omega^n)}(dx)\, K^n(x, dy), \\
\widehat{\xi}^n(\widehat{\omega}^n) &= \eta(y).
\end{aligned}$$

Next, we show that this construction satisfies the assertions (i) and (ii) of the lemma. Note first that the restrictions of the measure $\widehat{\mathbf{P}}^n$ on $(X \times Y, \mathcal{X} \otimes \mathcal{Y})$ and $(Y, \mathcal{Y})$ are the measures R^n and μ respectively. This and (A) imply

$$\mathcal{L}(\widehat{\xi}^n(\widehat{\omega}^n) \mid \widehat{\mathbf{P}}^n) = \mathcal{L}(\eta(y) \mid \widehat{\mathbf{P}}^n) = \mathcal{L}(\eta(y) \mid \mu) = \mathcal{L}(\xi \mid P).$$

To verify the property (i) we show that ($\widehat{\mathbf{P}}^n$-a.s.)

$$\xi^n(\widehat{\omega}^n) = \eta^n(y). \tag{2.127}$$

In fact, for $\widehat{\omega}^n = (\omega^n, x, y)$ we have $\xi^n(\widehat{\omega}^n) = \xi^n(\omega^n)$, and hence

$$\begin{aligned}
\widehat{\mathbf{P}}^n(\xi^n(\widehat{\omega}^n) = x) &= \int_{\Omega^n}\int_X I_{\{\xi^n(\widehat{\omega}^n)=x\}}\, \mathbf{P}^n(d\omega^n)\, \delta_{\xi^n(\omega^n)}(dx) \\
&= \int_{\Omega^n} \mathbf{P}^n(d\omega^n) \left[\int_X I_{\{\xi^n(\widehat{\omega}^n)=x\}}\, \delta_{\xi^n(\omega^n)}(dx)\right] \\
&= 1, \qquad (2.128)\\
\widehat{\mathbf{P}}^n(\eta^n(y) = x) &= \int_X\int_Y I_{\{\eta^n(y)=x\}}\, R^n(dx, dy) \\
&= \int_Y \mu(dy) \int_X I_{\{\eta^n(y)=x\}}\, \delta_{\eta^n(y)}(dx) \\
&= 1. \qquad (2.129)
\end{aligned}$$

Finally, (2.128) and (2.129) imply (2.127) and along with (C)

$$\widehat{\mathbf{P}}^n(\rho(\xi^n, \widehat{\xi}^n) \geq \varepsilon) = \widehat{\mathbf{P}}^n(\rho(\eta^n(y), \eta(y)) \geq \varepsilon) \to 0, \quad n \to \infty,$$

and the lemma follows.

2.11 Differentiable Experiments and Stochastic Expansions for Likelihood

1. The definition of λ-convergence of statistical experiments assumes the following two properties:

$$\mathcal{L}(\lambda^n \mid P^n) \to \mathcal{L}(\lambda \mid P)$$

and

$$Z_\alpha^n - Z_\alpha(\lambda^n) \xrightarrow{P^n} 0.$$

In this section we restrict ourselves to the main ideas of two constructive methods of verifying the λ-convergence that are widely applicable, especially for the case when the "LAN property" is proved. The first approach is the method of stochastic expansions which mostly works for the case of direct product of experiments that corresponds to the models with large number of *independent* observations.

The second method is the method of "differentiable experiments" based on the idea of using the smoothness properties of the considered distribution families. Contrary to the first method, this one "works" for more general situations (than just the case of direct product of experiments).

Theorem 2.24 below is an illustration of the joint application of these two approaches.

2. Stochastic expansion for likelihood. Let $\mathbb{E}^n = \mathcal{E}_1^n \times \ldots \times \mathcal{E}_n^n$ where

$$\mathcal{E}_i^n = (\Omega_i^n, \mathcal{F}_i^n; P_i^n, \widetilde{P}_i^n), \quad i = 1, \ldots, n,$$

are binary experiments. Thus $\mathbb{E}^n = (\Omega^n, \mathcal{F}^n; \mathbf{P}^n, \widetilde{\mathbf{P}}^n)$ with $\Omega^n = \Omega_1^n \times \ldots \times \Omega_n^n$, $\mathcal{F}^n = \mathcal{F}_1^n \otimes \ldots \otimes \mathcal{F}_n^n$, $\mathbf{P}^n = P_1^n \times \ldots \times P_n^n$, $\widetilde{\mathbf{P}}^n = \widetilde{P}_1^n \times \ldots \times \widetilde{P}_n^n$. Put

$$Z^n = \frac{d\widetilde{\mathbf{P}}^n}{d\mathbf{P}^n} \qquad \left(= \prod_{i=1}^n Z_i^n \quad \text{with} \quad Z_i^n = \frac{d\widetilde{P}_i^n}{dP_i^n} \right)$$

and

$$g_i^n = 2\left(\sqrt{Z_i^n} - 1\right).$$

All the derivatives Z_i^n under consideration are treated as the Radon–Nikodym derivations in the Lebesgue decomposition

$$\widetilde{P}_i^n(A) = \int_A Z_i^n \, dP_i^n + \widetilde{P}_i^n(A \cap N_i^n)$$

with $A \in \mathcal{F}_i^n$, $N_i^n \in \mathcal{F}_i^n$ and $P_i^n(N_i^n) = 0$.

Theorem 2.23 *Let g_i^n, $i = 1, \ldots, n$ satisfy the conditions:*

(*A*) $\limsup_n \sum_{i=1}^n E_i^n (g_i^n)^2 < \infty$;

(*B*) $\lim_n \sum_{i=1}^n E_i^n (g_i^n)^2 I_{\{|g_i^n| > \varepsilon\}} = 0, \quad \forall \varepsilon > 0.$

Then for Z^n the following expansion is fulfilled

$$Z^n = \exp\left\{ \sum_{i=1}^n [g_i^n - E_i^n g_i^n] - \frac{1}{2} \sum_{i=1}^n E_i^n (g_i^n)^2 \right\} \exp(\Psi^n) \tag{2.130}$$

with

$$\Psi^n = -\sum_{i=1}^n \widetilde{P}_i^n(N_i^n) + R^n$$

and

$$\mathbf{P}^n(|R^n| > \varepsilon) \to 0, \quad n \to \infty, \ \forall \varepsilon > 0.$$

Proof. It is based on the Taylor expansion and is given, for example, in Strasser (1985, Section 74). □

Remark 2.16 Let h_i^n, $i = 1, \dots, n$, $n \geq 1$, be such that h_i^n are $\mathcal{F}_i^n$-measurable, satisfy the conditions (A) and (B) (with g_i^n replaced by h_i^n) and

$$\lim_n \sum_{i=1}^n \int (g_i^n - h_i^n)^2 \, dP_i^n = 0,$$

or, equivalently,

$$\lim_n \sum_{i=1}^n \int \left[\sqrt{Z_i^n} - 1 - \frac{1}{2} h_i^n\right]^2 dP_i^n = 0. \tag{2.131}$$

Then the expansion (2.130) remains valid after exchanging g_i^n by h_i^n, Strasser (1985, Section 74).

3. Differentiable experiments. Let $\mathcal{E} = (\Omega, \mathcal{F}; P_\theta, \theta \in R; \nu)$ be a dominated statistical experiment with some σ-finite measure ν. Denote

$$z_\theta = \frac{dP_\theta}{d\nu}, \quad \theta \in R.$$

Definition 2.15 The experiment $\mathcal{E}$ is *differentiable* (in L_2) at the point $\theta = 0$ if the mapping $\theta \to \sqrt{z_\theta}$ (with values in $L_2(\Omega, \mathcal{F}, \nu)$) is Frechet differentiable, i.e. there exists an element $h \in L_2(\Omega, \mathcal{F}, \nu)$ such that

$$\theta^{-2} \int \left[\sqrt{z_\theta} - \sqrt{z_0} - \frac{\theta}{2} h\right]^2 d\nu \to 0, \quad \theta \to 0. \tag{2.132}$$

We formulate now some consequences following directly from this definition:

(a) $h(\omega) = 0$ (ν-a.s.) on the set $\{\omega : z_0(\omega) = 0\}$;

(b) $h(\omega)$ is of the form

$$h(\omega) = g(\omega)\sqrt{z_0},$$

with $g \in L_2(\Omega, \mathcal{F}, \nu)$;

(c) $\int g(\omega)\, P_0(d\omega) = 0$;

(d) $\lim_{\theta \to 0} \theta^{-2} P_\theta(N_\theta) = 0$;

where N_θ is the set from the Lebesgue decomposition

$$P_\theta(A) = \int_A Z_\theta \, dP_0 + P_\theta(A \cap N_\theta), \quad A \in \mathcal{F},$$

with $Z_\theta = z_\theta / z_0$.

To prove (a) note that if $\theta > 0$ and $A_1 = \{\omega : z_0(\omega) = 0,\ h(\omega) < 0\}$ then

$$\theta^{-2} \int_{A_1} \left[\sqrt{z_\theta} - \sqrt{z_0} - \frac{\theta}{2} h\right]^2 d\nu = \theta^{-2} \int_{A_1} \left[z_\theta - \theta \sqrt{z_0}\, h + \frac{\theta^2}{4} h\right]^2 d\nu$$
$$\geq \frac{1}{4} \int_{A_1} h^2\, d\nu.$$

If now $\theta \to 0$ then we conclude from this and (2.132) that $\nu(A_1) = 0$. Similarly, $\nu(A_2) = 0$ with

$$A_2 = \{\omega : z_0(\omega) = 0, \quad h(\omega) > 0\}$$

and (a) follows.

To prove (d) notice that $N_\theta = \{z_0 = 0\}$ and hence

$$P_\theta(N_\theta) = \int_{N_\theta} [\sqrt{z_\theta} - \sqrt{z_0}]^2\, d\nu$$
$$\leq 2 \int_{N_\theta} \left[\sqrt{z_\theta} - \sqrt{z_0} - \frac{\theta}{2} h\right]^2 d\nu + 2 \int_{N_\theta} \left[\frac{\theta}{2} h\right]^2 d\nu.$$

Now (d) follows from (a) and (2.132).

To select (b), define g in the form

$$g(\omega) = \begin{cases} h(\omega)\sqrt{z_0(\omega)} & \text{if } z_0(\omega) > 0 \\ 0, & \text{if } z_0(\omega) = 0 \end{cases}.$$

Then (a) implies the equality $h(\omega) = g(\omega)\sqrt{z_0(\omega)}$.

It remains to prove (c). From (2.132)

$$\theta^{-2} \int_\Omega \left[\frac{dP_\theta}{dP_0} - 1 - \frac{\theta}{2} h\right]^2 dP_0 \to 0, \quad \theta \to 0. \tag{2.133}$$

Hence, because L_2-convergence implies L_1-convergence, we derive

$$\frac{1}{2} \int g dP_0 = \lim_{\theta \to 0} \theta^{-1} \int \left[\frac{dP_\theta}{dP_0} - 1\right] dP_0. \tag{2.134}$$

Now the Lebesgue decomposition $P_\theta(A) = \int_A Z_\theta\, dP_0 + P_\theta(A \cap N_\theta)$ leads us to the equality

$$\frac{1}{2} \int \left[\frac{dP_\theta}{dP_0} - 1\right] dP_0 = -\int \left[\frac{dP_\theta}{dP_0} - 1\right]^2 dP_0 + P_\theta(N_\theta). \tag{2.135}$$

Moreover,

$$\lim_{\theta \to 0} \theta^{-1} \left\{ \int \left[\frac{dP_\theta}{dP_0} - 1\right]^2 dP_0 + P_\theta(N_\theta) \right\} = 0. \tag{2.136}$$

Indeed, on one hand (d) yields

$$\lim_{\theta\to 0} \theta^{-1} P_\theta(N_\theta) = 0,$$

and on the other hand (2.133) implies

$$\lim_{\theta\to 0} \theta^{-2} \int \left[\frac{dP_\theta}{dP_0} - 1\right]^2 dP_0 = \int \left[\frac{1}{2} g\right]^2 dP_0 < \infty,$$

and, hence

$$\lim_{\theta\to 0} \theta^{-1} \int \left[\frac{dP_\theta}{dP_0} - 1\right]^2 dP_0 = 0.$$

The required equality $\int g\, dP_0 = 0$ in (c) follows directly from (2.134)–(2.136).

4. Starting with a differentiable experiment $\mathcal{E} = (\Omega, \mathcal{F}; P_\theta, \theta \in R; \nu)$ we construct the experiments

$$\mathcal{E}^n = (\Omega^n, \mathcal{F}^n; P_\alpha^n, \alpha \in R), \quad n \geq 1,$$

with

$$\Omega^n = \Omega \times \ldots \times \Omega, \quad \mathcal{F}^n = \mathcal{F} \otimes \ldots \otimes \mathcal{F}, \quad P_\alpha^n = P_{\alpha/\sqrt{n}} \times \ldots \times P_{\alpha/\sqrt{n}}.$$

Theorem 2.24 *If $\mathcal{E}$ is a differentiable experiment (in the sense of (2.132)) then the densities $Z_\alpha^n = dP_\alpha^n/dP_0^n$ obey the following stochastic expansions*

$$Z_\alpha^n = \exp\left\{\frac{\alpha}{\sqrt{n}} \sum_{i=1}^n h(\omega_i) - \frac{\alpha^2}{2} \|h\|_{P_0}^2\right\} \exp(\Psi_n) \tag{2.137}$$

with $\Psi_n \xrightarrow{P_0^n} 0$, $n \to \infty$.

Proof. It is based on the stochastic expansion from Theorem 2.23 and the assumption of differentiability of the experiment $\mathcal{E}$. The statement (2.133) with $\theta = \alpha/\sqrt{n}$ implies

$$\lim_n n \int \left[\sqrt{Z_{\alpha n^{-1/2}}} - 1 - \frac{\alpha}{\sqrt{n}} h\right]^2 P_0(d\omega) = 0.$$

Hence, the condition (2.131) holds with

$$Z_i^n = Z_{\alpha n^{-1/2}}(\omega_i), \quad h_i^n = \frac{\alpha}{\sqrt{n}} h(\omega_i), \quad P_i^n = P_0.$$

The conditions (A) and (B) of Theorem 2.23 are also satisfied (with g_i^n instead of h_i^n). Thus, according to the remark after Theorem 2.23, for any $\alpha \in R$

$$Z_\alpha^n = \exp\left\{\frac{\alpha}{\sqrt{n}} \sum_{i=1}^n \left[h(\omega_i) - E_0\, h(\omega_i)\right] - \frac{\alpha^2}{2} \|h\|_{P_0}^2\right\} \exp(\Psi^n), \tag{2.138}$$

with

$$\Psi^n = -n\,P_{\alpha n^{-1/2}}(N_{\alpha n^{-1/2}}) + R^n, \qquad P_0^n(|R^n| > \varepsilon) \to 0, \quad n \to \infty,\ \forall \varepsilon > 0.$$

Due to (c) and (d) it holds $\mathbf{E}_0\, h = 0$ which yields for each $\alpha \not\equiv 0$

$$n\,P_{\alpha/\sqrt{n}}(N_{\alpha/\sqrt{n}}) \to 0,\ n \to \infty.$$

Therefore, (2.138) implies the desired expansion (2.137). □

5. Suppose for simplicity $\|h\|^2_{P_0} = 1$ and let $\mathcal{E} = (R, \mathcal{B}(R); P_\alpha, \alpha \in R)$ be a Gaussian shift experiment such that

$$Z_\alpha(\lambda) = \frac{dP_\alpha}{dP_0}(\lambda) = \exp\left\{\alpha\lambda - \frac{\alpha^2}{2}\right\}.$$

If we denote

$$\lambda^n = \frac{1}{\sqrt{n}}\sum_{i=1}^{n} h(\omega_i),$$

then the Central Limit Theorem for the sums of i.i.d. random variables with finite second moment yields

$$\mathcal{L}(\lambda^n \mid P_0^n) \to \mathcal{L}(\lambda \mid P_0) = \mathcal{N}(0,1)$$

and

$$Z_\alpha^n - Z_\alpha(\lambda^n) = Z_\alpha(\lambda^n)\,(\exp(\Psi^n) - 1) \xrightarrow{P_0^n} 0, \quad n \to \infty.$$

Therefore, the differentiablity condition for the experiment $\mathbb{E}$ directly implies the condition of λ-convergence for the corresponding "localized" experiments $\mathcal{E}^n$ to the Gaussian shift experiment $\mathcal{E}$.

2.12 Extended λ-Convergence

1. The definition of λ-convergence $\mathcal{E}^n \xrightarrow{\lambda} \mathcal{E}$ of statistical experiments $\mathcal{E}^n = (\Omega^n, \mathcal{F}^n;\ P_\alpha^n, \alpha \in \mathcal{A})$, $n \geq 1$, to a canonical experiment $\mathcal{E} = (X, \mathcal{X}; P_\alpha, \alpha \in \mathcal{A})$ is reduced to the following two conditions (see notation and Definition 2.6 in Section 2.4):

$$Z_\alpha^n - Z_\alpha(\lambda^n) \xrightarrow{P^n} 0, \quad \alpha \in \mathcal{A}, \tag{2.139}$$

and

$$\mathcal{L}(\lambda^n | P^n) \xrightarrow{w} \mathcal{L}(\lambda | P) \quad (= P), \tag{2.140}$$

where

$$Z_\alpha^n(\omega) = \frac{dP_\alpha^n}{dP^n}(\omega), \quad \omega \in \Omega^n,$$

and

$$Z_\alpha(x) = \frac{dP_\alpha}{dP}(x), \quad x \in X.$$

In this situation the "limit" experiment $\mathcal{E}$ is completely determined by the measure P and the process $\mathcal{Z} = (Z_\alpha(x),\, x \in X,\, \alpha \in \mathcal{A})$. To emphasize this fact we will denote it by $\mathcal{E} = \mathcal{E}(\mathcal{Z}, P)$. In such situations it is natural to indicate the λ-convergence in the form

$$\mathcal{E}^n \xrightarrow{\lambda} \mathcal{E}(\mathcal{Z}, P).$$

However, there are many statistics problems where (2.139) holds but the condition (2.140) is not satisfied and instead, tightness of the family of distributions $(\mathcal{L}(\lambda^n|P^n), n \geq 1)$ takes place. Thus, there is a possibility to speak at least about λ-convergence along some subsequence (n') : $\mathcal{E}^{n'} \xrightarrow{\lambda} \mathcal{E}'$ (instead of the whole sequence $\mathcal{E}^n$). The notion of *extended* λ-convergence introduced below corresponds precisely to this situation.

2. Let $(\mathcal{E}(\nu))_{\nu \in N}$ be a family of canonical experiments $\mathcal{E}(\nu) = (X, \mathcal{X}; P_\alpha(\nu), \alpha \in \mathcal{A})$ given on a common measurable space $(X, \mathcal{X})$. This family moreover plays the role of "the set of accumulation points of the sequence of experiments $\mathcal{E}^n$, $n \geq 1$". The experiments $\mathcal{E}(\nu), \nu \in N$, are assumed to be *weakly equivalent* in the sense that they have a common likelihood ratio

$$Z_\alpha(x) = \frac{dP_\alpha(\nu)}{dP(\nu)}(x), \quad x \in X, \tag{2.141}$$

where $P(\nu) = P_{\alpha_0}(\nu)$ and α_0 is some (common for all $\nu \in N$) point from the set $\mathcal{A}$. The condition (2.141) means in fact that the family of experiments $(\mathcal{E}(\nu), \nu \in N)$ is determined by the field $\mathcal{Z} = (Z_\alpha(x),\, x \in X,\, \alpha \in \mathcal{A})$ and by the collection of measures $\mathcal{P} = (P(\nu),\, \nu \in N)$. We shall use the notation $\boldsymbol{\mathcal{E}}(\mathcal{Z}, \mathcal{P})$ for the family with this property.

Definition 2.16 A sequence of experiments $\mathcal{E}^n$, $n \geq 1$, follows the conditions of *extended λ-convergence* of to a family $\boldsymbol{\mathcal{E}}(\mathcal{Z}, \mathcal{P})$ (notation $\mathcal{E}^n \xrightarrow{\lambda} \boldsymbol{\mathcal{E}}(\mathcal{Z}, \mathcal{P})$) if there exist $\mathcal{F}^n/\mathcal{X}$-measurable random elements $\lambda^n = \lambda^n(\omega)$, $n \geq 1$, such that

(i) for any $\alpha \in \mathcal{A}$

$$Z_\alpha^n - Z_\alpha(\lambda^n) \xrightarrow{P^n} 0,$$

with $P^n = P_{\alpha_0}^n$, $Z_\alpha^n = dP_\alpha^n / dP^n$, $n \geq 1$;

(ii) the family of distributions $(\mathcal{L}(\lambda^n \mid P^n), n \geq 1)$ is *tight* and the set of all its *accumulation* points (i.e. the set of all limits of weakly converging subsequences) belongs to the set $\mathcal{P} = (P(\nu), \nu \in N)$.

Remark 2.17 If the set $\mathcal{P}$ consists of *one* point, say P, then the extended λ-convergence simply coincides with the λ-convergence.

Remark 2.18 The condition of extended λ-convergence can be treated in the following manner. Let Σ denote the original sequence of experiments $\mathcal{E}^n$, $\Sigma = (\mathcal{E}^n, n \geq 1)$, and let $\mathfrak{A}_\lambda(\Sigma)$ be the set of λ-limit points for Σ, i.e. the set of experiments $\mathcal{E}$ for which λ-convergence $\mathcal{E}^{n'} \xrightarrow{\lambda} \mathcal{E}$ is fulfilled along some subsequence $(n') \subseteq (n)$. Then the condition $\mathcal{E}^n \xrightarrow{\lambda} \boldsymbol{\mathcal{E}}(\mathcal{Z}, \mathcal{P})$ implies the inclusion $\mathfrak{A}_\lambda(\Sigma) \subseteq \boldsymbol{\mathcal{E}}(\mathcal{Z}, \mathcal{P})$.

Remark 2.19 If the condition (i) in Definition 2.16 is fulfilled not only for any $\alpha \in \mathcal{A}$ but *uniformly* on K, $K \in \mathcal{K}$, with some collection $\mathcal{K}$ of subsets in $\mathcal{A}$, i.e.

$$\sup_{\alpha \in K} P^n(|Z_\alpha^n - Z_\alpha(\lambda^n)| > \varepsilon) \to 0, \quad n \to \infty,$$

for any $\varepsilon > 0$, then we say that *$\mathcal{K}$-uniform extended λ-convergence* takes place (notation $\mathcal{E}^n \xrightarrow{\lambda(\mathcal{K})} \boldsymbol{\mathcal{E}}(\mathcal{Z}, \mathcal{P})$).

In Section 2.5 we introduced the notion of regularity for a "limit" experiment $\mathcal{E}$ which was used, for example, for the proof of the main assertions of that section. The corresponding generalization of this notion to the case of a family $\boldsymbol{\mathcal{E}}(\mathcal{Z}, \mathcal{P})$ of weakly equivalent "limit" experiments $\mathcal{E}(\nu)$, $\nu \in N$, is given in the next definition. It is convenient to introduce an additional notation P_ν for the measures $P(\nu)$ from the family $\mathcal{P}$ and stand E_ν for integration with respect to P_ν.

Definition 2.17 The family $\boldsymbol{\mathcal{E}}(\mathcal{Z}, \mathcal{P})$ is called *regular* if the following conditions are satisfied:

(i) for each $\varepsilon > 0$ and any $\alpha \in \mathcal{A}$

$$\sup_{\nu \in N} E_\nu Z_\alpha I(Z_\alpha > h) \to 0, \quad h \to \infty;$$

(ii) for each $\varepsilon > 0$ and any $\alpha \in \mathcal{A}$

$$\sup_{\nu \in N} P_\nu(D^\delta Z_\alpha > \varepsilon) \to 0, \quad \delta \to 0,$$

with

$$D^\delta Z_\alpha(x) = \{y : \sup_{\{\rho(x,y) \leq \delta\}} |Z_\alpha(x) - Z_\alpha(y)|\};$$

(iii) the family of measures $\mathcal{P} = (P_\nu, \nu \in N)$ is compact for the weak convergence.

Remark 2.20 Note that the conditions (i),(ii) follow from (iii) and from the continuity of the field $\mathcal{Z} = (Z_\alpha(x),\ x \in X, \alpha \in \mathcal{A})$ in $x \in X$ and $\alpha \in \mathcal{A}$ (see Lemma 2.3 in Section 2.6).

Definition 2.18 A regular family of experiments $\boldsymbol{\mathcal{E}}(\mathcal{Z},\mathcal{P})$ is *$\mathcal{K}$-regular* for some collection $\mathcal{K}$ of subsets of the parameter set $\mathcal{A}$ if the conditions (i) and (ii) are satisfied uniformly in $\alpha \in K$ for each $K \in \mathcal{K}$ (i.e. after replacing $\sup_{\nu \in N}$ by $\sup_{\nu \in N, \alpha \in K}$).

The family $\boldsymbol{\mathcal{E}}(\mathcal{Z},\mathcal{P})$ introduced above, describes the set $\mathfrak{A}_\lambda(\Sigma)$ of λ-limits of subsequences of experiments $(\mathcal{E}^{n'})$, $(n') \subseteq (n)$, see Remark 2. Let now $\mathfrak{A}_w(\Sigma)$ be the set of all *weak* limits of subsequences $(\mathcal{E}^{n''})$, $(n'') \subseteq (n)$, i.e. the set of all *weak accumulation* points of the sequence Σ. The following result claims the coincidence of these two accumulation sets if one does not distinguish equivalent experiments in the set $\mathfrak{A}_w(\Sigma)$.

Theorem 2.25 *Let the condition of the extended λ-convergence $\mathcal{E}^n \xrightarrow{\lambda} \boldsymbol{\mathcal{E}}(\mathcal{Z},\mathcal{P})$ hold and let the family $\boldsymbol{\mathcal{E}}(\mathcal{Z},\mathcal{P})$ be regular. Then*

$$\mathfrak{A}_w(\Sigma) = \mathfrak{A}_\lambda(\Sigma).$$

Proof. Let $\mathcal{E} \in \mathfrak{A}_\lambda(\Sigma)$ and let a subsequence (n') be such that $\mathcal{E}^{n'} \xrightarrow{\lambda} \mathcal{E}$. The condition of the extended λ-convergence implies $\mathcal{E} \in \boldsymbol{\mathcal{E}}(\mathcal{Z},\mathcal{P})$ and the regularity of the family $\boldsymbol{\mathcal{E}}(\mathcal{Z},\mathcal{P})$ yields the regularity of the experiment $\mathcal{E}$. Therefore (Theorem 2.6 in Section 2.5) the condition of weak convergence $\mathcal{E}^{n'} \xrightarrow{w} \mathcal{E}$ is also satisfied and $\mathfrak{A}_\lambda(\Sigma) \subseteq \mathfrak{A}_w(\Sigma)$.

Next we prove the inverse inclusion. Let $\mathcal{E} \in \mathfrak{A}_w(\Sigma)$ and $\mathcal{E}^{n'} \xrightarrow{w} \mathcal{E}$. The condition (ii) in Definition 2.16 provides the tightness of the family of distributions $\mathcal{L}(\lambda^{n'}|P^{n'})$, and hence, there exists a $(n'') \subseteq (n')$ such that the weak convergence $\mathcal{L}(\lambda^{n''} \mid P^{n''}) \xrightarrow{w} P_\nu$, $\nu \in N$ holds. Along with the condition (i) of Definition 2.16 this leads to λ-convergence $\mathcal{E}^{n''} \xrightarrow{\lambda} \mathcal{E}_\nu = \mathcal{E}(\mathcal{Z}, P_\nu)$. Keeping in mind what was proved above we obtain the weak convergence $\mathcal{E}^{n''} \xrightarrow{w} \mathcal{E}$ and thus the experiments $\mathcal{E}$ and $\mathcal{E}_\nu$ are equivalent. Therefore, the experiment $\mathcal{E}$ (more precisely, its equivalent version E_ν) belongs to $\mathfrak{A}_\lambda(\Sigma)$. The theorem is proved. □

3. The following result for the minimax risks follows from the previous theorem and from Theorem 2.3 in Section 2.2.

Theorem 2.26 *Let the conditions of Theorem 2.25 hold. Then for any lower semicontinuous loss function W $(W \in \mathcal{W}_{lc})$*

$$\begin{aligned}\liminf_n \mathcal{R}(\mathcal{E}^n, W) &\geq \inf_{\mathcal{E} \in \mathfrak{A}_w(\Sigma)} \mathcal{R}(\mathcal{E}, W)\\ &= \inf_{\mathcal{E} \in \mathfrak{A}_\lambda(\Sigma)} \mathcal{R}(\mathcal{E}, W)\\ &\geq \inf_{\mathcal{E} \in \boldsymbol{\mathcal{E}}(\mathcal{Z},\mathcal{P})} \mathcal{R}(\mathcal{E}, W).\end{aligned}$$

If, moreover, $\mathcal{K}$-uniform extended λ-convergence holds and the family $\mathfrak{E}(\mathcal{Z},\mathcal{P})$ is $\mathcal{K}$-regular then for any continuous bounded loss function W $(W \in \mathcal{W}_c^b)$

$$\sup_{K\in\mathcal{K}} \limsup_n \mathcal{R}(\mathcal{E}_K^n, W) = \sup_{\mathcal{E}\in\mathfrak{A}_\lambda(\Sigma)} \mathcal{R}(\mathcal{E}, W) \leq \sup_{\mathcal{E}\in\mathfrak{E}(\mathcal{Z},\mathcal{P})} \mathcal{R}(\mathcal{E}, W).$$

2.13 Contiguity of Statistical Experiments. I

1. Above we considered the questions of convergence of statistical experiments $\mathcal{E}^n$, $n \geq 1$, to some limit experiment $\mathcal{E}$. In this situation one can say that the experiments $\mathcal{E}^n$, $n \geq 1$, "draw together" with the experiment $\mathcal{E}$. However, for many statistical problems one cannot guarantee the existence of a "limit" (in one or another sense) experiment $\mathcal{E}$ but often a new sequence of "accompanying" experiments $\widetilde{\mathcal{E}}^n$, $n \geq 1$, with a "simple structure" can be found such that the "distance" between $\mathcal{E}^n$ and $\widetilde{\mathcal{E}}^n$ tends to zero as n tends to infinity. Therefore, the sequence $\widetilde{\mathcal{E}}^n$, $n \geq 1$, can be considered as an approximation for the "original" sequence $\mathcal{E}^n$, $n \geq 1$, that gives us a possibility to solve some statistical problems for the experiments $\mathcal{E}^n$ using statistical conclusions for $\widetilde{\mathcal{E}}^n$. The concept of contiguity considered in this and next section follows Le Cam (1986) and this notion serve to describe accurately the idea of "accompanying" experiments.

Definition 2.19 A sequence of experiments $(\mathcal{E}^n)_{n\geq 1}$ is called *δ-contiguous* (or *δ-closed*) to a sequence $(\widetilde{\mathcal{E}}^n)_{n\geq 1}$ if the deficiency $\delta(\widetilde{\mathcal{E}}^n, \mathcal{E}^n) \to 0$, $n \to \infty$.

Definition 2.20 Sequences of experiments $(\mathcal{E}^n)_{n\geq 1}$ and $(\widetilde{\mathcal{E}}^n)_{n\geq 1}$ are *Δ-contiguous* (or *Δ-closed*) if Δ-distance $\Delta(\widetilde{\mathcal{E}}^n, \mathcal{E}^n) \to 0$, $n \to \infty$.

Directly from these definitions and from Theorem 2.13 of Section 1.7 we get (under the assumptions and notation of this theorem with $\|W_\alpha\| \leq 1$) that if $\delta(\widetilde{\mathcal{E}}^n, \mathcal{E}^n) = \delta_n$, then

$$\mathcal{R}(\widetilde{\mathcal{E}}^n, W) \leq \mathcal{R}(\mathcal{E}^n, W) + \delta_n$$

and if $\Delta(\widetilde{\mathcal{E}}^n, \mathcal{E}^n) = \Delta_n$ then

$$|\mathcal{R}(\widetilde{\mathcal{E}}^n, W) - \mathcal{R}(\mathcal{E}^n, W)| \leq \Delta_n. \tag{2.142}$$

2. In the same manner as Δ-convergence on *finite* parameter sets was used for the definition of the weak convergence of experiments, the following definitions are natural in the considered case.

Let K be a subset of the parameter set $\mathcal{A}$ $(K \subseteq \mathcal{A})$ and let $\mathcal{E}_K^n, \widetilde{\mathcal{E}}_K^n$ be the restrictions of experiments $\mathcal{E}^n, \widetilde{\mathcal{E}}^n$ on K:

$$\mathcal{E}_K^n = (\Omega^n, \mathcal{F}^n; P_\alpha^n, \alpha \in K), \qquad \widetilde{\mathcal{E}}_K^n = (\widetilde{\Omega}^n, \widetilde{\mathcal{F}}^n; \widetilde{P}_\alpha^n, \alpha \in K), \quad n \geq 1.$$

Let also $S(\mathcal{A})$ be the set of *all* finite subsets from $\mathcal{A}$.

Definition 2.21 The experiments $(\mathcal{E}^n)_{n\geq 1}$ are *weakly contiguous* to the experiments $(\widetilde{\mathcal{E}}^n)_{n\geq 1}$ if $\delta(\widetilde{\mathcal{E}}^n_\sigma, \mathcal{E}^n_\sigma) \to 0$ as $n \to \infty$ for any finite parameter set $\sigma \in S(\mathcal{A})$. The experiments $(\mathcal{E}^n)_{n\geq 1}$ and $(\widetilde{\mathcal{E}}^n)_{n\geq 1}$ are *weakly mutually contiguous* if $\Delta(\widetilde{\mathcal{E}}^n_\sigma, \mathcal{E}^n_\sigma) \to 0$, $n \to \infty$ for any $\sigma \in S(\mathcal{A})$.

Denote by μ^n_σ and $\widetilde{\mu}^n_\sigma$ the standard measures (see Section 1.1) corresponding to experiments $\mathcal{E}^n_\sigma$, $\widetilde{\mathcal{E}}^n_\sigma$ with a finite parameter set σ containing $\|\sigma\|$ points. By Theorem 1.9 of Section 1.7 the condition of weak mutual contiguity of the experiments $(\mathcal{E}^n)_{n\geq 1}$ and $(\widetilde{\mathcal{E}}^n)_{n\geq 1}$ is equivalent to the fact that the Lévy-Prokhorov distance $L(\cdot,\cdot)$ (see Shiryaev (1995, III.7)) between μ^n_σ and $\widetilde{\mu}^n_\sigma$ fulfills

$$L\left(\frac{1}{\|\sigma\|}\mu^n_\sigma, \frac{1}{\|\sigma\|}\widetilde{\mu}^n_\sigma\right) \to 0, \qquad n \to \infty, \quad \sigma \in S(\mathcal{A}).$$

If $\mathcal{E}^n \xrightarrow{w} \mathcal{E}$ and $W \in \mathcal{W}^b_c$ (i.e. W is continuous and bounded) then from Theorem 2.3 of Section 2.2)

$$\lim_n \mathcal{R}(\mathcal{E}^n_\sigma, W) = \mathcal{R}(\mathcal{E}_\sigma, W), \qquad \sigma \in S(\mathcal{A}).$$

From this, because $\sup_{\sigma\in S(\mathcal{A})} \mathcal{R}(\mathcal{E}_\sigma, W) = \mathcal{R}(\mathcal{E}, W)$ (see Section 1.7), we find that

$$\liminf_n \mathcal{R}(\mathcal{E}^n, W) \geq \mathcal{R}(\mathcal{E}, W). \tag{2.143}$$

Let now $(\mathcal{E}^n)_{n\geq 1}$ and $(\widetilde{\mathcal{E}}^n)_{n\geq 1}$ be weakly mutually contiguous sequences of experiments. Then from (2.142)

$$\lim_n |\mathcal{R}(\widetilde{\mathcal{E}}^n_\sigma, W) - \mathcal{R}(\mathcal{E}^n_\sigma, W)| = 0, \tag{2.144}$$

which is a generalization of the property (2.143).

Consider now the question of the generalization of (2.143) to the case of weakly mutually contiguous experiments. Define for a sequence of experiments $\Sigma = (\mathcal{E}^n)_{n\geq 1}$,

$$\mathcal{R}_*(\Sigma, W) = \sup_{\sigma\in S(\mathcal{A})} \liminf_n \mathcal{R}(\mathcal{E}^n_\sigma, W)$$

and

$$\mathcal{R}^*(\Sigma, W) = \sup_{\sigma\in S(\mathcal{A})} \limsup_n \mathcal{R}(\mathcal{E}^n_\sigma, W).$$

Theorem 2.27 *Let experiments $\Sigma = (\mathcal{E}^n)_{n\geq 1}$ and $\widetilde{\Sigma} = (\widetilde{\mathcal{E}}^n)_{n\geq 1}$ be weakly mutually contiguous and $W \in \mathcal{W}^b_c$. Then*

$$\liminf_n \mathcal{R}(\mathcal{E}^n, W) \geq \mathcal{R}_*(\widetilde{\Sigma}, W), \tag{2.145}$$

$$\limsup_n \mathcal{R}(\mathcal{E}^n, W) \geq \mathcal{R}^*(\widetilde{\Sigma}, W). \tag{2.146}$$

Proof. By virtue of (2.144) for any $\sigma \in S(\mathcal{A})$

$$\liminf_n \mathcal{R}(\mathcal{E}^n_\sigma, W) = \liminf_n \mathcal{R}(\widetilde{\mathcal{E}}^n_\sigma, W). \tag{2.147}$$

Hence

$$\mathcal{R}_*(\Sigma, W) = \mathcal{R}_*(\widetilde{\Sigma}, W) \tag{2.148}$$

and similarly

$$\mathcal{R}^*(\Sigma, W) = \mathcal{R}^*(\widetilde{\Sigma}, W).$$

Now Fatou's lemma implies

$$\begin{aligned} \liminf_n \mathcal{R}(\mathcal{E}^n, W) &\geq \liminf_n \sup_{\sigma \in S(\mathcal{A})} \mathcal{R}(\mathcal{E}^n_\sigma, W) \\ &\geq \sup_{\sigma \in S(\mathcal{A})} \liminf_n \mathcal{R}(\mathcal{E}^n_\sigma, W) = \mathcal{R}_*(\Sigma, W), \end{aligned}$$

which together with (2.148) implies (2.145). The inequality (2.146) is proved in the same way. □

3. The assertion (2.147) of Theorem 2.27 can be expressed in another form using the result of the lemma given below. Denote, similarly to the previous section, by $\mathfrak{A}_w(\widetilde{\Sigma})$ the set of all weak accumulation points for a sequence of experiments $\widetilde{\Sigma} = (\widetilde{\mathcal{E}}^n)_{n \geq 1}$ (i.e. $\widetilde{\mathcal{E}} \in \mathfrak{A}_w(\widetilde{\Sigma})$ if there exists a subsequence $(n') \subseteq (n)$ such that $\widetilde{\mathcal{E}}^{n'} \xrightarrow{w} \widetilde{\mathcal{E}}$).

Lemma 2.6 *For any loss function* $W \in \mathcal{W}^b_c$

$$\mathcal{R}^*(\widetilde{\Sigma}, W) = \sup_{\widetilde{\mathcal{E}} \in \mathfrak{A}_w(\widetilde{\Sigma})} \mathcal{R}(\widetilde{\mathcal{E}}, W). \tag{2.149}$$

Proof. Since the set of experiments $\mathfrak{A}_w(\widetilde{\Sigma})$ is compact for the topology of the weak convergence, Strasser (1985, Section 59), there is such an experiment $\mathcal{E}^*$ in $\mathfrak{A}_w(\widetilde{\Sigma})$ that

$$\sup_{\widetilde{\mathcal{E}} \in \mathfrak{A}_w(\widetilde{\Sigma})} \mathcal{R}(\widetilde{\mathcal{E}}, W) = \mathcal{R}(\mathcal{E}^*, W).$$

Let $\widetilde{\mathcal{E}}^{n'} \xrightarrow{w} \mathcal{E}^*$, $(n') \subseteq (n)$. Then for any $\sigma \in S(\mathcal{A})$

$$\limsup_n \mathcal{R}(\widetilde{\mathcal{E}}^n_\sigma, W) \geq \lim_{n'} \mathcal{R}(\widetilde{\mathcal{E}}^{n'}_\sigma, W) = \mathcal{R}(\mathcal{E}^*_\sigma, W).$$

Thus

$$\begin{aligned}\mathcal{R}^*(\widetilde{\Sigma}, W) &= \sup_{\sigma\in S(\mathcal{A})} \limsup_n \mathcal{R}(\widetilde{\mathcal{E}}^n_\sigma, W) \\ &\geq \sup_{\sigma\in S(\mathcal{A})} \mathcal{R}(\mathcal{E}^*_\sigma, W) = \mathcal{R}(\mathcal{E}^*, W) \\ &= \sup_{\widetilde{\mathcal{E}}\in\mathfrak{A}_w(\widetilde{\Sigma})} \mathcal{R}(\widetilde{\mathcal{E}}, W). \end{aligned} \tag{2.150}$$

Now we prove the inverse inequality. Let $\sigma \in S(\mathcal{A})$ and let a subsequence $(n') \subseteq (n)$ be such that

$$\limsup_n \mathcal{R}(\widetilde{\mathcal{E}}^n_\sigma, W) = \lim_{n'} \mathcal{R}(\widetilde{\mathcal{E}}^{n'}_\sigma, W).$$

One can extract from the subsequence (n') another subsequence (n'') such that the weak convergence holds:

$$\begin{aligned}\mathcal{R}^*(\widetilde{\Sigma}, W) &\leq \sup_{\sigma\in S(\mathcal{A})} \sup_{\widetilde{\mathcal{E}}\in\mathfrak{A}_w(\widetilde{\Sigma})} \mathcal{R}(\widetilde{\mathcal{E}}_\sigma, W) \\ &= \sup_{\widetilde{\mathcal{E}}\in\mathfrak{A}_w(\widetilde{\Sigma})} \sup_{\sigma\in S(\mathcal{A})} \mathcal{R}(\widetilde{\mathcal{E}}_\sigma, W) \\ &= \sup_{\widetilde{\mathcal{E}}\in\mathfrak{A}_w(\widetilde{\Sigma})} \mathcal{R}(\widetilde{\mathcal{E}}, W). \end{aligned} \tag{2.151}$$

The required equality (2.149) follows from (2.150) and (2.151). □

Theorem 2.28 *Let experiments* $\Sigma = (\mathcal{E}^n)_{n\geq 1}$ *and* $\widetilde{\Sigma} = (\widetilde{\mathcal{E}}^n)_{n\geq 1}$ *be weakly mutually contiguous and* $W \in \mathcal{W}^b_c$ *. Then*

$$\limsup_n \mathcal{R}(\mathcal{E}^n, W) \geq \sup_{\widetilde{\mathcal{E}}\in\mathfrak{A}_w(\widetilde{\Sigma})} \mathcal{R}(\widetilde{\mathcal{E}}, W),$$

where $\mathfrak{A}_w(\widetilde{\Sigma})$ *is the set of all weak accumulation points of the sequence of experiments* $\widetilde{\Sigma} = (\widetilde{\mathcal{E}}^n)_{n\geq 1}$ *.*

2.14 Contiguity of Statistical Experiments. II

1. In the previous section we considered the notion of contiguity which (analogously to the notion of convergence of experiments) can be called the "strong contiguity" (Definition 2.20) and the "weak contiguity" (Definition 2.21). Now we introduce the definition of contiguous experiments based on the concept of λ-convergence. For this we suppose one fairly special condition on an approximating sequence of experiments $(\widetilde{\mathcal{E}}^n)_{n\geq 1}$ which makes the problem under consideration similar to the case of λ-convergence. Notice that although this assumption is rather restrictive, it is satisfied for a wide class of statistical models (see, for example, Sections 3.6 and 3.7).

Let a sequence of experiments $\widetilde{\mathcal{E}}^n = (X, \mathcal{X}; \widetilde{P}^n_\alpha, \alpha \in \mathcal{A})$, $n \geq 1$, be defined on a Polish space $(X, \mathcal{X}; \rho)$ such that $\widetilde{P}^n_\alpha \ll \widetilde{P}^n_{\alpha_0}$ for all $\alpha \in \mathcal{A}$, with some fixed point α_0 in $\mathcal{A}$. For simplicity we denote $\widetilde{P}^n = \widetilde{P}^n_{\alpha_0}$. This particularly means that the experiments $\widetilde{\mathcal{E}}^n, n \geq 1$, are defined on the common probability space and they are canonical in terms of Section 1.1.

The special assumption with respect to $\widetilde{\mathcal{E}}^n$, $n \geq 1$, mentioned above is as follows: the likelihood $d\widetilde{P}^n_\alpha / d\widetilde{P}^n(x)$ do not depend on n, i.e. there exists a function $\widetilde{Z}_\alpha(x), x \in X, \alpha \in \mathcal{A}$, such that $\widetilde{P}^n$-a.s. for all $n \geq 1$

$$\frac{d\widetilde{P}^n_\alpha}{d\widetilde{P}^n}(x) = \widetilde{Z}_\alpha(x), \qquad x \in X, \quad \alpha \in \mathcal{A}. \tag{2.152}$$

Experiments $\widetilde{\mathcal{E}}^n$, $n \geq 1$, obeying this condition will be called *weakly equivalent.* For such situations the experiments $\widetilde{\mathcal{E}}^n$, $n \geq 1$, are really determined by the measures $\widetilde{P}^n$, $n \geq 1$, and by the process $\widetilde{Z} = (\widetilde{Z}_\alpha(x),\ x \in X,\ \alpha \in \mathcal{A})$. (Under the notation of Section 2.12, $\widetilde{\mathcal{E}}^n = \mathcal{E}(\widetilde{Z}, \widetilde{P}^n)$, $n \geq 1$). If we denote $\widetilde{\Sigma} = (\widetilde{\mathcal{E}}^n)_{n\geq 1}$, $\widetilde{\Xi} = (\widetilde{P}^n)_{n \geq 1}$, then it is natural to write $\widetilde{\Sigma} = (\widetilde{\mathcal{E}}^n)_{n \geq 1} = \mathcal{E}(\widetilde{Z}, \widetilde{\Xi})$.

Definition 2.22 We say that experiments $(\mathcal{E}^n)_{n\geq 1}$ with $\mathcal{E}^n = (\Omega^n, \mathcal{F}^n; P^n_\alpha, \alpha \in \mathcal{A})$, $n \geq 1$, are λ-*contiguous* to weakly equivalent experiments $(\widetilde{\mathcal{E}}^n)_{n \geq 1} = \mathcal{E}(\widetilde{Z}, \widetilde{\Xi})$ if the following two conditions hold:

(i) there exist $\mathcal{F}^n / \mathcal{X}$-measurable random elements $\lambda^n = \lambda^n(\omega)$ such that for any $\alpha \in \mathcal{A}$

$$Z^n_\alpha - \widetilde{Z}_\alpha(\lambda^n) \xrightarrow{P^n} 0,$$

with $Z^n_\alpha = dP^n_\alpha / dP^n$, $P^n = P^n_{\alpha_0}$;

(ii) the Lévy-Prokhorov distance

$$L(P^n_{\lambda^n}, \widetilde{P}^n) \to 0, \quad n \to \infty,$$

where $P^n_{\lambda^n} = \mathcal{L}(\lambda^n \mid P^n)$ is the distribution of λ^n with respect to P^n.

Remark 2.21 Note that if the experiments $\widetilde{\mathcal{E}}^n$, $n \geq 1$, coincide with one another ($\widetilde{\mathcal{E}}^n = \mathcal{E}$, $n \geq 1$) then this definition is reduced to the definition of λ-convergence $\mathcal{E}^n \xrightarrow{\lambda} \mathcal{E}$.

Remark 2.22 For a number of models (see, for example, Section 3.7 below) the condition (ii) is satisfied automatically because the measures $\widetilde{P}^n$ entering in the construction of the approximating experiments $\widetilde{\mathcal{E}}^n$ are chosen equal to $P^n_{\lambda^n}$.

2. As for the case of λ-convergence, in order to obtain the results on minimax risks under condition of λ-contiguity we need also some regularity conditions on the approximating sequence of experiments $(\widetilde{\mathcal{E}}^n)_{n \geq 1}$:

(*R*) (i) Experiments $\widetilde{\mathcal{E}}^n$, $n \geq 1$, are weakly equivalent, i.e. the condition (2.152) holds;

(ii) for any $\alpha \in \mathcal{A}$

$$\sup_n \widetilde{\mathcal{E}}^n \widetilde{Z}_\alpha I(\widetilde{Z}_\alpha > h) \to 0, \quad h \to \infty;$$

(iii) for any $\varepsilon > 0$

$$\sup_n \widetilde{P}^n(D^\delta \widetilde{Z}_\alpha > \varepsilon) \to 0, \quad \delta \to 0,$$

with

$$D^\delta \widetilde{Z}_\alpha(x) = \{y : \sup_{\rho(x,y) \leq \delta\}} |\widetilde{Z}_\alpha(x) - \widetilde{Z}_\alpha(y)|, \quad x \in X;$$

(iv) the family of measures $\widetilde{\Xi} = (\widetilde{P}^n)_{n\geq 1}$ is tight.

Analogously to Lemma 2.3 from Section 2.6 one can show that the conditions (i) and (ii) follow from (iv) and from the continuity of the function $\widetilde{Z}_\alpha(x)$ in the pair $x \in X$, $\alpha \in \mathcal{A}$. (Here we assume, as in the referred lemma, that $\mathcal{A}$ is a metric space.)

The condition (*R*) permits us to characterize easily the set of accumulation points $\mathfrak{A}_w(\widetilde{\Sigma})$ for the sequence of experiments $\widetilde{\Sigma} = (\widetilde{\mathcal{E}}^n)_{n\geq 1}$ for the topology of weak convergence. Indeed, due to Theorem 2.30 below, there is a one-to-one correspondence between the limit points of the sequence of measures $\widetilde{\Xi} = (\widetilde{P}^n)_{n\geq 1}$ and the points (equivalent classes) from $\mathfrak{A}_w(\widetilde{\Sigma})$.

First we state the result describing the relation between the notions of λ-contiguity and the notion of the extended λ-convergence introduced in Section 2.12.

Theorem 2.29 *Let a sequence of experiments $\Sigma = (\mathcal{E}^n)_{n\geq 1}$ be λ-contiguous to a sequence $\widetilde{\Sigma} = (\widetilde{\mathcal{E}}^n)_{n\geq 1}$ satisfying the condition (R), let $\widetilde{\mathcal{P}} = \mathfrak{A}_w(\widetilde{\Xi})$ be the set of all limit points for the sequence of measures $\widetilde{\Xi} = (\widetilde{P}^n)_{n\geq 1}$, and let $\mathcal{E}(\widetilde{Z}, \widetilde{\mathcal{P}}) = (\mathcal{E}(\widetilde{Z}, \widetilde{P}),\ \widetilde{P} \in \widetilde{\mathcal{P}})$ be the family of experiments corresponding to the process $\widetilde{Z}$ and to the family of measures $\widetilde{\mathcal{P}}$ with experiments $\widetilde{\mathcal{E}} = \mathcal{E}(\widetilde{Z}, \widetilde{P}) = (X, \mathcal{X}; \widetilde{P}_\alpha, \alpha \in \mathcal{A})$, $\widetilde{P} \in \mathcal{P}$, determined by the condition*

$$\frac{d\widetilde{P}_\alpha}{d\widetilde{P}}(x) = \widetilde{Z}_\alpha(x), \quad x \in X, \alpha \in \mathcal{A}.$$

Then the extended λ-convergence holds $\left(\mathcal{E}^n \xrightarrow{\lambda} \mathcal{E}(\widetilde{Z}, \widetilde{\mathcal{P}})\right)$ and the family of experiments $\mathcal{E}(\widetilde{Z}, \widetilde{\mathcal{P}})$ is regular in the sense of Definition 2.17 of Section 2.12.

Proof. First we verify that the family of experiments $\mathcal{E}(\widetilde{Z}, \widetilde{\mathcal{P}})$ is correctly constructed and obeys the regularity condition. From the conditions (*R*.iii) – (*R*.iv) it follows that for any $\alpha \in \mathcal{A}$

$$\sup_{\widetilde{P}\in\widetilde{\mathcal{P}}} \widetilde{E}\,\widetilde{Z}_\alpha I(\widetilde{Z}_\alpha > h) \to 0, \quad h \to \infty,$$
$$\sup_{\widetilde{P}\in\widetilde{\mathcal{P}}} \widetilde{P}(D^\delta \widetilde{Z}_\alpha > \varepsilon) \to 0, \quad \delta \to 0. \tag{2.153}$$

The relation (2.153) which means the uniform integrability of the function $\widetilde{Z}_\alpha$ under measures $\widetilde{P} \in \widetilde{\mathcal{P}}$ allows along with the equalities $\widetilde{E}^n \widetilde{Z}_\alpha = 1$, $n \geq 1$, to assert that $\widetilde{E}\widetilde{Z}_\alpha = 1$ for any $\alpha \in \mathcal{A}$ and for all $\widetilde{P} \in \widetilde{\mathcal{P}}$; hence the experiments $\widetilde{\mathcal{E}} = \mathcal{E}(\widetilde{Z}, \widetilde{P})$ are defined correctly and the family $\mathcal{E}(\widetilde{Z}, \widetilde{\mathcal{P}})$ is regular. The condition of the extended λ-convergence follows now directly from the condition of λ-contiguity. □

The result of this theorem also helps to describe the set of all weak limits (more precisely, all equivalent classes) and the set of all λ-limits for the original sequence of experiments $\Sigma = (\mathcal{E}^n)_{n\geq 1}$.

Theorem 2.30 *Let a sequence of experiments $\Sigma = (\mathcal{E}^n)_{n\geq 1}$ be λ-contiguous to a sequence $\widetilde{\Sigma} = (\widetilde{\mathcal{E}}^n)_{n\geq 1}$ obeying the condition (R). Then*

$$\mathfrak{A}_w(\Sigma) = \mathfrak{A}_\lambda(\Sigma) = \mathcal{E}(\widetilde{Z}, \widetilde{\mathcal{P}}) \tag{2.154}$$

and the sequences of experiments $\Sigma = (\mathcal{E}^n)_{n\geq 1}$ and $\widetilde{\Sigma} = (\widetilde{\mathcal{E}}^n)_{n\geq 1}$ are weakly mutually contiguous.

Proof. The assertion (2.154) was proved above. It only remains to check that Σ and $\widetilde{\Sigma}$ are weakly mutually contiguous. Let σ be an arbitrary finite set in $\mathcal{A}$. Then the condition of λ-contiguity implies rapprochement in the strong sense of the finite experiments $\mathcal{E}^n_\sigma$ and $\widetilde{\mathcal{E}}^n_\sigma$,

$$\Delta(\mathcal{E}^n_\sigma, \widetilde{\mathcal{E}}^n_\sigma) \to 0, \quad n \to \infty. \tag{2.155}$$

Indeed, because of the condition $(R.\text{iv})$ about the tightness of the family $\widetilde{\Xi} = (\widetilde{P}^n)_{n\geq 1}$, it suffices to prove the assertion (2.155) for any subsequence $(n') \subseteq (n)$ such that $\widetilde{P}^{n'} \xrightarrow{w} \widetilde{P}$. But with it, it can be concluded from above that the λ-convergence of the experiments $\mathcal{E}^{n'}$ and $\widetilde{\mathcal{E}}^{n'}$ to $\mathcal{E}(\widetilde{Z}, \widetilde{P})$ holds and (2.155) follows from Theorem 2.6, Section 2.5. □

Remark 2.23 Theorems 2.29 and 2.30 imply that all the asymptotic minimax assertions satisfied for the case of extended λ-convergence and for weakly mutually contiguous experiments remain valid under assumptions of λ-contiguity and the condition (R).

3. In Section 2.6 we considered the uniform versions of λ-convergence of statistical experiments which permitted us to obtain significantly more precise assertions about convergence of minimax risks. A similar situation arises for the models with contiguous experiments: uniform versions of λ-contiguity and of the conditions (R) permit to establish the "strong" contiguity of sequences of experiments (i.e. on

Δ-contiguity) leading to the corresponding "strong" results on rapprochement of risks. The exposition of these questions is similar to that of in Section 2.6 for the case of λ-convergence and in Section 2.12 for the uniform versions of extended λ-convergence.

Let some class $\mathcal{K}$ of subsets be fixed in the space $\mathcal{A}$.

Definition 2.23 A sequence of experiments $(\mathcal{E}^n)_{n\geq 1}$ with $\mathcal{E}^n = (\Omega^n, \mathcal{F}^n; P_\alpha^n, \alpha \in \mathcal{A})$, $n \geq 1$, is *K-uniformly* λ*-contiguous* to a sequence of weakly equivalent experiments $(\widetilde{\mathcal{E}}^n)_{n\geq 1}$ if the following conditions hold:

(i) there exist $\mathcal{F}^n/\mathcal{X}$-measurable random elements $\lambda^n = \lambda^n(\omega)$ such that

$$\sup_{\alpha\in K} P^n(|Z_\alpha^n - \widetilde{Z}_\alpha(\lambda^n)| > \varepsilon) \to 0, \qquad n \to \infty, \quad \forall \varepsilon > 0;$$

(ii) the Lévy-Prokhorov distance

$$L(P_{\lambda^n}^n, \widetilde{P}^n) \to 0, \quad n \to \infty,$$

where $P_{\lambda^n}^n = \mathcal{L}(\lambda^n \mid P^n)$ is the distribution of λ^n with respect to P^n.

If $\mathcal{K}$ is a collection of sets K in $\mathcal{A}$ for which the conditions (i) and (ii) are satisfied then we say that $\mathcal{K}$*-uniform* λ*-contiguity* of the sequence $(\mathcal{E}^n)_{n\geq 1}$ to the sequence $(\widetilde{\mathcal{E}}^n)_{n\geq 1}$ holds.

Now we give a *uniform* analog of the condition (R):

$(R\mathcal{K})$ (i) Experiments $\widetilde{\mathcal{E}}^n$, $n \geq 1$, are weakly equivalent, i.e. the condition (2.152) is satisfied;

(ii) for any $K \in \mathcal{K}$

$$\sup_n \sup_{\alpha\in K} \widetilde{\mathcal{E}}^n \widetilde{Z}_\alpha I(\widetilde{Z}_\alpha > h) \to 0, \quad h \to \infty;$$

(iii) for any $\varepsilon > 0$

$$\sup_n \sup_{\alpha\in K} \widetilde{P}^n(D^\delta \widetilde{Z}_\alpha > \varepsilon) \to 0, \quad \delta \to 0;$$

(iv) the sequence of measures $\widetilde{\Xi} = (\widetilde{P}^n)_{n\geq 1}$ is tight.

Remark 2.24 The conditions (ii) and (iii) follow from (iv) and the continuity of $\widetilde{Z}_\alpha(x)$ in both $x \in X$, $\alpha \in \mathcal{A}$. (Compare with Lemma 2.3 in Section 2.6.)

The uniform versions of the condition of λ-contiguity and of the condition (R) easily yield the uniform version of Theorem 2.29.

Theorem 2.31 *Let a sequence of experiments* $\Sigma = (\mathcal{E}^n)_{n\geq 1}$ *be* $\mathcal{K}$*-uniformly* λ*-contiguous to a sequence* $\widetilde{\Sigma} = (\widetilde{\mathcal{E}}^n)_{n\geq 1}$ *satisfying the condition* $(R\mathcal{K})$, *let* $\widetilde{\mathcal{P}} = \mathfrak{A}_w(\widetilde{\Xi})$ *be the set of weak limit points of the sequence of measures* $\widetilde{\Xi} = (\widetilde{P}^n)_{n\geq 1}$ *and let* $\mathcal{E}(\widetilde{Z}, \widetilde{\mathcal{P}}) = (\mathcal{E}(\widetilde{Z}, \widetilde{P}), \widetilde{P} \in \widetilde{\mathcal{P}})$ *be the family of experiments corresponding* *to the process* $\widetilde{\mathcal{Z}}$ *and to the family of measures* $\widetilde{\mathcal{P}}$. *Then the* $\mathcal{K}$*-uniform extended* λ*-convergence*

$$\mathcal{E}^n \xrightarrow{\lambda(\mathcal{K})} \mathcal{E}(\widetilde{\mathcal{Z}},\widetilde{\mathcal{P}})$$

holds and the family of experiments $\mathcal{E}(\widetilde{\mathcal{Z}},\widetilde{\mathcal{P}})$ *is* $\mathcal{K}$*-regular in the sense of Definition 2.18 from Section 2.12.*

Therefore, all the results of Section 2.12 related to the case of extended λ-convergence can be prolonged to the considered case. Moreover, one can also state an assertion about rapprochement of Δ-distances between experiments $\mathcal{E}^n$ and $\widetilde{\mathcal{E}}^n$.

Theorem 2.32 *Let a sequence of experiments* $\Sigma = (\mathcal{E}^n)_{n\geq 1}$ *be* $\mathcal{K}$*-uniform* λ*-contiguous to a sequence* $\widetilde{\Sigma} = (\widetilde{\mathcal{E}}^n)_{n\geq 1}$ *satisfying* $(R\mathcal{K})$. *Then for any* $K \in \mathcal{K}$

$$\Delta(\mathcal{E}^n_K, \widetilde{\mathcal{E}}^n_K) \to 0, \quad n \to \infty, \tag{2.156}$$

with the restrictions $\mathcal{E}^n_K$, $\widetilde{\mathcal{E}}^n_K$ *of the experiments* $\mathcal{E}^n$, $\widetilde{\mathcal{E}}^n$ *on the parameter set* K.

Proof. Let us fix arbitrary $K \in \mathcal{K}$. By Theorem 2.31 the sets of experiments $\Sigma = (\mathcal{E}^n)_{n\geq 1}$ and $\widetilde{\Sigma} = (\widetilde{\mathcal{E}}^n)_{n\geq 1}$ are relatively compact in the topology of $\lambda(K)$-convergence. If $\mathcal{E}$ is any limit point for the sequence $(\mathcal{E}^n)_{n\geq 1}$, i.e. $\mathcal{E}^{n'} \xrightarrow{\lambda(\mathcal{K})} \mathcal{E}$, $(n') \subseteq (n)$, then the conditions of the theorem immediately imply $\widetilde{\mathcal{E}}^{n'} \xrightarrow{\lambda(\mathcal{K})} \mathcal{E}$. This and Theorem 2.9 in Section 2.6 yield (2.156). □

Theorems 2.31 and 2.32 suggest the following approach to solve asymptotic statistical problems for experiments $\mathcal{E}^n$, $n \geq 1$, if the condition of $\mathcal{K}$-uniform contiguity is fulfilled and also the condition $(R\mathcal{K})$ holds.

First the set $\widetilde{\mathcal{P}}$ *of weak limit points of the sequence of measures* $\widetilde{\Xi} = (\widetilde{P}^n)_{n\geq 1}$ *is described, which immediately gives us the description of the set of limits points for the sequences of experiments* $\Sigma = (\mathcal{E}^n)_{n\geq 1}$ *and* $\widetilde{\Sigma} = (\widetilde{\mathcal{E}}^n)_{n\geq 1}$ *due to Theorem 2.31. Then for each measure* $\widetilde{P} \in \widetilde{\mathcal{P}}$, *more precisely, for the corresponding experiment* $\widetilde{\mathcal{E}} = \mathcal{E}(\widetilde{\mathcal{Z}},\widetilde{P})$, *the statistical problem under consideration is solved, and then the conclusions obtained are extended to the experiments* $\mathcal{E}^n$, $n \geq 1$, *by Theorem 2.32.*

Chapter 3

(γ, Γ)-Models. Convergence to (γ, Γ)-Models

3.1 Definition of (γ, Γ)-Models. Examples

1. The results of the previous chapters lead to the following conclusion: one or another type of convergence $\mathcal{E}^n \to \mathcal{E}$ of statistical experiments permits us to obtain the results of the form "$\liminf_n \mathcal{R}(\mathcal{E}^n) \geq \mathcal{R}(\mathcal{E})$" or "$\liminf_n \mathcal{R}(\mathcal{E}^n) = \mathcal{R}(\mathcal{E})$". Hence, if convergence of experiments $\mathcal{E}^n$ to some experiment $\mathcal{E}$ is stated then we get at least *lower bounds* for minimax risks which become effective for practical situations if one can calculate or estimate the value of the risk $\mathcal{R}(\mathcal{E})$ for the limit experiment.

In general, the calculation of $\mathcal{R}(\mathcal{E})$ is a fairly complicated problem, even for experiments with comparatively simple structure. In this chapter we concentrate on the case of *finite-dimensional* parameters, assuming experiments to be described by so-called (γ, Γ)-models which appear not only in many classical statistical problems, but also in statistics of random processes.

We shall suppose some σ-finite measure $\nu = \nu(d\omega)$ to be given on a space $(\Omega, \mathcal{F})$, as well as random elements $\lambda(\omega) = (\gamma(\omega), \Gamma(\omega))$ with $\gamma(\omega)$ having its values in R^d, and $\Gamma(\omega)$ belonging to the set of non-negative symmetric matrices $\mathcal{M}(R^d)$ of order $d \times d$. Let also $q = q(x)$, $x \in R^d$, be a density function with respect to the Lebesgue measure in R^d, i.e.

$$q(x) \geq 0, \qquad \int_{R^d} q(x) = 1.$$

Using the elements ν, q, γ and Γ one can introduce for each $\alpha \in R^d$ the probability measure P_α on $(\Omega, \mathcal{F})$ with

$$\frac{dP_\alpha}{d\nu}(\omega) = q(\gamma(\omega) - \Gamma(\omega)\alpha). \tag{3.1}$$

Of course, for the correctness of this definition one has to assume for each $\alpha \in R^d$

$$\int_\Omega q(\gamma(\omega) - \Gamma(\omega)\alpha)\, \nu(d\omega) = 1. \tag{3.2}$$

Definition 3.1 A statistical experiment $\mathcal{E} = (\Omega, \mathcal{F}; P_\alpha, \alpha \in R^d)$ with the measures P_α satisfying the properties (3.1) and (3.2) is called a (γ,Γ)-*model* or, sometimes, in order to emphasize all the entering components, a $(\gamma,\Gamma;q,\nu)$-model.

In the case of $q(x) > 0$, the property (3.1) is equivalent to the following representation

$$\frac{dP_\alpha}{dP_0}(\omega) = \frac{q(\gamma(\omega) - \Gamma(\omega)\alpha)}{q(\gamma(\omega))}. \tag{3.3}$$

Therefore, if $q(x) > 0$ then the properties (3.3) and (3.2) can be used for the definition of a (γ,Γ)-model.

Let a statistical experiment $\mathcal{E}$ be a (γ,Γ)-model and let $\mathcal{E}_c = (X, \mathfrak{X}; P'_\alpha, \alpha \in R^d)$ be the corresponding canonical experiment where $X = (R^d, \mathcal{M}^d)$, $\mathfrak{X}$ is the Borel σ-field on X, and the measures P_α are such that

$$P'_\alpha(A \times B) \equiv P_\alpha(\gamma(\omega) \in A,\, \Gamma(\omega) \in B) \qquad \left(= P_\alpha(\gamma^{-1}(A) \cap \Gamma^{-1}(B))\right).$$

Moreover, with

$$\mu(A \times B) \equiv \nu(\gamma(\omega) \in A,\, \Gamma(\omega) \in B) \qquad \left(= \nu(\gamma^{-1}(A) \cap \Gamma^{-1}(B))\right).$$

we obtain

$$\frac{dP'_\alpha}{d\mu}(\gamma,\Gamma) = q(\gamma - \Gamma\alpha),$$

where $\gamma \in R^d$, $\Gamma \in \mathcal{M}^d$ and $\int q(\gamma - \Gamma\alpha)\,\mu(d\omega) = 1$, $\alpha \in R^d$.

Due to Section 1.3 the experiments $\mathcal{E}_c$ and $\mathcal{E}$ are strongly and weakly equivalent, i.e. $\Delta(\mathcal{E}_c, \mathcal{E}) = 0$.

Denoting

$$z_\alpha(x) = \frac{dP'_\alpha}{d\mu}(x), \qquad x \in X = (R^d, \mathcal{M}^d),$$

we get

$$\frac{dP_\alpha}{d\nu}(\omega) = z_\alpha(\lambda(\omega)),$$

with $\lambda(\omega) = (\gamma(\omega), \Gamma(\omega))$.

Let $Z_\alpha(x) = dP'_\alpha/dP'_0(x)$ be the Radon-Nikodym derivative of the measure P'_α under the measure P'_0. Recall, Shiryaev (1980, Chapter 7, Section 6), that $Z_\alpha(x)$ can be taken in the form

$$Z_\alpha(x) = \frac{z_\alpha(x)}{z_0(x)}\, I_{\{z_0(x)>0\}}$$

or

$$Z_\alpha(x) = \frac{z_\alpha(x)}{z_0(x)}.$$

Obviously both of these versions coincide P_0'-a.s. It is also clear that

$$\mathbf{Z}_\alpha(\omega) \equiv \frac{dP_\alpha}{dP_0}(\omega) = Z_\alpha(\lambda(\omega)) \qquad (P_0 - \text{a.s.}),$$

where $Z_\alpha(x)$ is any version of $dP_\alpha'/dP_0'(x)$.

Let us consider a number of examples of (γ,Γ)-models.

Example 3.1 Let P be a probability measure on R^d with some density $q = q(x)$ with respect to the Lebesgue measure:

$$\frac{dP}{dx}(x) = q(x).$$

Let the family $(P_\alpha, \alpha \in R^d)$ of probability measures be defined such that

$$P_\alpha(A) = P(A - \alpha),$$

with $A - \alpha = \{y - \alpha : y \in A\}$, $A \in \mathcal{B}(R^d)$. Evidently

$$\frac{dP_\alpha}{dx}(x) = q(x - \alpha).$$

Therefore, the experiment $\mathcal{E} = (R^d, \mathcal{B}(R^d); P_\alpha, \alpha \in R^d)$, called *a shift experiment*, is an example of (γ,Γ)-model with $\gamma = x$, $\Gamma = 1_d$ the unit $d \times d$-matrix, $\mu(dx) = dx$, $x \in R^d$.

If the density $q(x)$ is positive for all $x \in R^d$, then similarly to (3.3)

$$\frac{dP_\alpha}{dP_0}(x) = \frac{q(x-\alpha)}{q(x)}. \tag{3.4}$$

Example 3.2 Let $d = 1$, $\gamma = x$, $\Gamma = 1$, $\mu(dx) = (1 + \sin x)dx$ and

$$q(x) = \frac{1}{2\pi} I_{[-\pi,\pi]}(x).$$

It is clear that $\int q(x)\,dx = 1$, $\int q(x-\alpha)\,\mu(dx) = 1$ for any $\alpha \in R^d$. Put

$$\frac{dP_\alpha}{d\mu}(x) = q(x - \alpha).$$

The experiment $\mathcal{E} = (R, \mathcal{B}(R); P_\alpha, \alpha \in R)$ so constructed is a (γ,Γ)-model but it is *not* a shift experiment (like in the preceding example).

Example 3.3 Let $d = 1$, $\gamma = x$, $\Gamma = 1$, $\mu(dx) = dx$ and

$$q(x) = pe^{-px} I_{\{x \geq 0\}}.$$

Let the family of measures $(P_\alpha, \alpha \in R)$ be defined by

$$\frac{dP_\alpha}{d\mu}(x) = q(x-\alpha) = pe^{-p(x-\alpha)} I_{\{x \geq \alpha\}}.$$

The corresponding experiment $\mathcal{E} = (R, \mathcal{B}(R); P_\alpha, \alpha \in R)$, called an *exponential* experiment on R, is a particular case of the shift experiment from Example 3.2.

Example 3.4 Let, similarly to Example 3.1, a measure P on R^d have a density $q(x)$ with respect to the Lebesgue measure, and let measures P_α be defined by the equalities

$$\frac{dP_\alpha}{dx}(x) = q(x - \Gamma\alpha),$$

with some fixed matrix Γ from $\mathcal{M}^d$. Then the experiment $\mathcal{E} = (R^d, \mathcal{B}(R^d); P_\alpha, \alpha \in R^d)$ is the (γ,Γ)-model with a nonrandom element Γ. Clearly, the reparametrization $\alpha' = \Gamma\alpha$ reduces this case to the case of a shift experiment. For the positive density $q(x)$ we have, similarly to (3.4),

$$\frac{dP_\alpha}{dP_0}(x) = \frac{q(x - \Gamma\alpha)}{q(x)}.$$

The following example generalizes the previous one when the element Γ is *random.*

Example 3.5 Let independent random elements γ, Γ be defined on a probability space $(\Omega, \mathcal{F}, P)$ with values respectively in $R^d, \mathcal{M}^d$, and let P^γ, P^Γ be the distributions of these random elements such that

$$P(\gamma \in A, \Gamma \in B) = P^\gamma(A)\, P^\Gamma(B) \tag{3.5}$$

for all $A \in \mathcal{B}(R^d)$, $B \in \mathcal{B}(\mathcal{M}^d)$.

Let also $q(x)$ be the density of the distribution P^γ with respect to the Lebesgue measure,

$$\frac{dP^\gamma}{dx}(x) = q(x), \tag{3.6}$$

and $q(x) > 0$ for all $x \in R^d$.

Define on $(\Omega, \mathcal{F})$ the experiment $\mathcal{E} = (\Omega, \mathcal{F}; P_\alpha, \alpha \in R^d)$ for which the measures P_α are determined by the equalities

$$\frac{dP_\alpha}{dP}(\omega) = \frac{q(\gamma(\omega) - \Gamma(\omega)\alpha)}{q(\gamma(\omega))}. \tag{3.7}$$

We now verify that P_α are probability measures and the experiment $\mathcal{E}$ was defined accurately. Independence of γ and Γ implies

$$\begin{aligned} P_\alpha(\Omega) &= \int \frac{dP_\alpha}{dP}(\omega)\, P(d\omega) \\ &= \int \frac{q(\gamma(\omega) - \Gamma(\omega)\alpha)}{q(\gamma(\omega))}\, P(d\omega) \\ &= \iint \frac{q(x - M\alpha)}{q(x)}\, P^\gamma(dx)\, P^\Gamma(dM) \\ &= \int \left[\int \frac{q(x - M\alpha)}{q(x)}\, q(x)\, dx \right] P^\Gamma(dM) = 1, \end{aligned} \tag{3.8}$$

since for any $\alpha \in R^d$ and $M \in \mathcal{M}^d$ it holds

$$\int q(x - M\alpha)\, dx = \int q(x)\, dx = 1.$$

Hence the experiment $\mathcal{E}$ is defined accurately.

Experiments obeying (3.5)-(3.7) will be called *mixed* (γ,Γ)*-models.*

Remark 3.1 The typical example of mixed (γ,Γ)-models is given by mixed normal (or mixed Gaussian) experiments (see Example 3.7 below) for which the density q coincides with the standard Gaussian density φ. Different definitions of such experiments are given in various sources. In particular, often the equalities (3.5), (3.7) (with $q = \varphi$) and 3.1.19 are often taken for this definition; Jeganathan (1980), Swensen (1980), Le Cam and Yang (1990). In that case, the condition (3.6) describing the distribution of the vector γ under the measure $P = P_0$ follows (for $q = \varphi$) from the three ones mentioned above. Here the natural question arises , of whether one can define in the general case, mixed (γ,Γ)-models using conditions of the type of (3.5), (3.7) and (3.8). The answer to this question is negative (as follows from Example 3.2) even for the case of $\Gamma \equiv 1$. Nevertheless, it is not difficult to show that (3.6) is a consequence of the conditions (3.5), (3.7) and (3.8) if the Fourier transform of the density q is a positive function. (Notice that for the case of Example 3.2 the Fourier transform $\widehat{q}(\lambda)$ is equal to zero for all non-zero integer λ).

We now present give some useful properties of mixed (γ,Γ)-models.

Lemma 3.1 *Let an experiment $\mathcal{E} = (\Omega, \mathcal{F}; P_\alpha, \alpha \in R^d)$ be a mixed (γ,Γ)-model, i.e. the conditions (3.5) - (3.7) are fulfilled. Then*

(i) the measure P_0 coincides with the measure P;

(ii) the distribution of the element Γ under the measure P_α does not depend on $\alpha \in R^d$ and coincides with P^Γ;

(iii) the distribution of the random vector $\gamma - \Gamma\alpha$ under the measure P_α does not depend on $\alpha \in R^d$ and coincides with P^γ;

(iv) the distribution P_α^γ *of the random vector* γ *under the measure* P_α *has the density* $p_\alpha^\gamma(x)$ *with respect to the Lebesgue measure of the form*

$$p_\alpha^\gamma(x) = \int q(\gamma - M\alpha)\, P^\Gamma(dM).$$

Remark 3.2 The last expression means that the P_α-distribution of γ is the mixture of the distributions corresponding to the experiments with the shift parameter for different values of the scale factor Γ (it is the property that justifies the name "mixed" for the experiment $\mathcal{E}$).

Proof. The first assertion follows directly from (3.7). To prove (ii) notice that as in (3.8) for any $B \in \mathcal{B}(\mathcal{M}^d)$

$$\begin{aligned}
P_\alpha(\Gamma \in B) &= \int I_B(\Gamma)\,\frac{dP_\alpha}{dP}(\omega)\,P(d\omega)\\
&= \int I_B(\Gamma)\,\frac{q(\gamma(\omega)-\Gamma(\omega)\alpha)}{q(\gamma(\omega))}\,P(d\omega)\\
&= \iint I_B(M)\,\frac{q(x-M\alpha)}{q(x)}\,P^\gamma(dx)\,P^\Gamma(dM)\\
&= \int \left[\int \frac{q(x-M\alpha)}{q(x)}\,q(x)dx\right] I_B(M)\,P^\Gamma(dM)\\
&= \int I_B(M)\,P^\Gamma(dM) = P^\Gamma(B),
\end{aligned}$$

which yields (ii).

Similarly, for each $A \in \mathcal{B}(R^d)$

$$\begin{aligned}
P_\alpha(\gamma - \Gamma\alpha \in A) &= \int I_A(\gamma-\Gamma\alpha)\,\frac{dP_\alpha}{dP}(\omega)\,P(d\omega)\\
&= \int I_A(\gamma-\Gamma\alpha)\,\frac{q(\gamma(\omega)-\Gamma(\omega)\alpha)}{q(\gamma(\omega))}\,P(d\omega)\\
&= \iint I_A(x-M\alpha)\,\frac{q(x-M\alpha)}{q(x)}\,P^\gamma(dx)\,P^\Gamma(dM)\\
&= \iint I_A(x-M\alpha)\,\frac{q(x-M\alpha)}{q(x)}\,q(x)\,dx\,P^\Gamma(dM)\\
&= \int I_A(x)\,q(x)\,dx \int P^\Gamma(dM)\\
&= \int I_A(x)\,q(x)\,dx = P^\gamma(A),
\end{aligned}$$

which proves (iii). □

Remark 3.3 Shift experiments, as well as experiments obtained by a reparametrization from these ones (as in Example 3.4), are particular cases of mixed models. Thus, the assertions of Lemma 3.1 are valid for this situation.

2. It is worth indicating an important subclass of (γ,Γ)-models for which the density q coincides with the standard normal density $\varphi=\varphi(x)$ in R^d,

$$\varphi(x)=(2\pi)^{-d/2}e^{-|x|^2/2},\qquad x\in R^d.$$

Due to (3.3) such (γ,Γ)-models have a likelihood of the form

$$\frac{dP_\alpha}{dP_0}(\omega)=\frac{\varphi(\gamma(\omega)-\Gamma(\omega)\alpha)}{\varphi(\gamma(\omega))}=\exp\left\{(\alpha,\Gamma(\omega)\gamma(\omega))-\frac{1}{2}\left(\alpha,\Gamma^2(\omega)\alpha\right)\right\}. \tag{3.9}$$

Here $(\cdot,\cdot)$ means the scalar product in R^d. We shall use also the notation $(\gamma,\Gamma;\varphi)$-models for such experiments. All the examples considered below are related to $(\gamma,\Gamma;\varphi)$-models.

Example 3.6 Let $(\Omega,\mathcal{F},(\mathcal{F}_t)_{t\geq 0},P)$ be a probability space with the filtration of σ-fields $\mathcal{F}=(\mathcal{F}_t)_{t\geq 0}$. Let $W=(W_t,\mathcal{F}_t)_{t\geq 0}$ be a standard Wiener process on this space and let $Y=(Y_t,\mathcal{F}_t)_{t\geq 0}$ be a random process with the stochastic differential (see Liptser and Shiryaev (1974, Chapter 4))

$$dY_t=\alpha\, a(Y,t)\,dt+dW_t,\qquad t\leq T, \tag{3.10}$$

where $a(Y,t)$ is an anticipative functional. Let P_α be the distribution of the process Y with the differential (3.10) in the space $\mathbf{C}_T$ of continuous functions on $[0,T]$ with the Borel σ-field $\mathcal{B}(\mathbf{C}_T)$. Then P_0 is simply the Wiener measure on $(\mathbf{C}_T,\mathcal{B}(\mathbf{C}_T))$. Under well-known conditions (see, for example, Liptser and Shiryaev (1974, Chapter 7, Section 2)) the density $Z_\alpha^T(Y)=\dfrac{dP_\alpha}{dP_0}(Y)$ of the measure P_α with respect to the Wiener measure P_0 is of the form

$$Z_\alpha^T(Y)=\exp\left\{\alpha\int_0^T a(Y,s)\,dY_s-\frac{\alpha^2}{2}\int_0^T a^2(Y,s)\,ds\right\}.$$

Define

$$\Gamma_T^2(Y)=\int_0^T a^2(Y,s)\,ds,\quad \gamma_T(Y)=\Gamma_T^{-1}(Y)\int_0^T a(Y,s)\,dY_s.$$

Then

$$Z_\alpha^T(Y)=\exp\left\{\alpha\,\Gamma_T(Y)\gamma_T(Y)-\frac{\alpha^2}{2}\Gamma_T^2(Y)\right\}.$$

Due to (3.9) this implies that the experiment $\mathcal{E}=(\mathbf{C}_T,\mathcal{B}(\mathbf{C}_T);P_\alpha,\alpha\in R)$ is a $(\gamma,\Gamma;\varphi)$-model.

The experiment $\mathcal{E}$ is strongly equivalent to the experiment

$$\mathcal{E}_T=(R\times R_+,\mathcal{B}(R\times R_+);P'_\alpha,\alpha\in R^d),$$

where P'_α is the distribution (with respect to P_α) of the pair of generally dependent, random variables $\lambda_T=(\gamma_T,\Gamma_T)$ defined by (3.11).

It is worth mentioning that the fixed moment T in the above consideration can be replaced by a Markov time $T = T(Y)$. In particular, if $T = T_h(Y)$ with $T_h(Y) = \inf\{t : \int_0^t a^2(Y,s)\,ds \geq h\}$ for some constant $h > 0$, then

$$\Gamma_{T_h}(Y) = \sqrt{h}, \qquad \gamma_{T_h}(Y) = h^{-1/2} \int_0^{T_h} a(Y,s)\,dY_s.$$

With this it follows that (Liptser and Shiryaev (1974, Chapter 17, Section 2))

$$\mathcal{L}(\gamma_{T_h}(Y) \mid P_0) = \mathcal{N}(0,1).$$

This implies that the experiment $\mathcal{E}_{T_h}$ is a normal (Gaussian) shift (see Example 1.2 in Section 1.1).

Example 3.7 The famous properties of local asymptotic normality (LAN) and mixed LAN (see Section 3.6 below) are related to normal and mixed normal experiments which are also particular cases of (γ,Γ)-models. Let $\xi \sim \mathcal{N}(0,1)$ be a Gaussian random vector in R^d with independent components, let Γ be a non-random matrix from $\mathcal{M}^d$ and $\alpha \in R^d$. We consider the following three cases that lead to normal experiments $(\mathcal{E}_1, \mathcal{E}_2, \mathcal{E}_3)$, generated by the elements γ and Γ:

1) $\quad \gamma = \Gamma\alpha + \xi;$

2) $\quad \gamma = \Gamma\alpha + \Gamma^{1/2}\xi;$

3) $\quad \gamma = \alpha + \Gamma^{-1/2}\xi.$

If P'_α is the distribution of γ for a given α, then $P'_\alpha \sim P'_0$ and the corresponding density $Z_\alpha(\gamma) = \dfrac{dP'_\alpha}{dP'_0}(\gamma)$ fulfills:

1*) $\quad Z_\alpha(\gamma) = \exp\left\{(\alpha, \Gamma\gamma) - \tfrac{1}{2}(\alpha, \Gamma^2\alpha)\right\} \quad \left(= \dfrac{d\mathcal{N}_{(\Gamma\alpha,1)}}{d\mathcal{N}_{(0,1)}}\right);$

2*) $\quad Z_\alpha(\gamma) = \exp\left\{(\alpha, \gamma) - \tfrac{1}{2}(\alpha, \Gamma\alpha)\right\} \quad \left(= \dfrac{d\mathcal{N}_{(\Gamma\alpha,\Gamma)}}{d\mathcal{N}_{(0,\Gamma)}}\right);$

3*) $\quad Z_\alpha(\gamma) = \exp\left\{(\alpha, \Gamma\gamma) - \tfrac{1}{2}(\alpha, \Gamma\alpha)\right\} \quad \left(= \dfrac{d\mathcal{N}_{(\alpha,\Gamma^{-1})}}{d\mathcal{N}_{(0,\Gamma^{-1})}}\right).$

For the case of 1)

$$Z_\alpha(\gamma) = \frac{\varphi(\gamma - \Gamma\alpha)}{\varphi(\gamma)},$$

with $\varphi(\gamma) = (2\pi)^{-d/2} e^{-|\gamma|^2/2}$, $\gamma \in R^d$.

In the case of 2)

$$Z_\alpha(\gamma) = \frac{\varphi(\Gamma^{-1/2}\gamma - \Gamma^{1/2}\alpha)}{\varphi(\Gamma^{-1/2}\gamma)} = \frac{\varphi(\widetilde{\gamma} - \widetilde{\Gamma}\alpha)}{\varphi(\widetilde{\gamma})}$$

with $\widetilde{\gamma} = \Gamma^{-1/2}\gamma$, $\widetilde{\Gamma} = \Gamma^{1/2};$.

In the case of 3)

$$Z_\alpha(\gamma) = \frac{\varphi(\Gamma^{1/2}\gamma - \Gamma^{1/2}\alpha)}{\varphi(\Gamma^{1/2}\gamma)} = \frac{\varphi(\widetilde{\gamma} - \widetilde{\Gamma}\alpha)}{\varphi(\widetilde{\gamma})}$$

with $\widetilde{\gamma} = \Gamma^{1/2}\gamma$, $\widetilde{\Gamma} = \Gamma^{1/2}$.

Hence, for all three cases, the corresponding experiments are $(\gamma, \Gamma; \varphi)$-models.

Let us consider now *mixed normal* (γ,Γ)*-models* for which the matrix Γ is *random*. This kind of experiments leads to mixed (γ,Γ)-models (see Example 3.5) with the density q coinciding with the Gaussian density φ.

According to the definition in Example 3.5 mixed normal (γ,Γ)-models are described by the following conditions:

(1) random elements γ and Γ are independent under the measure $P = P_0$;

(2) the element γ is standard normal under the measure P_0;

(3) $\displaystyle \frac{dP_\alpha}{dP_0}(\omega) = \frac{\varphi(\gamma(\omega) - \Gamma(\omega)\alpha)}{\varphi(\gamma(\omega))} = \exp\left\{(\alpha, \Gamma(\omega)\gamma(\omega)) - \frac{1}{2}\left(\alpha, \Gamma^2(\omega)\alpha\right)\right\}.$

Moreover, due to Lemma 3.1

(4) the distribution of the random element Γ under the measure P_α does not depend on the value of parameter α, $\mathcal{L}(\Gamma \mid P_\alpha) = \mathcal{L}(\Gamma \mid P_0) = P^\Gamma$;

(5) the distribution of the random vector $\gamma - \Gamma\alpha$ under P_α does not depend on the parameter α and it is standard normal:

$$\mathcal{L}(\gamma - \Gamma\alpha \mid P_\alpha) = \mathcal{L}(\gamma \mid P_0) = \mathcal{N}(0,1).$$

(6) the distribution $\mathcal{L}(\gamma \mid P_\alpha)$ of the random vector γ under P_α is a mixture of normal distributions and has, with respect to the Lebesgue measure, the density $p_\alpha^\gamma(x)$ of the form

$$p_\alpha^\gamma(x) = \int \varphi(x - M\alpha)\, P^\Gamma(dM).$$

The following (martingale) example is a generalization of Example 3.6.

Example 3.8 Let $M = (M_t)_{t\geq 0}$ be a d-dimensional continuous square integrable martingale given on a probability space $(\Omega, \mathcal{F}, \mathbf{\mathcal{F}} = (\mathcal{F}_t)_{t\geq 0}, P)$ endowed with a filtration of σ-field $\mathbf{\mathcal{F}} = (\mathcal{F}_t)_{t\geq 0}$. Let also $\tau = \tau(\omega)$ be a Markov time. It is well known (see, for example, Liptser and Shiryaev (1978, Chapter 7)) that if

$$E\exp\{(\alpha, \langle M\rangle_\tau\alpha)/2\} < \infty, \tag{3.11}$$

where $\langle M\rangle$ is the quadratic characteristic of M, then the measure P_α defined on $(\Omega, \mathcal{F})$ by the equality

$$\frac{dP_\alpha}{dP_0} = \exp\left\{(\alpha, M_\tau) - \frac{1}{2}(\alpha, \langle M\rangle_\tau\alpha)\right\},$$

is a probability measure.

With γ and Γ defined by

$$\Gamma^2 = \langle M \rangle_T, \qquad \gamma = \Gamma^{-1} M_T,$$

the experiment

$$\mathcal{E} = (\Omega, \mathcal{F}_T; P_\alpha, \alpha \in R^d)$$

is a $(\gamma, \Gamma; \varphi)$-model.

3.2 Generalized Bayes Approach for (γ,Γ)-Models. Lemma About Averaging

1. Let an experiment $\mathcal{E} = (\Omega, \mathcal{F};\ P_\alpha, \alpha \in R^d)$ be a $(\gamma, \Gamma; q, \nu)$-model. This means that the densities of measures P_α with respect to the dominating σ-finite measure ν are of the form $dP_\alpha/d\nu = q(\gamma - \Gamma\alpha)$, the same as for shift experiments (see Example 3.1 and 3.4 in Section 3.1). However, there is a number of results describing the structure of Bayes, minimax and regular (equivariant) decisions (estimators) for shift experiments and these results are based on the generalized Bayes approach (see, for example Blyth (1951), Boll (1955), Strasser(1985)). In this and the next sections we show that these results can be extended from the case of shift experiments to the case of arbitrary (γ,Γ)-models.

2. Before formulating the main results we recall some definitions related to the Bayes method in the statistical decision theory, and also explain the motivation and the essence of the generalized Bayes approach.

Let π be some measure on the parameter space $\mathcal{A} = R^d$ often called an a priori distribution of the parameter α. Let then $(\mathbf{A}, \mathcal{A})$ be a *decision space* and let $W = (W(a,\alpha), a \in \mathbf{A}, \alpha \in \mathcal{A})$ be a *loss function.* If ρ is a decision for the experiment $\mathcal{E}$, then the Bayes (or π-Bayes) *risk of the decision* ρ in the experiment $\mathcal{E}$ is the value $R_\pi(\mathcal{E}, W, \rho)$ of the form

$$R_\pi(\mathcal{E}, W, \rho) = \int R_\alpha(\mathcal{E}, W, \rho)\, \pi(d\alpha), \tag{3.12}$$

where $R_\alpha(\mathcal{E}, W, \rho)$ is the risk of the decision ρ at a point $\alpha \in \mathcal{A}$ (see Section 1.3).

The Bayes (or π-Bayes) risk in the experiment $\mathcal{E}$ for the loss function W and for an a priori distribution π is the value $\mathcal{R}_\pi(\mathcal{E}, W)$ with

$$\mathcal{R}_\pi(\mathcal{E}, W) = \inf_\rho R_\pi(\mathcal{E}, W, \rho),$$

where inf is taken over the class of all admissible decisions.

A decision ρ^* for which $R_\pi(\mathcal{E}, W, \rho^*) = \mathcal{R}_\pi(\mathcal{E}, W)$, is called *Bayes* or *$\pi$-Bayes.* It is important to mention that under the Bayes approach one can compare any two decisions ρ_1 and ρ_2: a decision ρ_1 is *preferable* to ρ_2 if $R_\pi(\mathcal{E}, W, \rho_1) <$

$R_\pi(\mathcal{E}, W, \rho_2)$. For the minimax approach the problem of comparison becomes more complicated because one has to compare two functions $R_\alpha(\mathcal{E}, W, \rho_1)$ and $R_\alpha(\mathcal{E}, W, \rho_2)$. Therefore, the problem of searching the Bayes risk is simpler than the similar problem for minimax framework.

Under the Bayes approach it is convenient to introduce, for each a priori distribution π, the measure P_π on the product space $(\mathcal{A}, \mathcal{B}(\mathcal{A})) \times (\Omega, \mathcal{F})$ which is called the *Bayes measure* and defined as the *semi-direct* product of the measure π and the measures P_α,

$$P_\pi = \pi \times P_\alpha,$$

i.e.

$$P_\pi(A \times B) = \int_A P_\alpha(B)\, \pi(d\alpha), \qquad A \in \mathcal{B}(\mathcal{A}), \quad B \in \mathcal{F}.$$

Then the parameter α is naturally treated as a random element having under the measure P_π the distribution π and each measure P_α is obtained from the measure P_π under the condition that the value of the parameter is $\alpha \in \mathcal{A}$.

If ρ is a non-randomized decision for the experiment $\mathcal{E}$, i.e. $\rho = \rho(\omega)$ is a measurable mapping from Ω to $\mathbf{A}$ then

$$R_\alpha(\mathcal{E}, W, \rho) = E_\alpha W(\rho, \alpha) = \int W(\rho(\omega), \alpha)\, P_\alpha(d\omega)$$

and the expression (3.12) for the Bayes risk $R_\pi(\mathcal{E}, W, \rho)$ can be written in the form

$$\begin{aligned} R_\pi(\mathcal{E}, W, \rho) &= \int E_\alpha W(\rho, \alpha)\, \pi(d\alpha) \\ &= \iint W(\rho(\omega), \alpha)\, P_\alpha(d\omega)\, \pi(d\alpha) \\ &= E_\pi W(\rho, \alpha), \end{aligned}$$

where E_π is the integration under the measure P_π.

Similarly, in the case of a randomized decision ρ, i.e. a Markov kernel $\rho = \rho(\omega, da)$ from $(\Omega, \mathcal{F})$ to $(\mathbf{A}, \mathcal{A})$ (see Section 1.3) we have

$$R_\alpha(\mathcal{E}, W, \rho) = \iint W(a, \alpha)\, \rho(\omega, da)\, P_\alpha(d\omega)$$

and

$$\begin{aligned} R_\pi(\mathcal{E}, W, \rho) &= \iiint W(a, \alpha)\, \rho(\omega, da)\, P_\alpha(d\omega)\, \pi(d\alpha) \\ &= E_\pi \int W(a, \alpha)\, \rho(\omega, da). \end{aligned}$$

Hence, for both cases the value of the Bayes risk is determined by the distribution of the decision ρ under the Bayes measure P_π and the Bayes risk of a decision ρ can be treated as the mean value of the loss $W(\rho, \alpha)$ under the Bayes measure P_π.

As we have already noticed, the Bayes approach is usually simpler than the minimax one and often an explicit expression for the Bayes decision ρ^* can be found. However, the Bayes approach also has the essential disadvantage of lacking the reasons for choosing one or another a priori distribution. Nevertheless, investigation of the structure of Bayes risks is useful at least because the Bayes risk is always not bigger than the minimax one. Thus, if the value of the Bayes risk for some a priori distribution is known, then it automatically establishes the *lower* bound for the minimax risk. Moreover, applying the classical minimax theorem one can state (Strasser (1985, Chapter 8)) that the *upper* bound of Bayes risks over all possible a priori distributions (and even over all distributions with finite supports) is just the value of the minimax risk.

For a number of statistical problems one can find a priori distribution (called "the worst" or "least favorable") maximizing the corresponding Bayes risk, which therefore coincides with the minimax one. The corresponding Bayes decision is usually minimax. The problem here is that the set of a priori distributions (probability measures on the parameter space) is not compact in the weak topology. That is why "the worst" a priori distribution may not be a *probability* measure but it is only a *weak limit* of some sequence of prior distributions. This is exactly the situation with shift experiments for which "the least favorable" prior distribution is the uniform distribution of the whole parameter space R^d and it is not a probability measure.

3. In order to describe π-Bayes risks and π-Bayes decisions for non-probability a priori distributions of the type of uniform distribution on R^d, the *generalized Bayes approach* is used and is developed as follows.

A sequence of probability distributions (π_n) which "approximates" in some sense the distribution π is considered, and the π-Bayes risk of decisions with respect to the a priori distribution π is treated as the limit of the π_n-Bayes risks. But here one must keep in mind that this limit does not depend on a particular choice of an approximating sequence (π_n).

Usually such sequences (π_n) are constructed in one of the following two ways:

(1) Let (K_n) be an increasing sequence of convex, compact sets in R^d, $K_n \uparrow R^d$ (for example, the sequence of balls with radius n and center at zero). Then each measure π_n is the *uniform* distribution on the set $K_n, n \geq 1$,

$$\pi_n(dx) = \frac{1}{\pi(K_n)} I_{\{x \in K_n\}}\, dx.$$

(2) Let Θ be a subset in R^d, and let μ be a probability measure on Θ with a density $p(x)$ with respect to the Lebesgue measure:

$$\int_\Theta p(x)\, dx = 1.$$

Let then θ_0 be an internal point in Θ treated as "a localization point" (see Section 3 of Chapter 2). If the measure μ is the distribution of the "original" parameter θ, then the measures π_n are the distributions of the "localized" parameter $\alpha = n(\theta - \theta_0)$. This means that

$$\pi_n(dx) = \frac{1}{n} p(\theta_0 + \frac{1}{n} x)\, dx.$$

For instance, if μ is the standard normal distribution on the whole space R^d, then the measures π_n are normal with the parameters $(0, n^2)$.

In what follows we do not indicate an explicit construction of an approximating sequence π_n but only assume that π_n converge to the uniform distribution π on the whole space R^d. This condition can be formulated in different ways and we choose the variant which is most convenient for the exposition below. First we explain the idea behind this condition. The main property of the measure π is its invariance. If $\pi - \alpha$ denotes the shift of the measure π by a vector $\alpha \in R^d$, i.e.

$$(\pi - \alpha)(A) = \pi(A + \alpha), \qquad A \in \mathcal{B}(R^d),$$

with $A + \alpha = \{\alpha + \beta,\ \beta \in A\}$, then $\pi - \alpha$ coincides with π for all α. In other words: for any measurable function $f(x)$

$$\int f(x)\, (\pi - \alpha)(dx) = \int f(x)\, \pi(dx) = \int f(x)\, dx.$$

Thus, the demand of convergence of the measures π_n to π means that, after normalization, the values of the integrals $\int f(x)\, (\pi_n - \alpha)(dx)$ converge to $\int f(x)\, dx$. In order to avoid introducing normalizing factors it is convenient to consider fractions of the type $\int h(x)\, (\pi_n - \alpha)(dx) / \int g(x)\, (\pi_n - \alpha)(dx)$ for any two functions h, g such that $|h| \le |g|$. The condition introduced below means convergence of these fractions to the value $\int h(x) dx / \int g(x) dx$, uniformly in $\alpha \in R^d$ and in functions h, g.

Let us first introduce the following notation. Let $\delta(r)$ be a function on R_+ monotonously decreasing to zero, $\delta(r) \downarrow 0$ as $r \to \infty$. Denote by $\Delta(\delta)$ the class of non-negative functions on R^d such that

$$\int_{|x| \ge r} g(x)\, dx \le \delta(r), \qquad r \ge 0,$$

i.e. the function $\delta(r)$ controls the rate with which the functions from the class $\Delta(\delta)$ decrease to zero as $|x| \to \infty$.

Now we formulate the condition on approximating measures π_n.

(II) For any function $\delta(r) \downarrow 0$ the following convergence under the measures π_n is fulfilled uniformly in functions $g(x)$ and $h(x)$ with $g \in \Delta(\delta)$ and $|h| \le g$:

$$\left| \frac{\int h(x)\, (\pi_n - \alpha)(dx)}{\int g(x)\, (\pi_n - \alpha)(dx)} - \frac{\int h(x) dx}{\int g(x) dx} \right| \xrightarrow{\pi_n} 0.$$

In this condition α is treated as a random parameter having the distribution π_n. Therefore the condition (Π) can be rewritten in the following form: for any $\varepsilon > 0$ and any function $\delta(r) \downarrow 0$

$$\pi_n \left\{ \alpha : \sup_{g\in\Delta(\delta),|h|\leq g} \left| \frac{\int h(x)\,(\pi_n - \alpha)(dx)}{\int g(x)\,(\pi_n-\alpha)(dx)} - \frac{\int h(x)dx}{\int g(x)dx} \right| > \varepsilon \right\} \to 0$$

as $n \to \infty$.

It is straightforward to check that the condition (Π) is true for the two above considered examples.

Example 3.9 Let $K_n = \{x \in R^d : |x| \leq n\}$ be the ball with radius n and let π_n be the uniform distribution on K_n,

$$\pi_n(dx) = \frac{1}{\pi(K_n)} I_{\{x\in K_n\}}\, dx,$$

where $\pi(K_n)$ is the Lebesgue measure of the ball $K_n, n \geq 1$. Let also $\Pi_n = \pi(K_n)\,\pi_n$, i.e. Π_n is the restriction of the Lebesgue measure π on K_n. It is clear that for any $\alpha \in R^d$

$$\Pi_n - \alpha = \pi(K_n)(\pi_n - \alpha).$$

Let $g(x)$ and $h(x)$ be functions on R^d such that $g \in \Delta(\delta)$ and $|h| \leq g$. Then

$$\frac{\int h(x)\,(\pi_n-\alpha)(dx)}{\int g(x)\,(\pi_n-\alpha)(dx)} = \frac{\int h(x)\,(\Pi_n-\alpha)(dx)}{\int g(x)\,(\Pi_n-\alpha)(dx)}.$$

Since $|h| \leq g$, it follows $\int |h(x)|(\Pi_n - \alpha)(dx) \leq \int g(x)(\Pi_n-\alpha)(dx)$, for proving the condition (Π) it suffices to verify that

$$\int g(x)\,(\Pi_n - \alpha)(dx) \xrightarrow{\pi_n} \int g(x)\,dx \tag{3.13}$$

uniformly in $g \in \Delta(\delta)$. Note that

$$\int g(x)(\Pi_n - \alpha)(dx) = \int_{\{x:\,|x-\alpha|\leq n\}} g(x)\,dx.$$

If $\alpha \in R^d$ and $|\alpha| \leq n - \sqrt{n}$, then the inclusion $\{x : |x-\alpha| \leq n\} \subseteq \{x : |x| \leq \sqrt{n}\}$ implies

$$\left| \int g(x)\,(\Pi_n-\alpha)(dx) - \int g(x)\,dx \right| \leq \int_{\{x:\,|x|\geq\sqrt{n}\}} g(x)\,dx \leq \delta(\sqrt{n}).$$

Therefore,

$$\pi_n\left\{ \left| \int g(x)\,(\Pi_n-\alpha)(dx) - \int g(x)\,dx \right| \geq \delta(\sqrt{n}) \right\} \leq \pi_n(|\alpha| \geq n - \sqrt{n}) \to 0,$$

as $n \to \infty$. This proves (3.13) and the condition (Π).

Example 3.10 Let θ_0 be an internal point of a set Θ in R^d and let $p(x)$ be the uniform continuous density on Θ, i.e.

$$p(x) \geq 0, \quad x \in R^d, \tag{3.14}$$

$$\int_\Theta p(x)\, dx = 1, \tag{3.15}$$

$$w(p,\varepsilon) = \sup_{x\in\Theta, |\alpha|\leq\varepsilon} |p(x) - p(x+\alpha)| \to 0, \quad \varepsilon \to 0. \tag{3.16}$$

Let the measures π_n be defined by the condition

$$\pi_n(dx) = \frac{1}{n} p(\theta_0 + \frac{1}{n}x)dx.$$

Next, let a function $\delta(r) \downarrow 0$ be fixed and let $g \in \Delta(\delta)$ and $|h| \leq g$. Then

$$\frac{\int h(x)\,(\pi_n - \alpha)(dx)}{\int g(x)\,(\pi_n - \alpha)(dx)} = \frac{\int h(x)\, p(\theta_0 + \frac{1}{n}(x+\alpha))dx}{\int g(x)\, p(\theta_0 + \frac{1}{n}(x+\alpha))dx} = \frac{\int h(x)\, p(\theta + \frac{1}{n}x)dx}{\int g(x)\, p(\theta + \frac{1}{n}x)dx}$$

where θ denotes $\theta_0 + \alpha/n$. We now show that the following does hold uniformly in $g \in \Delta(\delta)$ and $\theta \in \Theta$:

$$\int g(x)\, p\left(\theta + \frac{1}{n}x\right)\, dx - p(\theta) \int g(x)\, dx \to 0, \quad n \to \infty. \tag{3.17}$$

Indeed, (3.14)-(3.16) easily imply that $\|p\| = \sup\{|p(\theta)|, \theta \in \Theta\} < \infty$, and

$$\begin{aligned} &\left| \int g(x)\, p\left(\theta + \frac{1}{n}x\right)\, dx - p(\theta) \int g(x)\, dx \right| \\ &\leq 2\|p\| \int_{|x|\geq r} g(x)\, dx + \int_{|x|\leq r} g(x)\, \left| p\left(\theta + \frac{1}{n}x\right) - p(\theta) \right|\, dx \\ &\leq 2\,\|p\|\, \delta(r) + w(p, r/n) \int g(x)\, dx \\ &\leq 2\,\|p\|\, \delta(r) + w(p, r/n)\, \delta(0). \end{aligned}$$

But the last term tends to zero as n and then r tend to infinity which yields (3.17).

A similar assertion holds with $g(x)$ substituted by $h(x)$. Moreover,

$$\begin{aligned} \pi_n\{\theta : p(\theta) < a\} &= \pi_n\{\alpha : p(\theta_0 + \alpha/n) < a\} \\ &= \frac{1}{n} \int_{\{p(\theta_0 + x/n) < a\}} p(\theta_0 + x/n)\, dx \\ &= \int_{\{p(y) < a\}} p(y)\, dy \to 0, \quad a \to 0. \end{aligned}$$

Hence, from (3.17)

$$\left| \frac{\int h(x)\, p(\theta_0 + (x+\alpha)/n)\, dx}{\int g(x)\, p(\theta_0 + (x+\alpha)/n)\, dx} - \frac{p(\theta_0 + \alpha/n) \int h(x)\, dx}{p(\theta_0 + \alpha/n) \int g(x)\, dx} \right| \xrightarrow{\pi_n} 0$$

and the condition (II) follows.

4. Now we formulate the main assertion of this section which describes, for an arbitrary (γ,Γ)-model, the distribution of the random vector $\xi = \gamma - \Gamma\alpha$ under the generalized Bayes measure P_π (which is treated as the limit of the distributions under the Bayes measures P_{π_n}).

To simplify our notation we will write from now on P_n instead of P_{π_n}.

Lemma 3.2 *("About Averaging") Let a sequence of a priori distributions π_n satisfy the condition* (II) *and let an experiment $\mathcal{E} = (\Omega, \mathcal{F}; P_\alpha, \alpha \in R^d)$ be a $(\gamma,\Gamma;\, q,\nu)$-model such that the following condition holds:*

(*T*) *the variables $\xi = \gamma - \Gamma\alpha$ and Γ^{-1} are tight under the Bayes measures P_n, i.e.*

$$\limsup_n P_n(|\gamma - \Gamma\alpha| > r) + P_n(\|\Gamma^{-1}\| > r) \to 0, \quad r \to \infty.$$

If Q_n is the a posteriori distribution of the vector $\xi = \gamma - \Gamma\alpha$, i.e. the conditional distributions of ξ with respect to the σ-field $\mathcal{F}$ and the measure P_n

$$Q_n = \mathcal{L}(\gamma - \Gamma\alpha \mid \mathcal{F}; P_n),$$

then

$$\|Q_n - Q\| \xrightarrow{P_n} 0,$$

where Q is the non-random measure on R^d with density $q(x)$, $Q(dx) = q(x)dx$, and $\|Q_n - Q\|$ is the variation distance between the random measure Q_n and measure Q.

Proof. We have to prove that the following property holds uniformly in functions $f(x)$ on R^d with $\|f\| = \sup\{|f(x)|, x \in R^d\} \le 1$:

$$\left|\int f(x)\,Q_n(dx) - \int f(x)\,Q(dx)\right| \xrightarrow{P_n} 0. \tag{3.18}$$

Standard algebra lead us to the following representation for the integral with respect to the conditional distribution Q_n:

$$\int f(x)\,Q_n(dx) = E_n[f(\gamma - \Gamma\alpha) \mid \mathcal{F}] = \frac{\int f(\gamma - \Gamma y)\,q(\gamma - \Gamma y)\,\pi_n(dy)}{\int q(\gamma - \Gamma y)\,\pi_n(dy)}. \tag{3.19}$$

Define $h(x) = f(x)q(x)$. Then from (3.19)

$$\begin{aligned}\int f(x)\,Q_n(dx) &= \frac{\int h(\gamma - \Gamma y)\,\pi_n(dy)}{\int q(\gamma - \Gamma y)\,\pi_n(dy)}\\ &= \frac{\int h(\gamma - \Gamma\alpha - \Gamma(y-\alpha))\,\pi_n(dy)}{\int q(\gamma - \Gamma\alpha) - \Gamma(y-\alpha))\,\pi_n(dy)}\\ &= \frac{\int h(\xi - \Gamma x)\,(\pi_n - \alpha)(dx)}{\int q(\xi - \Gamma x)\,(\pi_n - \alpha)(dx)}.\end{aligned}$$

Since $\int q(x)dx = 1$, the assertion (3.18) can be rewritten in the following form:

$$\left| \frac{\int h(\xi - \Gamma x)(\pi_n - \alpha)(dx)}{\int q(\xi - \Gamma x)(\pi_n - \alpha)(dx)} - \frac{\int h(x)\, dx}{\int q(x)\, dx} \right| \xrightarrow{P_n} 0. \tag{3.20}$$

In this form the assertion of the lemma is very similar to the condition (Π). We now show that the condition (Π) really implies (3.20). By definition $|h| = |f\, q| \leq \|f\|\, q \leq q$, and in order to apply the condition (Π) we only have to verify that the functions of the form $g(x) = q(\xi - \Gamma x)$ belong to some class $\Delta(\delta)$ for each for fixed values of ξ, Γ.

Suppose for a moment that the following inequalities are fulfilled for a constant $A > 0$ and all $n \geq 1$ with P_n-probability one:

$$|\xi| = |\gamma - \Gamma\alpha| \leq A, \quad \|\Gamma^{-1}\| \leq A. \tag{3.21}$$

Denote $R(r) = Ar + A^2$, $r > 0$. Then for $|x| \geq R(r)$

$$|\Gamma x| \geq |x|/\|\Gamma^{-1}\| \geq r + A$$

and

$$|\Gamma x - \xi| \geq r + A - r = r.$$

Therefore,

$$\int_{|x| \geq R(r)} q(\xi - \Gamma x)\, dx \leq |\det \Gamma^{-1}| \int_{|y| \geq r} q(y)\, dy \leq A^d\, \delta_0(r),$$

with

$$\delta_0(r) = \int_{|y| \geq r} q(y)\, dy \to 0, \quad r \to \infty.$$

These relations mean that for any ξ, Γ satisfying (3.21), the function $g(x) = q(\xi - \Gamma x)$ belongs to the class $\Delta(\delta)$ with the function $\delta(r)$ defined by the equality

$$\delta(R(r)) = A^d\, \delta_0(r).$$

Thus, under (3.21) the assertion (3.19) as well as the whole lemma follow directly from the condition (Π).

In the general case the variables $\xi = \gamma - \Gamma\alpha$ and Γ^{-1} are not bounded but due to the condition (T) one can choose a constant A in such a way that the probability

$$P_n(|\gamma - \Gamma\alpha| > A) + P_n(\|\Gamma^{-1}\| > A)$$

is arbitrary small for all $n \geq 1$. Taking into account the statements proved previously, this implies the assertion of the lemma. □

Remark 3.4 Often (see examples below) a stronger condition on the considered (γ, Γ)-model $\mathcal{E}$ can be checked:

(T^*) The variable $\gamma - \Gamma\alpha$ and Γ^{-1} are tight, uniformly in $\alpha \in R^d$, under the measure P_α, i.e.

$$\sup_\alpha P_\alpha(|\gamma - \Gamma\alpha| > r) + P_\alpha(\|\Gamma^{-1}\| > r) \to 0, \quad r \to \infty.$$

The condition (T) evidently follows from (T^*) for arbitrary choice of a priori measures π_n, $n \geq 1$.

5. Before turning to the corollaries of the Lemma "About Averaging" we show that the required condition (T) (or the stronger condition (T^*)) is satisfied for the typical examples of (γ, Γ)-models.

For shift models (Examples 3.1 and 3.4 from Section 3.1) and also for mixed (γ, Γ)-models (Example 3.5 there) the distribution of the variables $\gamma - \Gamma\alpha$ and Γ^{-1} under measures P_α is the same for all $\alpha \in R^d$ due to Lemma 3.1 from Section 3.1. Hence

$$\begin{aligned} &\sup_\alpha P_\alpha(|\gamma - \Gamma\alpha| > r) + P_\alpha(\|\Gamma^{-1}\| > r) \\ &\qquad = P_0(|\gamma| > r) + P_0(\|\Gamma^{-1}\| > r) \to 0, \quad r \to \infty, \end{aligned}$$

i.e. the condition (T^*) is satisfied and Lemma "About Averaging" applies.

Now we consider the less trivial case of diffusion and martingale (γ, Γ)-models (see Examples 3.6 and 3.8 from Section 3.1). We at once consider the general case of a martingale $(\gamma, \Gamma; \varphi)$-model defined by an R^d-valued continuous square-integrable martingale $M = (M_t, 0 \leq t \leq \infty)$ with a quadratic characteristic $\langle M \rangle = (\langle M \rangle_t, 0 \leq t \leq \infty)$ with values in $\mathcal{M}^d$.

Let M satisfy the condition

$$a\, 1_d \leq \langle M \rangle_\infty \leq A\, 1_d, \tag{3.22}$$

with some positive constants a and A where 1_d is the unit $d \times d$-matrix and the notation $M \leq M'$ means that the matrix $M' - M$ is non-negative. The condition (3.22) obviously implies (3.11) which guarantees the correctness of the definition of the $(\gamma, \Gamma; \varphi)$-model with the likelihood of the form

$$\frac{dP_\alpha}{dP_0} = \exp\left\{(\alpha, M_\infty) - \frac{1}{2}(\alpha, \langle M \rangle_\infty \alpha)\right\},$$

and $\Gamma^2 = \langle M \rangle_\infty$, $\gamma = \Gamma^{-1} M_\infty = \langle M \rangle_\infty^{-1/2} M_\infty$.

Due to Girsanov's theorem, see e.g. Liptser and Shiryaev (1974, Chapter 7), the process $M - \alpha\langle M \rangle$ for any $\alpha \in R^d$ is a square integrable martingale under the measure P_α with the same quadratic characteristic $\langle M \rangle$, thus

$$E_\alpha (M - \langle M \rangle_\infty \alpha)(M - \langle M \rangle_\infty \alpha)^T = E_0\, M\, M^T = E_0\, \langle M \rangle_\infty,$$

where the symbol ^{T} means transposition. Rewriting this equality in the form

$$E_\alpha (\Gamma\gamma - \Gamma^2\alpha)(\Gamma\gamma - \Gamma^2\alpha)^T = E_0\, \Gamma^2,$$

we get from (3.22) for all $\alpha \in R^d$

$$E_\alpha (\gamma - \Gamma\alpha)(\gamma - \Gamma\alpha)^T \le \frac{A}{a} 1_d,$$

hence

$$E_\alpha |\gamma - \Gamma\alpha|^2 \le d \frac{A}{a}$$

and the condition (T^*) is a consequence of (3.22) for the considered class of martingale (γ, Γ)-models.

6. Now we consider some corollaries of Lemma "About Averaging" which are important for applications to statistical decision problems for (γ, Γ)-models. A straightforward application of this lemma leads to the following result:

Corollary 3.1 *Let* Y *be a Polish space and let* $f(x, y)$ *be a bounded measurable function of two variables* $x \in R^d$, $y \in Y$. *If an experiment* $\mathcal{E} = (\Omega, \mathcal{F}; P_\alpha, \alpha \in R^d)$ *is a* (γ, Γ)*-model and the conditions* (Π), (T) *are satisfied, then for any* Y*-valued* $\mathcal{F}$*-measurable random element* η *the following holds:*

$$E_n \left[f(\gamma - \Gamma\alpha, \eta) \mid \mathcal{F} \right] - F(\eta) \xrightarrow{P_n} 0, \tag{3.23}$$

with $F(\eta) = \int f(x, \eta)\, q(x)\, dx$, *and also*

$$E_n f(\gamma - \Gamma\alpha, \eta) - E_n F(\eta) \to 0, \quad n \to \infty. \tag{3.24}$$

Moreover, the convergence (3.23), (3.24) is uniform in all such functions f *with* $\|f\| \le a$ *for every constant* $a > 0$.

Proof. The desired result follows directly from the representation

$$E_n \left[f(\gamma - \Gamma\alpha, \eta) \mid \mathcal{F} \right] = \int f(x, \eta)\, Q_n(dx),$$

$$F(\eta) = \int f(x, \eta)\, Q(dx)$$

and from Lemma "About Averaging". □

Another evident consequence of that lemma is the independence of the variable $\gamma - \Gamma\alpha$ under the generalized Bayes measure P_π from "observations" γ and Γ since the conditional distribution of $\gamma - \Gamma\alpha$ with respect to γ, Γ does not depend on γ, Γ.

Corollary 3.2 *Let an experiment* $\mathcal{E} = (\Omega, \mathcal{F}; P_\alpha, \alpha \in R^d)$ *be a* (γ, Γ)*-model and let the conditions* (Π) *and* (T) *hold. Then the random variables* $\xi = \gamma - \Gamma\alpha$ *asymptotically (as* $n \to \infty$*) independent of the* σ*-field* $\mathcal{F}$ *under the measures* P_n.

Proof. Let η be an arbitrary bounded $\mathcal{F}$-measurable variable and let A be an arbitrary Borel subset in R^d. It suffices to check the following convergence (uniformly in all such η and A):

$$E_n\,\eta\, I_A(\gamma-\Gamma\alpha) - E_n\,\eta\, P_n(\gamma-\Gamma\alpha\in A)\to 0,\quad n\to\infty. \tag{3.25}$$

Notice first that by Lemma "About Averaging"

$$P_n(\gamma-\Gamma\alpha\in A)\to Q(A)=\int_A q(x)\,dx. \tag{3.26}$$

Now an application of Corollary 3.1 with $f(x,y)=y\,I_A(x)$ yields for

$$\begin{aligned} f(\gamma-\Gamma\alpha,\eta) &= \eta\, I_A(\gamma-\Gamma\alpha),\\ F(\eta) &= \int \eta\, I_A(x)\,q(x)\,dx = \eta\, Q(A), \end{aligned}$$

the following convergence

$$E_n\,\eta\, I_A(\gamma-\Gamma\alpha) - Q(A)\,E_n\,\eta\to 0,\quad n\to\infty,$$

which along with (3.26) implies (3.25) and the assertion of the corollary follows. □

7. The result given below (which also follows from Lemma "About Averaging") is called, for the case of a Gaussian shift experiment $\mathcal{E}$, the *Bernstein-von Mises phenomenon* and it describes the structure of the posteriori distribution of the parameter α for the generalized a priori distribution π.

Corollary 3.3 *Let an experiment $\mathcal{E}=(\Omega,\mathcal{F};P_\alpha,\alpha\in R^d)$ be a (γ,Γ)-model and let the conditions* (Π) *and* (T) *hold. If G_n is the posteriori distribution of the parameter α, i.e.*

$$G_n=\mathcal{L}(\alpha\mid\mathcal{F};P_n),$$

then

$$\|G_n-G\|\xrightarrow{P_n}0,$$

with the random measure G on R^d of the form

$$G(dx)=|\det\Gamma|\,q(\gamma-\Gamma x)\,dx.$$

Remark 3.5 If the density q coincides with the standard Gaussian density φ, i.e. $\mathcal{E}$ is a $(\gamma,\Gamma;\varphi)$-model, then the measure G is Gaussian for a fixed γ and Γ with mean $\Gamma^{-1}\gamma$ and covariance matrix Γ^{-2}.

Proof. Similarly to the proof of Lemma "About Averaging", it is necessary to verify that, uniformly in functions $h(x)$ on R^d with $\|h\|\le 1$,

$$\int h(x)\,G_n(dx)-\int h(x)\,G(dx)\xrightarrow{P_n}0,$$

or, in another way,

$$E_n\big[h(\alpha)\mid\gamma,\Gamma\big]-|\det\Gamma|\int h(x)\,q(\gamma-\Gamma x)\,dx\xrightarrow{P_n}0. \tag{3.27}$$

In order to prove this assertion, we again apply Corollary 3.1 with $\eta=(\gamma,\Gamma)$ and $f(x,\eta)=f(\gamma,\Gamma,x)=h(\Gamma^{-1}(\gamma-x))$. We have

$$\begin{aligned}
f(\gamma-\Gamma\alpha,\eta) &= h(\Gamma^{-1}(\gamma-(\gamma-\Gamma\alpha)))=h(\alpha),\\
F(\eta) &= \int f(x,\eta)\,q(x)\,dx\\
&= \int h(\Gamma^{-1}(\gamma-x))\,q(x)\,dx=|\det\Gamma|\int h(x)\,q(\gamma-\Gamma x)\,dx
\end{aligned}$$

and the assertion of Corollary 3.1 means exactly (3.27). □

8. The following result describes the class of all possible distributions (under the generalized Bayes measure) of the variables $\Gamma(\varkappa-\alpha)$ for arbitrary estimators $\varkappa$ of the parameter α. This result will be used later in order to describe the structure of regular estimators for (γ,Γ)-models (see Section 3.4 below).

Corollary 3.4 *Let an experiment $\mathcal{E}=(\Omega,\mathcal{F};P_\alpha,\alpha\in R^d)$ be a (γ,Γ)-model and let the conditions (Π) and (T) hold. Let then $\varkappa$ be an arbitrary non-randomized estimator of the parameter α, i.e. a measurable mapping from $(\Omega,\mathcal{F})$ to R^d, and let there exist a weak limit of the distributions of the variable $\Gamma(\varkappa-\alpha)$ under the Bayes measures P_n,*

$$\mathcal{L}(\Gamma(\varkappa-\alpha)\mid P_n)\xrightarrow{w}K.$$

Then K is the convolution of the measure Q (with $Q(dx)=q(x)dx$) and some another distribution H,

$$K=Q*H.$$

Proof. The weak convergence due to the Prokhorov theorem (see Billingsley (1968, Section 6)), implies tightness of the family of distributions of the variable $\Gamma(\varkappa-\alpha)$ under the measures $P_n, n\ge 1$, and the condition (T) means the same for $\gamma-\Gamma\alpha$. Hence, again by the Prokhorov theorem, for a properly selected subsequence (n'), the joint weak convergence of the distributions of the variables $\Gamma(\varkappa-\alpha)$ and $\gamma-\Gamma\alpha$ under the measures $P_{n'}$ holds, and, hence, the differences $\Gamma(\varkappa-\alpha)-(\gamma-\Gamma\alpha)=\Gamma\varkappa-\gamma$, satisfy

$$\mathcal{L}(\Gamma\varkappa-\gamma\mid P_{n'})\xrightarrow{w}H,$$

where H is a distribution in R^d.

Now we notice that the variable $\Gamma(\varkappa-\alpha)$ can be represented as the sum of the variables $\Gamma\varkappa-\gamma$ and $\gamma-\Gamma\alpha$, and the first of them is measurable with respect to the pair (γ,Γ), while the second one (due to Corollary 3.2) is asymptotically independent of (γ,Γ) and has the limit distribution Q. This implies that the distribution of $\Gamma(\varkappa-\alpha)$ under the measure $P_{n'}$ converges (as $n'\to\infty$) to the convolution $Q*H$ as required. □

Remark 3.6 If for an estimator $\varkappa$ the condition of weak convergence of the distributions $K_n=\mathcal{L}(\Gamma(\varkappa-\alpha)\mid P_n)$ to some fixed distribution K is not satisfied then the result of Corollary 3.4 can be clearly extended to each weak limit point (accumulation point) of the sequence $K_n, n\ge 1$: if

$$K_{n'}\xrightarrow{w}K$$

for some subsequence (n') then $K=Q*H$ for some distribution H.

Remark 3.7 A similar result can be obtained also for randomized estimators. But using the method based on the representation of a randomized estimator as a non-randomized one, perhaps, for an extended experiment (see Section 1.3 Chapter 1), one can reduce the case of randomized estimators to the case we considered previously (see Section 3.3 below for further details).

3.3 Lower Bound of Minimax Risk for (γ,Γ)-Models. Minimax Estimators

1. In this section we describe the structure of generalized Bayes and minimax estimators for (γ,Γ)-models with the help of Lemma "About Averaging" and its corollaries stated in Section 3.2.

As it is usual for estimation problems, a decision space now is assumed to be coinciding with the parameter space R^d, and a loss function $W(a,\alpha)$ is taken in the form

$$W(a,\alpha)=w(a-\alpha),\qquad a\in R^d,\quad \alpha\in R^d,$$

where w is a non-negative function on R^d (usually some additional conditions are imposed on w; see, for example, conditions (wq) or (wq^*) below).

Let $\mathcal{E}=(\Omega,\mathcal{F};P_\alpha,\alpha\in R^d)$ be a statistical experiment which is assumed to be a $(\gamma,\Gamma;q,\nu)$-model. We shall denote by $\varkappa=\varkappa(\omega)$ the Markov non-randomized estimators of the parameter α with values in R^d. The risk of an estimator $\varkappa$ at a point $\alpha\in R^d$ is defined by the expression

$$R_\alpha(\mathcal{E},W,\varkappa)=E_\alpha W(\varkappa-\alpha),$$

and the quantity

$$R(\mathcal{E},W,\varkappa)=\sup_\alpha R_\alpha(\mathcal{E},W,\varkappa)=\sup_\alpha E_\alpha W(\varkappa-\alpha)$$

determines the risk of an estimator $\varkappa$ in the experiment $\mathcal{E}$ corresponding to the minimax approach.

The minimax risk of estimation in the experiment $\mathcal{E}$ for the loss function w is the lower bound of the risks over all possible estimators,

$$\mathcal{R}(\mathcal{E},W) = \inf_{\varkappa} R(\mathcal{E},W,\varkappa) = \inf_{\varkappa} \sup_{\alpha} R_\alpha(\mathcal{E},W,\varkappa),$$

and an estimator $\varkappa^*$ is minimax if $R(\mathcal{E},W,\varkappa^*) = \mathcal{R}(\mathcal{E},W)$.

2. As we have already mentioned above, the investigation of minimax problem is usually carried out based on the generalized Bayes approach by making use of "the least favorable" which is the uniform distribution on the whole parameter space R^d. Therefore, we begin with the characterization of generalized Bayes decision for the estimation problem under consideration. In the first result we restrict ourselves to the case of a bounded loss function w.

Theorem 3.1 *Let an experiment $\mathcal{E} = (\Omega,\mathcal{F};P_\alpha,\alpha\in R^d)$ be a $(\gamma,\Gamma;q,\nu)$-model and a sequence of prior distribution (π_n) be such that the conditions (Π) and (T) from Section 2 are fulfilled, and let a loss function w be bounded and satisfy the following condition:*

(wq) *for every $y\in R^d$, $M\in\mathcal{M}^d$*

$$\int w(y+Mx)\,q(x)\,dx \geq \int w(Mx)\,q(x)\,dx.$$

Then $\varkappa_\pi = \Gamma^{-1}\gamma$ is the generalized Bayes estimator in the sense that for any other estimator $\varkappa$

$$R_\pi(\mathcal{E},W,\varkappa) \geq R_\pi(\mathcal{E},W,\varkappa_\pi),$$

with

$$R_\pi(\mathcal{E},W,\varkappa) = \limsup_n R_{\pi_n}(\mathcal{E},W,\varkappa) = \limsup_n \int R_\alpha(\mathcal{E},W,\varkappa)\,\pi_n(d\alpha).$$

Also

$$\mathcal{R}_\pi(\mathcal{E},W) = R(\mathcal{E},W,\varkappa_\pi) = E_\pi\,\Psi(\Gamma,w),$$

with

$$\Psi(\Gamma,w) = |\det\Gamma| \int_{R^d} w(x)\,q(\Gamma x)\,dx \qquad \left(= \int_{R^d} w(\Gamma^{-1}x)\,q(x)\,dx\right), \tag{3.28}$$

$$E_\pi\,\Psi(\Gamma,w) = \limsup_n E_n\,\Psi(\Gamma,w), \tag{3.29}$$

where E_n denotes the expectation under the Bayes measure P_n corresponding to the prior distribution π_n (see Section 3.2).

Remark 3.8 The following well-known and widespread condition called the *Anderson lemma* (see Anderson (1955), Ibragimov and Khasminskii (1979, Lemma II.11.1)) is sufficient for (wq):

(wq^*) The sets $\{x \in R^d : w(x) \le b\}$ and $\{x \in R^d : q(x) \ge b\}$ are convex and centrally symmetric for any $b > 0$.

Proof of Theorem 3.1. First we consider the case of a non-randomized estimator $\varkappa$. We apply Corollary 3.5 of Lemma "About Averaging" from Section 3.2 to $\eta = (\zeta, \Gamma) = (\varkappa - \Gamma^{-1}\gamma, \Gamma)$ and $f(x,\eta) = f(\zeta,\Gamma,\eta) = w(\zeta + \Gamma^{-1}x)$. We have

$$\begin{aligned} f(\gamma - \Gamma\alpha, \eta) &= w(\zeta + \Gamma^{-1}(\gamma - \Gamma\alpha)) = w(\varkappa - \alpha), \\ F(\eta) &= \int f(x,\eta)\, q(x)\, dx = \int w(\zeta + \Gamma^{-1}x)\, q(x)\, dx, \end{aligned}$$

and by Corollary 3.5

$$E_n\, w(\varkappa - \alpha) - E_n\, F(\eta) \to 0, \qquad n \to \infty.$$

Then, the condition (wq) ensures that

$$F(\eta) \ge \Psi(\Gamma, W) = \int w(\Gamma^{-1}x)\, q(x)\, dx,$$

which immediately implies

$$E_n\, w(\varkappa - \alpha) \ge E_\pi\, \Psi(\Gamma, w) = \limsup_n\, E_n\, \Psi(\Gamma, w).$$

Let now ρ be a *randomized* estimator. Because the decision space R^d is Polish and the experiment $\mathcal{E}$ is dominated we can restrict ourselves to the case of Markov decisions (see Section 1.5 of Chapter 1) supposing that ρ is a Markov kernel. Let

$$R_\alpha(\mathcal{E}, w; \rho) = E_\alpha \int w(y - \alpha)\, \rho(\omega, dy)$$

be the corresponding risk at the point α.

Due to Lemma 1.2 in Section 1.4 one can extend the original experiment $\mathcal{E} = (\Omega, \mathcal{F}; P_\alpha, \alpha \in R^d)$ to an experiment $\widetilde{\mathcal{E}} = (\widetilde{\Omega}, \widetilde{\mathcal{F}}; \widetilde{P}_\alpha, \alpha \in R^d)$ and define a new estimator $\widetilde{\varkappa} = \widetilde{\varkappa}(\widetilde{\omega})$ on it in such a way that

$$\widetilde{E}_\alpha\, w(\widetilde{\varkappa} - \alpha) = E_\alpha \int w(y - \alpha)\, \rho(\omega, dy).$$

Note (see Lemma 1.1 from Section 1.4) that the likelihood ratio (the derivatives of $\widetilde{P}_\alpha$ with respect to $\widetilde{\nu}$) for the experiment $\widetilde{\mathcal{E}}$ is of the same form as for the experiment $\mathcal{E}$, i.e. $\widetilde{\mathcal{E}}$ is the $(\gamma, \Gamma; q, \widetilde{\nu})$-model and, moreover, with the same γ, Γ, q, as for $\mathcal{E}$. Applying now Theorems 3.1 to the experiment $\widetilde{\mathcal{E}}$ and taking into account that $\widetilde{E}_\alpha \Psi(\Gamma, w) = E_\alpha \Psi(\Gamma, w)$, we obtain

$$R_\pi(\mathcal{E}, W, \rho) = R_\pi(\widetilde{\mathcal{E}}, W, \widetilde{\varkappa}) \ge \widetilde{E}_\pi\, \Psi(\Gamma, w) = E_\pi\, \Psi(\Gamma, w).$$

To complete the proof of the theorem it remains only to notice that for $\varkappa = \varkappa_\pi = \Gamma^{-1}\gamma$ it holds $\zeta = \varkappa_\pi - \Gamma^{-1}\gamma = 0$, and hence, $F(\eta) = \Psi(\Gamma, W)$ and $E_\pi\, w(\varkappa_\pi - \alpha) = E_\pi\, \Psi(\Gamma, w)$.

3. Now we consider the case of an *unbounded* loss function $w = w(x)$. We will assume the following condition to be satisfied instead of the condition (wq):

(wq') For any $y \in R^d, M \in \mathcal{M}^d$ and any number $c > 0$

$$\int w_c(y + Mx)\, q(x)\, dx \geq \int w_c(Mx)\, q(x)\, dx,$$

where $w_c(x) = \min\{c, w(x)\}$ is the trimming of the function $w(x)$ at the level $c > 0$.

It is easy to check that the condition (wq^*) is sufficient not only for (wq) but also for (wq').

Theorem 3.2 *Let an experiment $\mathcal{E} = (\Omega, \mathcal{F}; P_\alpha, \alpha \in R^d)$ be a $(\gamma, \Gamma; q, \nu)$-model, let a sequence of prior measures (π_n) obey the conditions (Π) and (T), and let a loss function w follow the condition (wq'). Let then the variables*

$$\Psi(\Gamma, w) = \int w(\Gamma^{-1}x)\, q(x)\, dx$$

be integrable under the measures P_α, uniformly in $\alpha \in R^d$, i.e.

$$\sup_\alpha E_\alpha\, \Psi(\Gamma, w)\, I_{\{\Psi(\Gamma,w) > h\}} \to 0, \qquad h \to \infty. \tag{3.30}$$

Then the estimator $\varkappa_\pi = \Gamma^{-1}\gamma$ is generalized Bayes and

$$\mathcal{R}_\pi(\mathcal{E}, w) = R_\pi(\mathcal{E}, w, \varkappa_\pi) = E_\pi \Psi(\Gamma, w),$$

with

$$E_\pi\, \Psi(\Gamma, w) = \limsup_n E_n\, \Psi(\Gamma, w).$$

Proof. Application of Theorem 3.1 with the loss function w_c provides

$$\mathcal{R}_\pi(\mathcal{E}, w) \geq \mathcal{R}_\pi(\mathcal{E}, w_c) = R_\pi(\mathcal{E}, w_c, \varkappa_\pi) = E_\pi\, \Psi(\Gamma, w_c). \tag{3.31}$$

Next, the condition of uniform integrability (3.30) implies the similar uniform integrability of the variables $\Psi(\Gamma, w)$ under the Bayes measures P_n, that is

$$\limsup_n E_n\, \Psi(\Gamma, w_c) \to \limsup_n E_n\, \Psi(\Gamma, w), \quad c \to \infty. \tag{3.32}$$

It follows from (3.31) and (3.32)

$$\mathcal{R}_\pi(\mathcal{E}, w) \geq E_\pi \Psi(\Gamma, w).$$

Since $R_\pi(\mathcal{E}, w, \varkappa_\pi) = E_\pi\, \Psi(\Gamma, w)$, the theorem is proved. □

Remark 3.9 The condition of uniform integrability (3.30) is obviously satisfied for a bounded loss function w. In the general situation one can indicate the following two conditions are sufficient for (3.30): $\Gamma \geq a\,1_d$ for some number $a > 0$; for every $A > 0$

$$\int \overline{w}(Ax)\, q(x)\, dx < \infty,$$

with $\overline{w}(x) = \sup\{w(y) : |y| \leq |x|\}$.

4. Now we return to the estimation problem under the minimax setup. The following fact is essential here: *the minimax risk is always greater or equal than the Bayes one for any a priori distribution.* Thus the value of generalized Bayes risk calculated above is at the same time the *lower bound* for the minimax risk of estimators. Moreover, this lower bound is accurate for mixed (γ, Γ)-models and the generalized Bayes estimator $\varkappa_\pi = \Gamma^{-1}\gamma$ is also minimax for such models.

Therefore, applying directly the results of Theorems 3.1 and 3.2 we can immediately state the following assertion related to the minimax risk of arbitrary estimators for (γ, Γ)-models.

Corollary 3.5 *Let an experiment $\mathcal{E} = (\Omega, \mathcal{F}; P_\alpha, \alpha \in R^d)$ be a $(\gamma, \Gamma; q, \nu)$-model, let a sequence of prior distributions (π_n) obey the conditions (Π) and (T) from Section 3.2, and let a loss function w satisfy the conditions (wq') and (3.30). Then the following lower bound of minimax risk of estimation for the experiment $\mathcal{E}$ and the loss function w holds:*

$$\mathcal{R}(\mathcal{E}, w) \geq E_\pi\, \Psi(\Gamma, w) = \limsup_n E_n\, \Psi(\Gamma, w),$$

with $\Psi(\Gamma, w)$ defined in (3.28).

5. Now we present some examples illustrating the previous results.

Example 3.11 Let Γ be a non-random matrix. It is clear that for any α

$$E_\alpha\, \Psi(\Gamma, w) = \int w(\Gamma^{-1}x)\, q(x)\, dx = E_0\, w(\Gamma^{-1}\gamma).$$

Hence for any estimator $\varkappa$

$$R(\mathcal{E}, w, \varkappa) \geq E_0\, w(\Gamma^{-1}\gamma) = \int w(\Gamma^{-1}x)\, q(x)\, dx.$$

Example 3.12 Let an experiment $\mathcal{E}$ be a mixed $(\gamma, \Gamma; q, \nu)$-model (Example 3.5 in Section 3.1. The case of shift experiments also enters here). Then, the condition (T) is fulfilled automatically (see Section 3.2). Since the distributions of Γ and $\gamma - \Gamma\alpha$ under the measure P_α do not depend on $\alpha \in R^d$ and $\mathcal{L}(\gamma \mid P_0) = Q$ with the measure Q of the form $Q(dx) = q(x)dx$ (Lemma 3.1 in Section 3.2), it holds for each $\alpha \in R^d$

$$E_\alpha \Psi(\Gamma, w) = E_0\, w(\Gamma^{-1}\gamma).$$

Hence, $E_\pi \Psi(\Gamma, w) = E_0 \Psi(\Gamma, w)$ and for any estimator $\varkappa$ we obtain, using independence of γ and Γ under the measure P_0, that

$$R(\mathcal{E}, w, \varkappa) \ge E_0\, w(\Gamma^{-1}\gamma) = \iint w(M^{-1}x)\, q(x)\, dx\, P^\Gamma(dM), \tag{3.33}$$

where P^Γ is the distribution of the element Γ. It is important to mention that under the above assumption we also have equality in (3.33) and the generalized Bayes estimator $\varkappa_\pi = \Gamma^{-1}\gamma$ is simultaneously minimax. Indeed, for any α

$$E_\alpha\, w(\varkappa_\pi - \alpha) = E_\alpha\, w(\Gamma^{-1}\gamma - \alpha) = E_\alpha\, w(\Gamma^{-1}(\gamma - \Gamma\alpha)) = E_0\, w(\Gamma^{-1}\gamma) = \mathcal{R}(\mathcal{E}, w),$$

and the assertion follows.

Example 3.13 Let $\mathcal{E} = (R_+, \mathcal{B}(R_+); P_\alpha, \alpha \in R_+)$ be the exponential experiment on R_+ (cf. Example 3.3 in Section 3.1), which is a particular case of a shift experiment. Then

$$E_\pi \Psi(\Gamma, w) = \int_0^\infty w(x)\, q(x)\, dx = p \int_0^\infty w(x)\, e^{-px}\, dx.$$

The function $q(x)$ does not satisfy the condition (wq^*). Nevertheless, if the function $w(x)$ is subconvex (i.e. satisfies (wq^*)), then it is easily seen that the condition (wq') holds and one can apply Theorem 2.

In accordance with Remark 3.4 and Theorems 3.1 and 3.2

$$R(\mathcal{E}, w, \varkappa) \ge p \int_0^\infty w(x)\, e^{-px} dx.$$

Example 3.14 Consider the (γ, Γ)-model described in Example 3.8 in Section 3.1. We will suppose there exist constants $a > 0$ and $A > a$ such that

$$a\,\mathbf{1}_d \le \langle M \rangle_\tau \le A\,\mathbf{1}_d. \tag{3.34}$$

This condition ensures the condition (3.10) which in turn implies that the experiment (3.11) is well defined and the condition (T) required in Theorems 3.1 and 3.2 (see Section 3.2) is fulfilled. An application of Corollary 3.5 leads to the following lower estimates for the considered (martingale) model:

$$R(\mathcal{E}, w, \varkappa) = \sup_\alpha E_\alpha\, w(\varkappa - \alpha) \ge \limsup_n E_n \left[\int w(\langle M \rangle_\tau^{-1/2} x) \varphi(x)\, dx \right].$$

3.4 Structure of Regular Estimators for Some (γ, Γ)-Models

1. The minimax assertions of the type of (3.31) from the previous section claim optimality of the *minimax* estimators in the class of *all* (randomized or non-randomized) estimators. In this section we consider another optimality criterion based on the following idea: one first fix some subclass of "natural" estimators in the class of all estimators and ther search "optimal" ones only within this subclass.

One of several possibility to realize this idea is to consider so-called regular estimators. It is the description of the structure of regular estimators that is undertaken in this section.

Definition 3.2 Let $\mathcal{E} = (\Omega, \mathcal{F}; P_\alpha, \alpha \in \mathcal{A})$ be a statistical experiment with a vector space $\mathcal{A}$. An estimator $\varkappa = \varkappa(\omega)$, having values in $\mathcal{A}$, is *regular* if for all $\alpha \in \mathcal{A}$

$$\mathcal{L}(\varkappa - \alpha \mid P_\alpha) = \mathcal{L}(\varkappa \mid P_0).$$

For experiments which are (γ, Γ)-models, the result of Corollary 3.4 from Section 2.2 suggests to consider the *normalized* losses $\Gamma(\varkappa - \alpha)$ instead of the usual losses $\varkappa - \alpha$ of an estimator $\varkappa$. This fact can be also explained as follows. For the case of a fixed non-random matrix Γ (see Example 3.4 in Section 3.1) the natural parameter in the experiment $\mathcal{E}$, being (γ, Γ)-model, is $\Gamma\alpha$, and after such reparametrization, the experiment $\mathcal{E}$ becomes a shift experiment and the quantity $\Gamma(\varkappa - \alpha)$ is exactly the estimation error for the new parameter $\Gamma\alpha$. Similar arguments, but in the conditional form, are valid for models of mixed type.

Theorem 3.3 *Let* $\mathcal{E} = (\Omega, \mathcal{F}; P_\alpha, \alpha \in R^d)$ *be a* (γ, Γ)-*model obeying the condition* (T^*) *from Section 2. If*

$$\mathcal{L}(\Gamma(\varkappa - \alpha) \mid P_\alpha) = \mathcal{L}(\Gamma\varkappa \mid P_0), \qquad \alpha \in R^d, \tag{3.35}$$

then

$$\mathcal{L}(\Gamma(\varkappa - \alpha) \mid P_\alpha) = \mathcal{L}(\Gamma\varkappa \mid P_0) = Q * H, \qquad \alpha \in R^d, \tag{3.36}$$

for some distribution H *in* R^d *and* $Q(dx) = q(x)\,dx$.

Proof. Note first that the condition (3.35) implies for any Bayes measure P_π the equality

$$\mathcal{L}(\Gamma(\varkappa - \alpha) \mid P_\pi) = \mathcal{L}(\Gamma\varkappa \mid P_0).$$

Hence a similar assertion is also true for the generalized Bayes measure P_π corresponding to the uniform prior distribution. Now the assertion of the theorem follows directly from Corollary 3.4 of Section 3.2. □

Remark 3.10 Sometimes it is useful to represent the result of this theorem in a slightly different form. Namely, for any estimator $\varkappa$ satisfying the condition (3.35), there exist mutually independent random vectors $\widetilde{\gamma}$ and ζ in R^d such that $\mathcal{L}(\widetilde{\gamma} \mid P_0) = Q$ and

$$\mathcal{L}(\Gamma(\varkappa - \alpha) \mid P_\alpha) = \mathcal{L}(\Gamma\varkappa \mid P_0) = \mathcal{L}(\widetilde{\gamma} + \zeta), \qquad \alpha \in R^d. \tag{3.37}$$

2. Here we apply the obtained result to describe distributions of regular estimators in the sense of Definition 3.2.

We begin with the case of shift experiments (see Examples 3.1 and 3.4 in Section 3.1).

Theorem 3.4 *Let an experiment $\mathcal{E} = (\Omega, \mathcal{F}; P_\alpha, \alpha \in R^d)$ be a (γ, Γ)-model with a non-random non-singular matrix Γ and $\mathcal{L}(\gamma \mid P_0) = Q$ where $Q(dx) = q(x)\,dx$. Then any regular estimator κ fulfills the following property:*

$$\mathcal{L}(\varkappa \mid P_0) = \mathcal{L}(\Gamma^{-1}\gamma \mid P_0) * H,$$

i.e. the P_0-distribution of $\varkappa$ is the convolution of the P_0-distribution of the element $\Gamma^{-1}\gamma$ and some distribution H on R^d.

Proof. The assertion follows directly from the representation (3.36) in Theorem 3.3 since the matrix Γ is non-random. □

3. For general (γ, Γ)-models the situation is somewhat more complicated and pithy results with respect to the distributions of regular estimators can be obtained only for the case when the distribution of Γ under P_α does not depend on $\alpha \in R^d$. In accordance with Remark 3.4 in Section 3.2 this case corresponds to mixed type (γ, Γ)-models. Therefore, we consider only such models and it is more convenient to formulate the assertions in the form (3.37).

Theorem 3.5 *Let an experiment $\mathcal{E} = (\Omega, \mathcal{F}; P_\alpha, \alpha \in R^d)$ be a mixed type (γ, Γ)-model. If an estimator $\varkappa$ is such that the distribution of the pair $(\varkappa - \alpha, \Gamma)$ under the measure P_α does not depend on $\alpha \in R^d$,*

$$\mathcal{L}(\varkappa - \alpha, \Gamma \mid P_\alpha) = \mathcal{L}(\varkappa, \Gamma \mid P_0), \qquad \alpha \in R^d, \tag{3.38}$$

then there is a random vector ζ in R^d, independent of γ and Γ, such that

$$\mathcal{L}(\varkappa - \alpha \mid P_\alpha) = \mathcal{L}(\varkappa \mid P_0) = \mathcal{L}(\Gamma^{-1}(\gamma + \zeta) \mid P_0), \qquad \alpha \in R^d. \tag{3.39}$$

Proof. The condition (3.38) implies that the distribution of the pair $(\varkappa - \alpha, \Gamma)$ under the *Bayes measure* P_π coincides with $\mathcal{L}(\varkappa, \Gamma \mid P_0)$ for any prior distribution π:

$$\mathcal{L}(\varkappa - \alpha, \Gamma \mid P_\pi) = \mathcal{L}(\varkappa, \Gamma \mid P_0),$$

that also yields that

$$\mathcal{L}(\Gamma(\varkappa-\alpha),\Gamma \mid P_\pi) = \mathcal{L}(\Gamma\varkappa,\Gamma \mid P_0).$$

Since the condition (T^*) is fulfilled automatically for mixed type (γ,Γ)-models (see Section 3.2), we conclude similarly to (3.37) that for some random vector ζ which is independent of γ and Γ, it holds

$$\mathcal{L}(\Gamma\varkappa,\Gamma \mid P_0) = \mathcal{L}(\gamma+\zeta,\Gamma \mid P_0).$$

(Here we have use that $\mathcal{L}(\gamma \mid P_0) = Q$ and hence, one can take $\tilde{\gamma} = \gamma$, see Remark 3.10.) This assertion implies (3.39) since γ and Γ are independent under the measure P_0. □

Remark 3.11 Sometimes the result (3.39) of the theorem is formulated in the following form, conditional with respect to Γ: P_0-a.s.

$$\mathcal{L}(\varkappa \mid \Gamma; P_0) = \mathcal{L}(\Gamma^{-1}\gamma \mid \Gamma; P_0) * H_\Gamma,$$

i.e. the conditional (under the measure P_0) distribution of $\varkappa$ "with respect to Γ" is (P_0-a.s.) the convolution of the conditional distribution of $\Gamma^{-1}\gamma$ and of some distribution H_Γ.

3.5 λ-Convergence of Statistical Experiments to (γ,Γ)-Models. Asymptotically Minimax Estimation

1. In this section we show how the combination of asymptotically minimax theorems under the condition of $\lambda(\mathcal{K})$-convergence (Section 2.7) to (γ,Γ)-models and the lower bounds of minimax risk for these models (Section 3.3) allows us to describe the structure of asymptotically minimax estimators of an unknown parameter.

We shall consider a sequence of statistical experiments (in "θ-model")

$$\mathbb{E}^n = (\Omega^n, \mathcal{F}^n, \mathbf{P}^n_\theta, \theta \in \Theta), \qquad n \geq 1,$$

where Θ is a parameter subset in R^d. Let θ_0 be an internal point in Θ (a localization point) and let θ be represented in the form

$$\theta = \theta_0 + \varphi(n,\theta_0)\alpha,$$

with $\varphi(n,\theta_0) \downarrow 0$, $n \to \infty$, $\alpha \in R^d$. In order to simplify the notation we shall sometimes write $\varphi(n)$ instead of $\varphi(n,\theta_0)$. Denoting $P^n_\alpha = \mathbf{P}^n_{\theta_0+\varphi(n)\alpha}$ we consider the reparametrizied sequence of experiments (in "α-model")

$$\mathcal{E}^n = (\Omega^n, \mathcal{F}^n; P^n_\alpha, \alpha \in \mathcal{A}^n), \qquad n \geq 1,$$

with $\mathcal{A}^n = \{\alpha \in R^d : \theta_0 + \varphi(n)\alpha \in \Theta\}$. Notice that the measures P^n_α depend generally on the localization point θ_0 : $P^n_\alpha = P^n_\alpha(\theta_0)$.

Let $\varkappa^n$ be an estimator of α in the experiment $\mathcal{E}^n$ and let

$$T^n = \theta_0 + \varphi(n)\varkappa^n$$

be the corresponding estimator of θ in the experiment $\mathbb{E}^n$. If $w = w(x)$, $x \in R^d$, is a loss function then

$$\mathbf{E}_\theta^n\, w\left(\varphi^{-1}(n)(T^n - \theta)\right) = E_\alpha^n\, w(\varkappa^n - \alpha)$$

and hence for any set K in R^d

$$\sup_{\varphi^{-1}(n)(\theta-\theta_0)\in K} \mathbf{E}_\theta^n\, w\left(\varphi^{-1}(n)(T^n - \theta)\right) = \sup_{\alpha\in K} E_\alpha^n\, w(\varkappa^n - \alpha).$$

Let $\mathcal{K} = \{K_b : K_b \subseteq R^d\}$ with $K_b = \{\alpha : |\alpha| \le b\}, b > 0,$. Suppose that the experiments $\mathcal{E}^n$ $\lambda(\mathcal{K})$-converge to some $\mathcal{K}$-regular experiment

$$\mathcal{E} = (X, \mathfrak{X}; P_\alpha, \alpha \in R^d),$$

which is a $(\gamma,\Gamma;q,\nu)$-model with the property $q(x) > 0$, $x \in R^d$. In other words, we suppose that there exist elements $\lambda^n = (\gamma^n, \Gamma^n) \in R^d \times \mathcal{M}^d$, defined on $(\Omega^n, \mathcal{F}^n)$ such that

$$\mathcal{L}(\gamma^n, \Gamma^n \mid P_0^n) \to \mathcal{L}(\gamma, \Gamma \mid P_0)$$

and for every $b > 0$ and $\varepsilon > 0$

$$\sup_{|\alpha|\le b} P_0^n \left\{|Z_\alpha^n - Z_\alpha(\gamma^n, \Gamma^n)| > \varepsilon\right\} \to 0, \qquad n \to \infty,$$

with

$$Z_\alpha^n = \frac{dP_\alpha^n}{dP_0^n}, \qquad Z_\alpha = \frac{dP_\alpha}{dP_0} = \frac{q(\gamma - \Gamma\alpha)}{q(\gamma)}.$$

If $w = w(x)$ is a bounded loss function satisfying the *condition* (wq) from Section 3.3, then by Theorem 2.10 in Section 2.7

$$\lim_n \mathcal{R}(\mathcal{E}^n, w; K) = \mathcal{R}(\mathcal{E}, w; K)$$

for any $K \in \mathcal{K}$, where

$$\begin{aligned} \mathcal{R}(\mathcal{E}^n, w; K) &= \inf_{\varkappa^n} \sup_{\alpha\in K} E_\alpha^n\, w(\varkappa^n - \alpha), \\ \mathcal{R}(\mathcal{E}, w; K) &= \inf_{\varkappa} \sup_{\alpha\in K} E_\alpha\, w(\varkappa - \alpha), \\ \lim_{b\to\infty} \lim_n \mathcal{R}(\mathcal{E}^n, w; K_b) &= \lim_{b\to\infty} \mathcal{R}(\mathcal{E}, w; K_b) = \mathcal{R}(\mathcal{E}, w). \end{aligned}$$

Now the *lower bound* of minimax risk for the experiment $\mathcal{E}$ from Corollary 3.5 in Section 3.3 yields

$$\lim_{b\to\infty} \lim_n \mathcal{R}(\mathcal{E}^n, w; K_b) \ge E_\pi\, \Psi(\Gamma, w), \tag{3.40}$$

with

$$\Psi(\Gamma, w) = \int w(\Gamma^{-1}x)\, q(x)\, dx,$$

$$E_\pi \Psi(\Gamma, w) = \limsup_{b\to\infty} \int E_\alpha\, \Psi(\Gamma, w)\, \pi_b(d\alpha),$$

where π_b is the *uniform* distribution on the ball K_b.

For arbitrary estimators $\varkappa^n$ of the parameter α we get from (3.40)

$$\lim_{b\to\infty} \lim_n \sup_{|\alpha|\le b} E_\alpha^n\, w(\varkappa^n - \alpha) \ge E_\pi\, \Psi(\Gamma, w).$$

For the original "θ-models", the bound (3.40) means that for any estimators T^n of the parameter θ

$$\lim_{b\to\infty} \lim_n \sup_{|\varphi^{-1}(n)(\theta-\theta_0)|\le b} \mathbf{E}_\theta^n\, w\left(\varphi^{-1}(n)(T^n - \theta)\right) \ge E_\pi\, \Psi(\Gamma, w). \tag{3.41}$$

If the experiment $\mathcal{E}$ is a mixed (γ, Γ)-model (in particular, a shift experiment), then due to Example 3.12 in Section 3.3, the assertion (3.41) implies

$$\lim_{b\to\infty} \lim_n \sup_{|\varphi^{-1}(n)(\theta-\theta_0)|\le b} \mathbf{E}_\theta^n w\left(\varphi^{-1}(n)(T^n - \theta)\right) \ge E_0\, w(\Gamma^{-1}\gamma).$$

Moreover, under the considered assumptions, the equality does hold here (i.e. $E_0\, w(\Gamma^{-1}\gamma)$ is the sharp lower bound of the risk of estimation). In other words, according to Example 3.12 the estimator $\varkappa_\pi = \Gamma^{-1}\gamma$ is *minimax* for the limit experiment. Therefore, (see the remark after Theorem 2.10 in Section 2.7) one can construct *asymptotically minimax* estimators for the experiments $\mathcal{E}^n$, $n \ge 1$, of the form $\widehat{\varkappa}^n = (\widehat{\Gamma}^n)^{-1}\widehat{\gamma}^n$, with $\widehat{\lambda}^n = (\widehat{\gamma}^n, \widehat{\Gamma}^n)$ defined in Section 2.5.

Moreover, the loss function $w = w(x)$ is continuous, by virtue of Theorem 2.13 in Section 2.7 the following estimators

$$\widetilde{T}^n = \theta_0 + \varphi(n)(\Gamma^n)^{-1}\gamma^n, \qquad n \ge 1,$$

will be *asymptotically minimax*, i.e.

$$\lim_{b\to\infty} \lim_n \sup_{|\varphi^{-1}(n)(\theta-\theta_0)|\le b} \mathbf{E}_\theta^n\, w\left(\varphi^{-1}(n)(\widetilde{T}^n - \theta)\right) = E_0\, w(\Gamma^{-1}\gamma).$$

3.6 Approximation of Statistical Experiments by $(\gamma, \Gamma; \varphi)$-Models. I. Discrete Time

1. In Section 2.11 we considered stochastic expansion of the likelihood ratio for the case of direct products $\mathcal{E}^n = \mathcal{E}_1^n \times \ldots \times \mathcal{E}_n^n$ of binary statistical experiments ("independent observations"), and these results really mean an approximation of $\mathcal{E}^n$ by experiments $\widetilde{\mathcal{E}}^n$ which are $(\gamma, \Gamma; \varphi)$-models (more precise, by Gaussian shifts). In the present section we consider the problem of approximating the experiments $\mathcal{E}^n$ by $(\gamma,\Gamma;\varphi)$-models without special assumptions of "direct product structure" for these experiments. In other words, we consider the case of "dependent observation" for the models in discrete time and in the framework of extended λ-convergence (see Section 2.12 of Chapter 2).

2. Let $\mathbb{E}^n = (\Omega^n, \mathcal{F}^n; \mathbf{P}^n_\theta, \theta \in \Theta)$, $n \geq 1$, be statistical experiments with a parameter set Θ which is a subset in R^d. Let θ_0 be an internal point in Θ and let $\mathcal{E}^n = (\Omega^n, \mathcal{F}^n; P^n_\alpha, \alpha \in R^d)$, $n \geq 1$, be the corresponding localized experiments with $P^n_\alpha = \mathbf{P}^n_{\theta_0+\varphi(n)\alpha}$. Here the localizing sequence $(\varphi(n), n \geq 1)$ is chosen by *contiguity* reason $(P^n_{\alpha,\theta_0}) \triangleleft (P^n_{0,\theta_0})$ (see Section 1.12 of Chapter 1). In what follows, we assume $(\varphi(n), n \geq 1)$ to be a *numerical* sequence, but nothing would change if $\varphi(n)$ were positive symmetric matrices of dimension $d \times d$ with $\|\varphi(n)\| \to 0$, $n \to \infty$.

In order to simplify the notation we define $P^n_\alpha = P^n_{\alpha,\theta_0}$ and let

$$Z^n_\alpha(\omega) = \frac{dP^n_\alpha}{dP^n_0}(\omega) \qquad \left(= \frac{d\mathbf{P}^n_{\theta_0+\varphi(n)\alpha}}{d\mathbf{P}^n_0}(\omega) \right).$$

Let then $X = R^d \times \mathcal{M}^d$ where $\mathcal{M}^d$ be the set of positive symmetric $d \times d$-matrices and $\mathcal{B}(X)$ is the Borel σ-field in X. Points from X will be denoted in the form (γ,Γ) with $\gamma \in R^d$, $\Gamma \in \mathcal{M}^d$, and let

$$\begin{aligned} \varphi(\gamma) &= (2\pi)^{-d/2} \exp\{-|\gamma|^2/2\}, \qquad \gamma \in R^d, \\ Z_\alpha(\gamma,\Gamma) &= \frac{\varphi(\gamma - \Gamma\alpha)}{\varphi(\gamma)} = \exp\left\{(\alpha, \Gamma\gamma) - \frac{1}{2}|\Gamma\alpha|^2\right\}. \end{aligned}$$

We have already noted that the approximation of the experiments $\mathcal{E}^n, n \geq 1$, by (γ,Γ)-models will be carried out in the framework of the *extended* λ-convergence (Section 2.12 of Chapter 2). Thus we suppose to be given a family $\mathcal{P}$ of measures on $(X, \mathcal{B}(X))$ such that for any measure $P \in \mathcal{P}$ and any $\alpha \in R^d$

$$\int_X \exp\left\{(\alpha, \Gamma\gamma) - \frac{1}{2}|\Gamma\alpha|^2\right\} P(d\gamma, d\Gamma) = 1.$$

Now we associate with the family of measures $\mathcal{P}$ and functions $\mathcal{Z} = (Z_\alpha(\gamma,\Gamma)$, $\alpha \in R^d)$, the family $\mathcal{E}(\mathcal{Z},\mathcal{P})$ of experiments

$$\mathcal{E} = (X, \mathcal{B}(X); P_\alpha, \alpha \in R^d), \tag{3.42}$$

with measures P_α defined by the equalities

$$\frac{dP_\alpha}{dP}(\gamma,\Gamma) = Z_\alpha(\gamma,\Gamma) = \exp\left\{(\alpha, \Gamma\gamma) - \frac{1}{2}|\Gamma\alpha|^2\right\}. \tag{3.43}$$

Definition 3.3 (Compare with Definition 2.16 from Section 2.12). We say that *local* (at a point θ_0) *extended λ-convergence* of experiments $\mathbb{E}^n$ holds for the family of experiments $\mathcal{E}(\mathcal{Z},\mathcal{P})$ defined by (3.42) and (3.43) if there exist for each

$n \geq 1$ a random vector $\gamma^n = \gamma^n(\omega) : \Omega^n \to R^d$ and a random matrix $\Gamma^n = \Gamma^n(\omega) : \Omega^n \to \mathcal{M}^d$ such that for any $b > 0$ and $\varepsilon > 0$

$$\sup_{|\alpha|\leq b} \mathbf{P}^n_{\theta_0}\{|Z^n_\alpha - Z_\alpha(\gamma^n,\Gamma^n)| > \varepsilon\} \to 0, \qquad n \to \infty, \tag{3.44}$$

and the set $\mathfrak{A}_w(\Sigma)$ of weak limit points of the sequence $\Sigma = (P^n, n \geq 1)$ of distributions $P^n = \mathcal{L}(\gamma^n,\Gamma^n \mid \mathbf{P}^n_{\theta_0})$ of elements $\lambda^n = (\gamma^n,\Gamma^n)$ with respect to the measures $\mathbf{P}^n_{\theta_0}$ belongs to the family of measures $\mathcal{P}$:

$$\mathfrak{A}_w(\Sigma) \subseteq \mathcal{P}. \tag{3.45}$$

Remark 3.12 The condition (3.44) is equivalent to the fact that for any bounded sequence of points (α_n) in R^d

$$Z^n_{\alpha_n} - Z_{\alpha_n}(\gamma^n,\Gamma^n) \xrightarrow{\mathbf{P}^n_{\theta_0}} 0.$$

If the functions Z_α are of specific form (3.43) (that is, the log-likelihood is a quadratic function of the parameter α) then the family $\mathbb{E}^n$, $n \geq 1$, is called *Locally Asymptotically Quadratic* (LAQ) at the point θ_0, (see Jeganathan (1987), or Fabian and Hannan (1987)) If, in addition, the family $\mathcal{E}(\mathcal{Z},\mathcal{P})$ consists of *one* experiment $\mathcal{E}$ which is moreover mixed normal (mixed Gaussian), (see Section 3.1), then the experiments $\mathbb{E}^n$ are called *Locally Asymptotically Mixed Normal* (LAMN) or *Mixed LAN.* Finally, if the limit experiment $\mathcal{E}$ is a *Gaussian shift*, then the family $(\mathbb{E}^n)_{n\geq 1}$ is called *Locally Asymptotically Normal (LAN).* For more details see e.g. Jeganathan (1980), Ibragimov and Khasminskii (1981), Fabian and Hannan (1987), Le Cam and Yang (1990).

3. Now we present some sufficient conditions which ensure the property of extended λ-convergence assuming for each $n \geq 1$ that $\mathbb{E}^n$ is a *filtered statistical experiment* (see Section 1.2 in Chapter 1) with $\mathcal{F}^n_1 \subseteq \mathcal{F}^n_2 \subseteq \ldots \subseteq \mathcal{F}^n$. The exposition follows Fabian and Hannan (1987).

It is convenient to treat n as the series index and k as the number of observation for the nth series so that $\mathcal{F}^n_k$ can be understood as the collection of events observed up to moment k (inclusively) in the nth series. We also denote $\mathbf{P}^n = \mathbf{P}^n_{\theta_0}$, $\mathbf{P}^n_{\theta,k} = \mathbf{P}^n_\theta \mid \mathcal{F}^n_k$, $\mathbf{P}^n_k = \mathbf{P}^n \mid \mathcal{F}^n_k$, $k,n \geq 1$.

Suppose that for each n we are given a Markov time N^n with respect to the filtration $\mathcal{F}^n$ which is treated as the number of observations in the nth series, so that the measures $\mathbf{P}^n_\theta$ coincide with their restrictions on σ-fields $\mathcal{F}^n_{N^n}$ "up to the time N^n":

$$\mathbf{P}^n_\theta = \mathbf{P}^n_\theta \mid \mathcal{F}^n_{N^n}.$$

In what follows we also assume that all the measures $\mathbf{P}^n_\theta$ are *locally absolutely continuous* with respect to $\mathbf{P}^n$ ($\mathbf{P}^n_\theta \overset{\text{loc}}{\ll} \mathbf{P}^n$), i.e. $\mathbf{P}^n_{\theta,k} \ll \mathbf{P}^n_k$, $\theta \in \Theta$, $k,n \geq 1$.

Define for $k, n \geq 1$

$$\begin{aligned} p_k^n(\theta) &= d\mathbf{P}_{\theta,k}^n / d\mathbf{P}^n, \\ g_k^n(\theta) &= p_k^n(\theta)/p_{k-1}^n(\theta), \\ p_0^n(\theta) &= 1. \end{aligned}$$

Then by (3.44) we have ($\mathbf{P}^n$-a.s.)

$$\frac{d\mathbf{P}_\theta^n}{d\mathbf{P}^n} = \prod_{k=1}^{N^n} g_k^n(\theta)$$

and, hence, it follows for the localized experiments $\mathcal{E}^n(\theta_0) = (\Omega^n, \mathcal{F}^n; P_{\alpha,\theta_0}^n, \alpha \in \mathcal{A}^n)$ with $P_{\alpha,\theta_0}^n = \mathbf{P}_{\theta_0+\varphi(n)\alpha}^n$, $\mathcal{A}^n = \{\alpha \in R^d : \theta_0 + \varphi(n)\alpha \in \Theta\}$ that

$$Z_\alpha^n = \frac{dP_\alpha^n}{dP_0^n} = \prod_{k=1}^{N^n} g_k^n(\theta_0 + \varphi(n)\alpha), \qquad \alpha \in \mathcal{A}^n.$$

Define

$$q_k^n(\alpha) = \left[g_k^n(\theta_0 + \varphi(n)\alpha)\right]^{1/2}.$$

Theorem 3.6 *(Fabian and Hannan (1987)) Assume that for all $k, n \geq 1$ there exist random vectors $\xi_k^n : \Omega^n \to R^d$ such that*

(i) $\sum_{k=1}^{N^n} \mathbf{E}^n[\xi_k^n \mid \mathcal{F}_{k-1}^n] \xrightarrow{\mathbf{P}^n} 0$;

(ii) *for any $\varepsilon > 0$*

$$\sum_{k=1}^{N^n} \mathbf{E}^n[|\xi_k^n|^2 I(|\xi_k^n| > \varepsilon) \mid \mathcal{F}_{k-1}^n] \xrightarrow{\mathbf{P}^n} 0;$$

(iii) *the sequence of the sums V^n with*

$$V^n = \sum_{k=1}^{N^n} \mathbf{E}^n[|\xi n_k|^2 \mid \mathcal{F}_{k-1}^n]$$

is tight under the measure $\mathbf{P}^n$, i.e.

$$\limsup_n \mathbf{P}^n(V^n > h) \to 0, \qquad h \to \infty;$$

(iv) *for any $b > 0$ and $\varepsilon > 0$*

$$\sup_{|\alpha| \leq b} \mathbf{P}^n(Q^n(\alpha) > \varepsilon) \to 0, \qquad n \to \infty,$$

with

$$Q^n(\alpha) = \sum_{k=1}^{N^n} \mathbf{E}^n \left[\left(q_k^n(\alpha) - 1 - \frac{1}{2}(\alpha, \xi_k^n) \right)^2 \middle| \mathcal{F}_{k-1}^n \right].$$

Then the property (3.44) holds with

$$(\Gamma^n)^2 = \sum_{k=1}^{N^n} \mathbf{E}^n \left[\xi_k^n (\xi_k^n)^T \mid \mathcal{F}_{k-1}^n\right], \tag{3.46}$$

$$\gamma^n = (\Gamma^n)^{-1} = \sum_{k=1}^{N^n} \xi_k^n. \tag{3.47}$$

Remark 3.13 If the variables $g_k^n(\theta)$ are differentiable in θ, then the values ξ_k^n are usually chosen in the following form:

$$\xi_k^n = \varphi(n)\, \nabla g_k^n(\theta) = \varphi(n)\, \partial g_k^n(\theta)/\partial\theta.$$

If moreover, the values $[g_k^n(\theta)]^{1/2}$ are differentiable in θ (in mean quadratic sense) then $\xi^n = (\xi_k^n, k \geq 1)$ is a martingale difference sequence, i.e. $\mathbf{E}^n \left[\xi_k^n \mid \mathcal{F}_{k-1}^n\right] = 0$ and the condition (i) of the theorem is satisfied automatically.

Remark 3.14 Theorem 3.6 generalizes the result of Theorem 2.23 in Section 2.11 of Chapter 2 to the case of dependent observations.

4. The conditions of Theorem 3.6 clearly imply the first condition in Definition 3.3 of local extended λ-convergence. It is natural to ask about sufficient conditions for the second property (3.45) in this definition. It turns out that just one more condition is required, namely, the condition that the matrices Γ^n are bounded away from zero.

Theorem 3.7 *Let the conditions* (i)-(iv) *of Theorem 3.6 hold as well as the condition*

(v) $\limsup_n \mathbf{P}^n(\|(\Gamma^n)^{-1}\| > h) \to 0$, $\quad h \to \infty$, *with* Γ^n *defined in (3.46),* $n \geq 1$.

Then the sequence Σ *of distributions* $P^n = \mathcal{L}(\gamma^n, \Gamma^n \mid \mathbf{P}^n)$, $n \geq 1$, *is tight. If* $\mathcal{P} = \mathfrak{A}_w(\Sigma)$ *is the set of weak limit points of the sequence* Σ, *then the locally (at the point* θ_0*) extended* λ*-convergence of experiments* $\mathbb{E}^n$ *to the family of experiments* $\mathcal{E}(\mathcal{Z},\mathcal{P})$ *holds where the latter family is defined by the formulas (3.42),(3.43), with* γ^n, Γ^n *from (3.46),(3.47).*

Proof. The first assertion of the theorem is an easy consequence of the conditions (i), (iii), (v). The second follows directly from the first one and Definition 3.3. □

Since the function $Z_\alpha(\gamma,\Gamma) = \exp\left\{(\alpha, \Gamma\gamma) - \frac{1}{2}|\Gamma\alpha|^2\right\}$ is continuous in α, the family of experiments $\mathcal{E}(\mathcal{Z},\mathcal{P})$ is uniformly regular (see Remark 2.18 in Section 2.12) and under the conditions of Theorem 3.7 the asymptotic minimax result from Section 2.12 apply to the case of extended λ-convergence. Along with the theorems on lower bounds of estimation risk for (γ,Γ)-models this allows us to obtain asymptotic lower bounds of local minimax estimation risks for the considered framework of depended observations.

We now present the corresponding results for the cases of convergence to (γ,Γ)-models considered in the examples from the previous section. Let a given loss function $w(x)$ in R^d satisfy the condition (wq') from Section 3.3 where q coincides with the standard Gaussian density φ. We begin with the case when the condition of local asymptotic normality is fulfilled.

Theorem 3.8 *Let the condition* (i),(ii),(iv) *of Theorem 3.6 hold and let there exist a non-random matrix $\Gamma \in \mathcal{M}^d$ such that*

$$\Gamma^n \xrightarrow{\mathbf{P}^n} \Gamma,$$

with the matrices Γ^n defined in (3.46). Then the LAN condition for experiments $\mathbf{E}^n$ is satisfied at the point θ_0, i.e. λ-convergence of localized experiments $\mathcal{E}^n$ to the Gaussian shift experiment $\mathcal{E}$ with the information matrix Γ^2 is fulfilled, and for any estimators T^n of the parameter θ

$$\lim_{b\to\infty} \liminf_{n} \sup_{|\varphi^{-1}(n)(\theta-\theta_0)|\le b} \mathbf{E}^n_\theta\, w(\varphi^{-1}(n)(T^n-\theta)) \ge E\,w(\Gamma^{-1}\xi), \tag{3.48}$$

with a standard normal d-vector $\xi \sim \mathcal{N}(0, 1_d)$.

Proof. The LAN property will be proved if we verify asymptotic normality of the vectors γ^n which in turn is a consequence of the General Central Limit Theorem for martingales (see Liptser and Shiryaev (1986), Jacod and Shiryaev (1987)). The lower bound (3.48) follows from the general asymptotic minimax result of Theorem 2.10 in Section 2.7 under the conditions of λ-convergence and from the result of Example 3.11 in Section 3.3 on lower bounds of minimax risk for (γ,Γ)-models with a shift-type parameter. □

The following result are related to the case of *mixed LAN*.

Theorem 3.9 *Let the conditions (i),(ii),(iv) of Theorem 3.6 hold and let there exist a random matrix Γ with values in $\mathcal{M}^d$ such that the matrices Γ^n defined in (3.46) weakly converge to Γ:*

$$\Gamma^n \xrightarrow{d(\mathbf{P}^n)} \Gamma.$$

Let also the following condition hold:

(vi) *there exist such stopping times τ^n with respect to $\mathcal{F}^n$, $n \ge 1$, such that*

$$\Gamma^n - \mathbf{E}^n\big[\Gamma^n \mid \mathcal{F}^n_{\tau^n}\big] \xrightarrow{\mathbf{P}^n} 0,$$
$$\sum_{k=1}^{\tau^n} \mathbf{E}^n\big[|\xi^n_k|^2 \mid \mathcal{F}^n_{k-1}\big] \xrightarrow{\mathbf{P}^n} 0. \tag{3.49}$$

Then the experiments $\mathbf{E}^n$ are locally mixed LAN at the point θ_0, i.e. the localized experiments $\mathcal{E}^n$ λ-converge to a mixed Gaussian experiment $\mathcal{E}$, and for any *estimators T^n of parameter θ*

$$\lim_{b\to\infty} \liminf_n \sup_{|\varphi^{-1}(n)(\theta-\theta_0)|\le b} \mathbf{E}^n w(\varphi^{-1}(n)(T^n-\theta)) \ge E(\Gamma^{-1}\xi), \tag{3.50}$$

with a standard normal d-vector $\xi \sim \mathcal{N}(0, 1_d)$.

Remark 3.15 The condition (vi) can be treated by the following manner. Due to (3.49) the matrices Γ^n can be "slightly corrected" (with respect to the measure $\mathbf{P}^n$) in a way that those corrected matrices will be measurable functions of the first τ^n observations and, due to (3.49), the number τ^n is small compared with the whole number of observations N^n.

Proof. The condition (vi) implies asymptotic independence of the elements Γ^n and γ^n from (3.46),(3.47) (see Spokoiny (1986, 1992a)). This, along with the asymptotic expansion from Theorem 3.6, yields the mixed LAN condition. The assertion (3.50) follows now from the result of Example 3.12 in Section 3.3 similarly to the previous theorem. □

The assertion of this theorem can be generalized to the case when the weak convergence of the matrices Γ^n to some limit cannot be checked, however, the condition (vi) of asymptotic independence of Γ^n and γ^n is fulfilled.

Theorem 3.10 *Let the conditions* (i)-(vi) *hold. If $\mathcal{P}$ is the set of weak limit points of the sequence of distributions $P^n = \mathcal{L}(\Gamma^n \mid \mathbf{P}^n)$, then the extended λ-convergence of localized experiments $\mathcal{E}^n$ to the family $\mathcal{E}(\mathcal{Z},\mathcal{P})$ of mixed Gaussian experiments defined by measures $P \in \mathcal{P}$ holds. The corresponding experiment $\mathcal{E}(P)$ from $\mathcal{E}(\mathcal{Z},\mathcal{P})$ is mixed Gaussian where P is the distribution of the random matrix Γ.*

For any sequence of estimators $(T^n)_{n\ge1}$ of the parameter θ

$$\lim_{b\to\infty} \liminf_n \sup_{|\varphi^{-1}(n)(\theta-\theta_0)|\le b} \mathbf{E}^n_\theta w(\varphi^{-1}(n)(T^n-\theta)) \ge \inf_{P\in\mathcal{P}} E w(\Gamma^{-1}\xi).$$

Proof. It is similar to the proof of Theorem 3.8 and 3.9 and we omit it. □

The last results concern with the case of LAQ when the localized experiments $\mathcal{E}^n$ are approximated by martingale (γ,Γ)-models (see Example 3.8 in Section 3.1).

Theorem 3.11 *(Spokoiny (1986, 1992a)) Let the conditions (i)–(v) hold and also*

(vii) *for any $\varepsilon > 0$*

$$\limsup_n \mathbf{P}^n(\sup_t |V^n_{t+\delta} - V^n_t| > \varepsilon) \to 0, \qquad n \to \infty,$$

with

$$V^n_t = \sum_{k\le t\varphi^{-1}(n)} \mathbf{E}^n \left[|\xi^n_k|^2 \mid \mathcal{F}^n_{k-1}\right];$$

(viii) *there exist numbers* $a, A > 0$, *such that*

$$\mathbf{P}^n\left(\|(\Gamma^n)^{-1}\| > a^{-1}\right) + \mathbf{P}^n(\|\Gamma^n\| > A) \to 0, \qquad n \to \infty.$$

If $\mathcal{P} = \mathfrak{A}_w(\Sigma)$ *is the set of weak limit points of the sequence of distributions* $\Sigma = (P^n, n \geq 1)$ *with* $P^n = \mathcal{L}(\gamma^n, \Gamma^n \mid \mathbf{P}^n)$, $n \geq 1$, *and the family of experiments* $\mathcal{E}(\mathcal{Z},\mathcal{P})$ *is defined by (3.42),(3.43), then it holds for any sequence of estimators* $(T^n)_{n\geq 1}$ *of parameter* θ

$$\lim_{b\to\infty} \liminf_{n} \sup_{|\varphi^{-1}(n)(\theta-\theta_0)|\leq b} \mathbf{E}_\theta^n\, w\left(\varphi^{-1}(n)(T^n-\theta)\right)$$
$$\geq \inf_{\mathcal{E}\in\mathcal{E}(\mathcal{Z},\mathcal{P})} \mathcal{R}(\mathcal{E},w) = \inf_{\mathcal{E}\in\mathcal{E}(\mathcal{Z},\mathcal{P})} E_\pi\, \Psi(\Gamma,w), \tag{3.51}$$

where the value $E_\pi\, \Psi(\Gamma,w)$ *is the same as in (3.28), (3.29).*

Proof. It can be reduced to verifying the fact that every element $\mathcal{E}$ from the limit family $\mathcal{E}(\mathcal{Z},\mathcal{P})$ is a martingale (γ,Γ)-model (in the sense of Example 3.8 in Section 3.1) which in turn is guaranteed by the conditions (i)-(v), (vii), (viii). Moreover, the condition (vii) provides the existence for each measure $P \in \mathcal{P}$ of a continuous square integrable martingale M such that the experiment $\mathcal{E}(P)$ is the corresponding to M martingale (γ,Γ)-model. Moreover, the condition (viii) implies $a \leq \Gamma \leq A$ which in turn implies the lower bound (3.51) (see Example 3.14 in Section 3.3). For more details see Spokoiny (1986, 1992a). □

3.7 Approximation of Statistical Experiments by $(\gamma,\Gamma;\varphi)$-Models. II. Continuous Time

1. Let $\mathbb{E}^T$ be a statistical experiment generated by the solutions X of the stochastic differential equation

$$dX_t = f_t(X,\theta)\,dt + dW_t, \qquad X_0 = 0. \tag{3.52}$$

(Notations and assumptions here are the same as in Example 1.8, Section 1.1). To simplify the exposition, we further assume that $\theta \in \Theta \subseteq R^1$. The case of arbitrary dimension d can be considered in the same manner.

We study the question of the approximating the experiments $\mathbb{E}^T$, as $T \to \infty$, by experiments generated by diffusion processes satisfying "linearized" stochastic differential equations. For the sake of unification with the previous exposition we set $T = n$.

Assume that the following condition holds

$$\mathbf{E}_W^n \exp\left\{\frac{1}{2}\int_0^n |f_t(X,\theta)|^2\,dt\right\} < \infty, \tag{3.53}$$

where $\mathbf{E}_W^n$ means expectation under the Wiener measure $\mathbf{P}_W^n$. It is well known (see Liptser and Shiryaev (1974, Chapter 7)) that under this condition the measure $\mathbf{P}_\theta^n$, corresponding to the process X which follows the stochastic differential equation (3.52), is absolutely continuous with respect to the Wiener measure $\mathbf{P}_W^n$, and

$$d\mathbf{P}_\theta^n/d\mathbf{P}_W^n(x) = \exp\left\{\int_0^n f_t(x,\theta)\,dx_t - \frac{1}{2}\int_0^n |f_t(x,\theta)|^2\,dt\right\},$$

where $x = (x_t, t \le n) \in \mathbf{C}_n$, and $\mathbf{C}_n = \mathbf{C}[0,n]$ is the space of continuous functions on $[0,n]$ with the Borel σ-field $\mathcal{B}_n$.

Let θ_0 be an internal point in Θ, and let

$$\mathcal{E}^n = \mathcal{E}^n(\theta_0) = (\mathbf{C}_n, \mathcal{B}_n; P_\alpha^n, \alpha \in \mathcal{A}^n)$$

be the experiment localized at the point θ_0 with $\mathcal{A}^n = \{\alpha \in R : \theta_0 + \varphi(n)\alpha \in \Theta\}$, $P_\alpha^n = \mathbf{P}_{\theta_0+\varphi(n)\alpha}^n$, where $\varphi(n) \to 0$ is a localizing sequence.

Together with these two experiments we consider experiments

$$\widetilde{\mathcal{E}}^n = \widetilde{\mathcal{E}}^n(\theta_0) = (\mathbf{C}_n, \mathcal{B}_n; \widetilde{P}_\alpha^n, \alpha \in \mathcal{A}^n),$$

generated by random processes satisfying linearized stochastic differential equation of the form

$$dX_t = [f_t(X,\theta_0) + \varphi(n)\,\nabla f_t(X,\theta_0)]\,dt + dW_t, \qquad X_0 = 0. \tag{3.54}$$

Here the function $f_t(X,\theta)$ is assumed to be differentiable in θ at the point θ_0 and

$$\nabla f_t(X,\theta_0) = \left.\frac{\partial f_t(X,\theta_0)}{\partial\theta}\right|_{\theta=\theta_0}.$$

Define for the experiments $\mathcal{E}^n(\theta_0)$ the following two statistics $\gamma^n = \gamma^n(x)$ and $\Gamma^n = \Gamma^n(x)$, $x \in \mathbf{C}_n$:

$$\begin{aligned}(\Gamma^n)^2 &= \varphi^2(n)\int_0^n |\nabla f_t(x,\theta_0)|^2\,dt,\\ \Gamma^n\gamma^n &= \varphi(n)\int_0^n \nabla f_t(x,\theta_0)\,[dx_t - f_t(x,\theta_0)\,dt].\end{aligned} \tag{3.55}$$

The value $(\Gamma^n)^2$ is called the *normalized conditional Fisher information* at the point θ_0.

Let also $\mathbf{P}^n = P_{\theta_0}^n$,

$$Z_\alpha^n(x) = \frac{dP_\alpha^n}{dP_0^n}(x) = \frac{dP_{\theta_0+\varphi(n)\alpha}^n}{dP_{\theta_0}^n}(x), \qquad \alpha \in \mathcal{A}^n, \quad x \in \mathbf{C}_n.$$

Lemma 3.3 *Let the condition (3.53) hold and also*

$$\limsup_n \mathbf{P}^n(\|\Gamma^n\| > h) \to 0, \qquad h \to \infty, \tag{3.56}$$

$$\limsup_n \mathbf{P}^n\left(\varphi^2(n)\int_0^n \varepsilon_t^2(x,\delta)\,dt > \varepsilon\right) \to 0, \qquad \delta \to 0, \tag{3.57}$$

for an arbitrary positive ε , *where*

$$\varepsilon_t(x,\delta) = \sup_{|\alpha|\le\delta} \left\{ |f_t(x,\theta_0+\alpha) - f_t(x,\theta_0) - \nabla f_t(x,\theta_0)\alpha| / |\alpha| \right\}.$$

Then for any $\alpha \in R$

$$Z_\alpha^n - \exp\left\{\alpha\Gamma^n\gamma^n - \frac{1}{2}|\Gamma^n\alpha|^2\right\} \xrightarrow{\mathbf{P}^n} 0 \tag{3.58}$$

and, moreover, this convergence is uniform in α *on compacts, i.e. for each* $b > 0$ *and any* $\varepsilon > 0$

$$\sup_{|\alpha|\le b} \mathbf{P}^n \left(\left| Z_\alpha^n - \exp\left\{\alpha\Gamma^n\gamma^n - \frac{1}{2}|\Gamma^n\alpha|^2\right\} \right| > \varepsilon \right) \to 0, \qquad n \to \infty.$$

Remark 3.16 Notice that the smoothness condition on the function $f_t(x,\theta)$ at the point θ_0 means convergence to zero of $\varepsilon_t(x,\delta)$ as $\delta \to 0$. Therefore, the condition (3.57) can be treated as a condition of differentiability of the function $f_t(x,\theta)$ in some averaged sense.

Proof. It is easily derived by standard methods of asymptotic statistics (see Ibragimov and Khasminskii (1981, Chapter 2)), and we omit it. □

2. Now we consider the experiments $\widetilde{\mathcal{E}}^n = \widetilde{\mathcal{E}}^n(\theta_0)$ corresponding to the linearized equation (3.54). We assume the function $f_t(x,\theta_0)$ to satisfy for any $n \ge 1$ and any constant $b > 0$ the following condition:

$$\mathbf{E}_W^n \exp\left\{ \int_0^n |f_t(X,\theta_0)|^2\,dt + b^2\|\Gamma^n\|^2 \right\} < \infty,$$

with Γ^n defined in (3.55). With it

$$\begin{aligned} &\mathbf{E}_W^n \exp\left\{ \frac{1}{2}\int_0^n |f_t(X,\theta_0) + \varphi(n)\,\nabla f_t(X,\theta_0)|^2\,dt \right\} \\ &\le \mathbf{E}_W^n \exp\left\{ \int_0^n |f_t(X,\theta_0)|^2\,dt + |\alpha|^2\|\Gamma^n\|^2 \right\} < \infty, \end{aligned}$$

and the measures $\widetilde{P}_\alpha^n$ corresponding to (3.54) allows the following representation:

$$\begin{aligned} \widetilde{Z}_\alpha^n(x) &= \frac{d\widetilde{P}_\alpha^n}{d\widetilde{P}_0^n}(x) \\ &= \exp\left\{ \varphi(n)\alpha \int_0^n \nabla f_t(x,\theta_0)\,[dx_t - f_t(x,\theta_0)\,dt] \right. \\ &\qquad \left. - \frac{1}{2}\varphi^2(n)\,\alpha^2 \int_0^n |\nabla f_t(x,\theta_0)|^2\,dt \right\} \\ &= \exp\left\{ \alpha\,\Gamma^n\gamma^n - \frac{1}{2}\,|\Gamma^n\alpha|^2 \right\}. \end{aligned} \tag{3.59}$$

Here γ^n, Γ^n are as in (3.55) and (3.55). It follows from (3.59) that the experiment $\widetilde{\mathcal{E}}^n$ is the $(\gamma,\Gamma;\varphi)$-model. Let $\overline{\mathcal{E}}^n = (X, \mathcal{B}(X); \overline{P}^n_\alpha, \alpha \in \mathcal{A}^n)$ be its standard representation (see Section 1.1) with $X = R^1 \times \mathcal{M}^1$, $\overline{P}^n_\alpha = \mathcal{L}(\gamma^n, \Gamma^n \mid \widetilde{P}^n_\alpha)$. Then for all $n \geq 1$

$$d\overline{P}^n_\alpha / d\overline{P}^n_0(\gamma,\Gamma) = Z_\alpha(\gamma,\Gamma) = \exp\left\{\alpha\Gamma\gamma - \frac{1}{2}|\Gamma\alpha|^2\right\},$$

hence all the experiments $\widetilde{\mathcal{E}}^n$ are weakly equivalent. This fact along with the asymptotic expansion (3.58) yields λ-contiguity (in the sense of Definition 2.22 from Section 2.14) of the sequence of experiments $(\mathcal{E}^n)$ to the sequence of experiments $(\overline{\mathcal{E}}^n)$ which are equivalent to the experiments $(\widetilde{\mathcal{E}}^n)$.

Theorem 3.12 *Let the conditions (3.53),(3.56),(3.57) hold and also*

$$\limsup_n \mathbf{P}^n(\|(\Gamma^n)^{-1}\| > h) \to 0, \qquad h \to \infty. \tag{3.60}$$

Then the sequence of experiments $(\mathcal{E}^n)$ is $\mathcal{K}$-uniformly λ-contiguous to the sequence $(\overline{\mathcal{E}}^n)$ where $\mathcal{K}$ is the class of compact sets in R^1. Moreover, the experiments $(\overline{\mathcal{E}}^n)$ satisfy the condition $(R\mathcal{K})$ from Section 14 of Chapter 2.

Proof. The assertion of λ-contiguity follows directly from Lemma 3.3. The condition $(R\mathcal{K})$ follows from the tightness condition of the family of distributions $\overline{P}^n_0 = \mathcal{L}(\gamma^n, \Gamma^n \mid \widetilde{P}^n_0) = \mathcal{L}(\gamma^n, \Gamma^n \mid \mathbf{P}^n)$. The condition (3.56) implies easily tightness of the family of distributions of the elements $\Gamma^n\gamma^n, \Gamma^n$ which along with (3.60) yield the required assertion. □

This result allows us to apply the asymptotic minimax theorems from Section 2.14 of Chapter 2 corresponding to the case of λ-contiguity which are similar to the results obtained under λ-convergence. Using lower bounds of minimax estimation risk for (γ,Γ)-models from Section 3.3 we now are able, analogously to the preceding section, to state lower bounds of asymptotic minimax estimation risk for the original model (3.52). The corresponding conditions and exposition usually repeat those used in Theorems 3.8 through 3.11 of Section 3.6. We present here without proof the results for two more interesting cases when the approximating (γ,Γ)-models are of mixed or martingale type.

Theorem 3.13 *Let the conditions of Theorem 3.12 hold and let the matrices Γ^n from (3.55) weakly converge to a random matrix Γ with values in $\mathcal{M}^d$:*

$$\Gamma^n \xrightarrow{d(\mathbf{P}^n)} \Gamma, \tag{3.61}$$

Assume additionally that there exists a sequence of stopping times τ^n, $n \geq 1$, with respect to the filtration of σ-field $\mathcal{F}^X$ generated by the process X such that

$$\Gamma^n - \mathbf{E}^n\left[\Gamma^n \mid \mathcal{F}^n_{\tau^n}\right] \xrightarrow{\mathbf{P}^n} 0,$$

$$\varphi^2(n)\int_0^{\tau_n} |\nabla f_t(x,\theta_0)|^2\,dt \xrightarrow{\mathbf{P}^n} 0.$$

Then the experiments $\mathbb{E}^n$ follows the mixed LAN condition at the point θ_0, i.e. the localized experiments $\mathcal{E}^n$ λ-converge to the Gaussian shift experiment $\mathcal{E}$. For any sequence of estimators $(T^n)_{n\geq 1}$ of the parameter θ

$$\lim_{b\to\infty} \liminf_{n} \sup_{|\varphi^{-1}(n)(\theta-\theta_0)|\leq b} \mathbf{E}_\theta^n \, w(\varphi^{-1}(n)(T^n-\theta)) \geq E\, w(\Gamma^{-1}\xi), \tag{3.62}$$

with a standard normal d-vector $\xi \sim \mathcal{N}(0, 1_d)$.

In the next result we do not require that the random matrices Γ^n weakly converge.

Theorem 3.14 *Let the conditions of Theorem 3.13 besides the condition (3.61) hold. If $\mathcal{P}$ is the set of weak limit points of the sequence of distributions $P^n = \mathcal{L}(\Gamma^n \mid \mathbf{P}^n)$, then the localized experiments $\mathcal{E}^n$ follows the condition of extended λ-convergence to to the family $\mathcal{E}(\mathcal{Z}, \mathcal{P})$ of mixed Gaussian experiments defined by measures $P \in \mathcal{P}$. Moreover, for any sequence of estimators $(T^n)_{n\geq 1}$ of the parameter θ*

$$\lim_{b\to\infty} \lim_{n} \sup_{|\varphi^{-1}(n)(\theta-\theta_0)|\leq b} \mathbf{E}_\theta^n \, w(\varphi^{-1}(n)(T^n-\theta)) \geq \inf_{P\in\mathcal{P}} E\, w(\Gamma^{-1}\xi).$$

In the last result we replace the condition (3.61) by another which which is typically sufficient for (3.61) but sometimes easier to verify, for further discussion see Spokoiny (1992a).

Theorem 3.15 *Let the conditions of Theorem 3.12 hold and also for any $\varepsilon > 0$ and some a, $A > 0$*

$$\mathbf{P}^n(\|(\Gamma^n)^{-1}\| > a^{-1}) \to 0, \qquad n \to \infty.$$

$$\mathbf{P}^n(\|\Gamma^n\| > A) \to 0, \qquad n \to \infty.$$

$$\limsup_{n} \mathbf{P}^n(\sup_t |V_{t+\delta}^n - V_t^n| > \varepsilon) \to 0, \qquad n \to \infty,$$

with

$$V_t^n = \varphi^2(n) \int_0^t |\nabla f_s(x, \theta_0)|^2 \, ds.$$

If $\mathcal{P}$ is the set of weak limit points of the sequence of distributions $\Xi = (P^n, n \geq 1)$ with $P^n = \mathcal{L}(\gamma^n, \Gamma^n \mid \mathbf{P}^n)$, $n \geq 1$, and the family of experiments $\mathcal{E}(\mathcal{Z}, \mathcal{P})$ is defined by (3.42),(3.43), then for any estimators T^n of the parameter θ

$$\lim_{b\to\infty} \liminf_{n} \sup_{|\varphi^{-1}(n)(\theta-\theta_0)|\leq b} \mathbf{E}_\theta^n \, w(\varphi^{-1}(n)(T^n-\theta))$$
$$\geq \inf_{\mathcal{E}\in\mathcal{E}(\mathcal{Z},\mathcal{P})} \mathcal{R}(\mathcal{E}, w) = \inf_{\mathcal{E}\in\mathcal{E}(\mathcal{Z},\mathcal{P})} E_\pi \, \Psi(\Gamma, w),$$

where the value $E_\pi \Psi(\Gamma, w)$ is defined by (3.28), (3.29).

Chapter 4

Local Convergence of Statistical Experiments and Global Estimation

4.1 Local λ-Convergence. Main Definitions

Let $\mathbb{E}^n = (\Omega^n, \mathcal{F}^n; \mathbf{P}_\theta^n, \theta \in \Theta)$, $n \geq 1$, be a sequence of statistical experiments. A typical situation in statistical problems is when "statistical data increase" as n grows. Usually, probability measures $\mathbf{P}_\theta^n, \theta \in \Theta$, $n \geq 1$, are such that for different θ' and θ'' from Θ the families $(\mathbf{P}_{\theta'}^n, n \geq 1)$ and $(\mathbf{P}_{\theta''}^n, n \geq 1)$ are asymptotically singular, i.e. there exist sets $A^n \in \mathcal{F}^n$ such that $\mathbf{P}_{\theta'}^n(A^n) \to 0$ but $\mathbf{P}_{\theta''}^n(A^n) \to 1$, $n \to \infty$. Therefore, one cannot expect that the experiments $\mathbb{E}^n, n \geq 1$, converge (in some reasonable sense) to some nontrivial experiment $\mathbb{E}$ which can be considered as an approximation of $\mathbb{E}^n$, at least for large n. However, various kinds of convergence considered in the preceding chapters apply in this situation if one does not operate with the experiments $\mathbb{E}^n, n \geq 1$, themselves but with their *localizations*. These localizations are introduced as follows (cf. Section 3.3 of Chapter 2).

Let θ_0 be some fixed point of the parametric set Θ which we treat as an *initial approximation* to the true value of the parameter θ. In other words, we suppose to be given some a priori information that the unknown value of the parameter θ is in some "vicinity" of θ_0. With respect to the parametric set Θ we assume that Θ is a subset of some Banach space $\mathcal{A}$.

We represent θ in the form

$$\theta = \theta_0 + \varphi(n, \theta_0)\alpha,$$

where $n \geq 1$, $\alpha \in \mathcal{A}$ and $(\varphi(n, \theta_0), n \geq 1)$ is some sequence of positive numbers tending monotonously to zero, $\varphi(n, \theta_0) \to 0$, $n \to \infty$. (A more general situation, when $(\varphi(n, \theta_0), n \geq 1)$ is a sequence of nondegenerated linear operators in $\mathcal{A}$ with decreasing norm, can be considered as well). This sequence $(\varphi(n, \theta_0), n \geq 1)$ is called a *localizing sequence* at the point θ_0.

Given experiments $\mathbb{E}^n, n \geq 1$, a localization point θ_0 and a localizing sequence $(\varphi(n, \theta_0), n \geq 1)$ we construct a new *localized* experiment

$$\mathcal{E}^n(\theta_0) = \left(\Omega^n, \mathcal{F}^n; P_{\alpha,\theta_0}^n, \alpha \in \mathcal{A}^n(\theta_0)\right),$$

with

$$P^n_{\alpha,\theta_0} = \mathbf{P}^n_{\theta_0+\varphi(n,\theta_0)\alpha}$$

and

$$\mathcal{A}^n(\theta_0) = \{\alpha \in \mathcal{A} : \theta_0 + \varphi(n,\theta_0)\alpha \in \Theta\}. \tag{4.1}$$

Let also

$$Z^n_\alpha(\theta_0;\omega) = \frac{dP^n_{\alpha,\theta_0}}{dP^n_{0,\theta_0}}(\omega) \qquad \left(= \frac{d\mathbf{P}^n_{\theta_0+\varphi(n,\theta_0)\alpha}}{d\mathbf{P}^n_{\theta_0}} \right)$$

be the Radon-Nykodim derivative of the measure P^n_{α,θ_0} with respect to the measure P^n_{0,θ_0}, where P^n_{0,θ_0} is the measure corresponding to the element $\alpha = 0$. The localizing sequence $(\varphi(n,\theta_0), n \geq 1)$ entering in the definition of the experiments $\mathcal{E}^n(\theta_0)$ is chosen usually in a way to provide the contiguity property for the family of measures $\left(P^n_{\alpha,\theta_0}, n \geq 1\right)$ to the family $\left(P^n_{0,\theta_0}, n \geq 1\right)$ (symbolically $\left(P^n_{\alpha,\theta_0}\right)_{n\geq 1} \triangleleft \left(P^n_{0,\theta_0}\right)_{n\geq 1}$). This can be explained in the following manner. If the distributions $\mathcal{L}\left(Z^n_\alpha(\theta_0) \mid P^n_{0,\theta_0}\right)$ converge weakly to the distribution of some random variable $Z_\alpha(\theta_0)$ with mathematical expectation equal to 1, then the following condition is necessary: $\left(P^n_{\alpha,\theta_0}\right)_{n\geq 1} \triangleleft \left(P^n_{0,\theta_0}\right)_{n\geq 1}$. (This statement is a part of "the first Le Cam's Lemma" , see Hajek and Sidak (1968); cf. Lemma 1.5 in Chapter 1).

In what follows we assume that the experiments $\mathcal{E}^n(\theta_0)$ converge to some canonical dominated experiment

$$\mathcal{E}(\theta_0) = (X, \mathcal{X}; P_{\alpha,\theta_0}\, \alpha \in \mathcal{A}),$$

where $(X, \mathcal{X})$ is a complete separable metric space with the Borel σ-field $\mathcal{X}$, and the family (P_{α,θ_0}) is dominated by the measure P_{0,θ_0}.

Contrary to the limit experiment $\mathcal{E}(\theta_0)$, the localized experiments $\mathcal{E}^n(\theta_0)$ have as the parametric set not the whole space $\mathcal{A}$ but only a subspace $\mathcal{A}^n(\theta_0)$, due to (4.1). However, if θ_0 is an internal point in Θ, then $\mathcal{A}^n(\theta_0) \uparrow \mathcal{A}$, $n \to \infty$, since $\varphi(n,\theta_0) \to 0$, $n \to \infty$. Taking this into account it is convenient to assume that $\mathcal{A}^n(\theta_0) = \mathcal{A}$ for all $n \geq 1$. This can be done without loss of generality setting $P^n_{\alpha,\theta_0} = P^n_{\alpha_0,\theta_0}$ for $\alpha \notin \mathcal{A}^n(\theta_0)$ with some $\alpha_0 \in \mathcal{A}^n(\theta_0)$.

Due to Section 2.4 of Chapter 2 the following definition seems to be natural.

Definition 4.1 We say that a sequence of experiments $\mathbb{E}^n$, $n \geq 1$, *λ-converges locally at a point* $\theta_0 \in \Theta$ to a canonical dominated experiment $\mathcal{E}(\theta_0)$ if $\mathcal{E}^n(\theta_0) \xrightarrow{\lambda} \mathcal{E}(\theta_0)$ for some localizing sequence $\varphi(n,\theta_0) \to 0$, $n \to \infty$. This means that there exist $\mathcal{F}^n/\mathcal{X}$-measurable random elements $\lambda^n_{\theta_0} : \Omega^n \to X$ such that

$$\mathcal{L}\left(\lambda^n_{\theta_0} \mid P^n_{0,\theta_0}\right) \to \mathcal{L}\left(\lambda \mid P_{0,\theta_0}\right) \qquad (= P_{0,\theta_0})$$

and

$$Z_\alpha^n(\theta_0;\omega) - Z_\alpha(\theta_0;\lambda_{\theta_0}^n(\omega)) \xrightarrow{P_{0,\theta_0}^n} 0,$$

where

$$Z_\alpha(\theta_0, x) = \frac{dP_{\alpha,\theta_0}}{dP_{0,\theta_0}}(x), \qquad x \in X,$$

and $\lambda = \lambda(x)$ is the canonical random element on X (i.e. $\lambda(x) = x$, $x \in X$).

Remark 4.1 Similar definitions can be given for other kinds of convergence of experiments considered in Chapter 2 (strong, weak, $\lambda(\mathcal{K})$-convergence). Below we use the following definition of local $\lambda(\mathcal{K})$-convergence.

Definition 4.2 Let $\mathcal{K} = \{K : K \subset \mathcal{A}\}$ be some class of subsets in $\mathcal{A}$. We say that experiments $\mathbb{E}^n$ *locally* $\lambda(\mathcal{K})$-*converge* at a point $\theta_0 \in \Theta$ to an experiment $\mathcal{E}(\theta_0)$ if $\lambda(\mathcal{K})$-convergence of localized experiments $\mathcal{E}^n(\theta_0)$ to $\mathcal{E}(\theta_0)$ holds, i.e. $\mathcal{E}^n(\theta_0) \xrightarrow{\lambda(\mathcal{K})} \mathcal{E}(\theta_0)$.

In these definitions the localization point θ_0 was fixed. However, it is worth noting that the localizing sequence $\varphi(n,\theta_0)$, $n \geq 1$, and the limit experiment $\mathcal{E}(\theta_0)$ may differ for different choices of θ_0.

Definition 4.3 We say that experiments $\mathbb{E}^n$ locally λ-converge to a family of experiments $\mathcal{E} = (\mathcal{E}(\theta_0), \theta_0 \in \Theta)$ if for each point $\theta_0 \in \Theta$ and for a proper choice of a localizing sequence $\varphi = \varphi(n,\theta_0)$ the local λ-convergence $\mathcal{E}^n(\theta_0) \xrightarrow{\lambda} \mathcal{E}(\theta_0)$ holds with corresponding random elements $\lambda_{\theta_0}^n$. In such situations we will write $\mathbb{E}^n \xrightarrow{(\lambda,\varphi)} \mathcal{E}$.

Remark 4.2 The *local* $\lambda(\mathcal{K})$-*convergence* can be defined in the same manner.

Let θ_0 be a localization point and let the true value of the parameter θ lie in some neighborhood of θ_0. If, moreover, local λ-convergence $\mathcal{E}^n(\theta_0) \xrightarrow{\lambda} \mathcal{E}(\theta_0)$ holds, then we can use the asymptotic minimax results from Chapter 2 to search the optimal estimators for θ. Of course, the corresponding estimators *depend* on the localization point θ_0. From the practical point of view this factor seems to be unsatisfactory since one cannot generally assume an "initial" approximation θ_0 to be known. This leads to the following important question: how can the well-developed "local" approach can be applied to "global" estimation, where a priori information about the true value of θ being near to θ_0 is not used? The main part of this chapter is devoted precisely to study this problem. The corresponding estimators constructed by local considerations, but operating globally, will be called *global estimators*.

4.2 Asymptotic Minimax Theorems under Local λ-Convergence

We begin by reformulating the main asymptotic minimax theorems from Chapter 2 for the situation when experiments $\mathbf{E}^n = (\Omega^n, \mathcal{F}^n; \mathbf{P}_\theta^n, \theta \in \Theta)$, $n \geq 1$, locally λ-converge to a family $\mathcal{E} = (\mathcal{E}(\theta_0), \theta_0 \in \Theta)$. Recall the notation from Section 4.1:

$$\mathcal{E}^n(\theta_0) = \left(\Omega^n, \mathcal{F}^n, P_{\alpha,\theta_0}^n, \alpha \in \mathcal{A}^n(\theta_0)\right),$$

with

$$P_{\alpha,\theta_0}^n = \mathbf{P}_{\theta_0+\varphi(n,\theta_0)\alpha}^n$$

and

$$\mathcal{A}^n(\theta_0) = \{\alpha \in \mathcal{A} : \theta_0 + \varphi(n,\theta_0)\alpha \in \Theta\}.$$

Since we deal with the estimation problem and it is natural to consider loss functions $W = (W_\alpha(a))$ of the following form

$$W_\alpha(a) = W(a - \alpha).$$

Under this assumption for $\theta = \theta_0 + \varphi(n,\theta_0)\alpha$ and $\widehat{\theta}^n = \theta_0 + \varphi(n,\theta_0)\widehat{\alpha}^n$

$$W\left(\varphi^{-1}(n,\theta_0)\left(\widehat{\theta}^n - \theta\right)\right) = W(\widehat{\alpha}^n - \alpha)$$

and if $K \subseteq \mathcal{A}$ and

$$U^n(\theta_0, K) = \left\{\theta \in \mathcal{A} : \varphi^{-1}(n,\theta_0)(\theta - \theta_0) \in K\right\},$$

then

$$\sup_{\theta \in U^n(\theta_0,K)} \mathbf{E}_\theta^n W\left(\varphi^{-1}(n,\theta_0)\left(\widehat{\theta}^n - \theta\right)\right) = \sup_{\alpha \in K} E_{\alpha,\theta_0}^n W(\widehat{\alpha}^n - \alpha). \tag{4.2}$$

It has already been noted in Section 2.3 of Chapter 2 that the relation (4.2) plays the key role in "transfering" the results from "α-models" to "θ-models". It is the relation which, combined with asymptotic minimax theorems for "α-models" gives us results of the type

$$\liminf_n \inf_{\widehat{\theta}^n} \sup_{\theta \in U^n(\theta_0,K)} \mathbf{E}_\theta^n W\left(\varphi^{-1}(n,\theta_0)\left(\widehat{\theta}^n - \theta\right)\right) \geq \inf_{\widehat{\alpha}} \sup_{\alpha \in K} E_{\alpha,\theta_0} W(\widehat{\alpha} - \alpha) \tag{4.3}$$

or shorter,

$$\liminf_n \mathcal{R}\left(\mathcal{E}^n(\theta_0), W, K\right) \geq \mathcal{R}\left(\mathcal{E}(\theta_0), W, K\right).$$

In order to state inequalities or (preferably) equalities of such form we have to impose, due to Chapter 2, certain assumptions on

(1) the kind of convergence of statistical experiments $\mathcal{E}^n(\theta_0)$, $n \geq 1$, to $\mathcal{E}(\theta_0)$;
(2) the loss function W;

(3) the classes of feasible estimators $\widehat{\theta}^n$ in (4.3).

We know that the assumption of weak convergence requires stronger assumptions on the smoothness of the loss function $W_\alpha(a)$ in a and a more involved class of decisions, including generalized randomized decisions from the classes $\mathfrak{B}(\mathcal{E}^n(\theta_0), \mathcal{A})$, $\mathfrak{B}(\mathcal{E}(\theta_0), \mathcal{A})$. In what follows we prefer to make *stronger* assumptions on the convergence of experiments, namely to assume *local* $\lambda(\mathcal{K})$-*convergence*. This kind of convergence simplifies further considerations since it allows to impose only very weak restrictions on loss functions and to consider only estimators determined by Markov kernels.

Let $\mathcal{K}$ be some class of subsets K in a Banach space $\mathcal{A}$. We shall suppose that all elements of $\mathcal{K}$ are of convex compact sets K and fulfilling the following property: if $K, K' \in \mathcal{K}$, then $K+K' \in \mathcal{K}$, where $K+K' = \{\alpha+\beta : \alpha \in K, \beta \in K'\}$. Moreover, the class $\mathcal{K}$ is assumed to be rich enough in the sense that for each point $\alpha \in \mathcal{A}$ there is a $K \in \mathcal{K}$ such that $\alpha \in K$.

In the case of a finite-dimensional space $\mathcal{A}$, the class $\mathcal{K}$ is assumed to be the class of all balls with the center at zero.

Loss functions W are supposed to be of the form $W_\alpha(a) = W(a-\alpha)$, $\alpha, a \in \mathcal{A}$, and bounded or, more general, satisfying, say, the condition $(W\mathcal{K})$ (see Section 1 in Chapter 2).

By $\mathcal{M}_0(\Omega, \mathcal{A})$ we denote the class of nonrandomized estimators $\varkappa = \varkappa(\omega)$ with values in $\mathcal{A}$. The class of Markov kernels $\varkappa = \varkappa(\omega, A)$, $A \in \mathcal{B}(\mathcal{A})$, is denoted by $\mathcal{M}(\Omega, \mathcal{A})$. If $\varkappa \in \mathcal{M}(\Omega, \mathcal{A})$, $\alpha \in \mathcal{A}$ and $c > 0$, then by $c(\varkappa - \alpha)$ we denote the Markov kernel $\varkappa(\omega, c^{-1}(A+\alpha))$ with $c^{-1}(A+\alpha) = \{\beta \in \mathcal{A} : c(\beta - \alpha) \in A\}$.

For an experiment $\mathcal{E} = (\Omega, \mathcal{F}, P_\alpha, \alpha \in \mathcal{A})$ we define

$$E_\alpha W(c(\varkappa - \alpha)) = E_\alpha \int_{\mathcal{A}} W(c(\beta - \alpha))\, \varkappa(\omega, d\beta)$$

and set, for $K \in \mathcal{K}$,

$$\mathcal{R}(\mathcal{E}, W; K) = \inf_{\varkappa \in \mathcal{M}(\Omega, \mathcal{A})} \sup_{\alpha \in K} E_\alpha W(\varkappa - \alpha).$$

In the case of $K = \mathcal{A}$ we use for this value the notation $\mathcal{R}(\mathcal{E}, W)$.

Below we suppose also that limit experiments $\mathcal{E}(\theta_0)$, $\theta_0 \in \Theta$, satisfy the following condition:

$$\sup_{K \in \mathcal{K}} \mathcal{R}(\mathcal{E}(\theta_0), W; K) = \mathcal{R}(\mathcal{E}(\theta_0), W) \tag{4.4}$$

which is certainly true if the loss function W is bounded and continuous (see Strasser (1985, Section 47)).

The parameter set Θ in $\mathcal{A}$ is supposed to satisfy the following condition, important for the local convergence framework:

(Θ) For each point $\theta_0 \in \Theta$ and any $K \in \mathcal{K}$ for n large enough, $n \geq n(\theta_0, K)$, the following inclusion holds

$$U^n(\theta_0, K) \subseteq \Theta$$

where

$$U^n(\theta_0, K) = \left\{\theta \in \mathcal{A} : \varphi^{-1}(n, \theta_0)(\theta - \theta_0) \in K\right\}.$$

In the case of a finite-dimensional space $\mathcal{A}$ this condition is reduced to openness of Θ. This condition is important for us since it guarantees the correctness of the definition of localized experiments $\mathcal{E}^n(\theta_0)$ and asymptotic statements below.

It follows from the previous assumptions and from Theorem 3.15, Section 3.7 7 that if $\mathcal{E}(\theta_0)$ is a regular experiment, then for any $K \in \mathcal{K}$

$$\begin{aligned} \liminf_n \inf_{\widehat{\theta}^n} \sup_{\theta \in U^n(\theta_0, K)} \mathbf{E}_\theta^n W\left(\varphi^{-1}(n, \theta_0)\left(\widehat{\theta}^n - \theta\right)\right) &= \inf_{\widehat{\alpha}} \sup_{\alpha \in K} E_{\alpha, \theta_0} W(\widehat{\alpha} - \alpha) \\ &= \mathcal{R}(\mathcal{E}(\theta_0), W; K), \end{aligned} \tag{4.5}$$

and, as corollary,

$$\sup_{K \in \mathcal{K}} \liminf_n \inf_{\widehat{\theta}^n} \sup_{\theta \in U^n(\theta_0, K)} \mathbf{E}_\theta^n W\left(\varphi^{-1}(n, \theta_0)\left(\widehat{\theta}^n - \theta\right)\right) = \mathcal{R}(\mathcal{E}(\theta_0), W). \tag{4.6}$$

In the case of a finite-dimensional space $\mathcal{A}$, when the class $\mathcal{K}$ consists of balls $B_b = \{\alpha : \|\alpha\| \leq b\}$, $b > 0$, one obtains from (4.6) the traditional version of the asymptotic minimax theorem:

$$\lim_{b \to \infty} \liminf_n \inf_{\widehat{\theta}^n} \sup_{\|\varphi^{-1}(n, \theta_0)(\theta - \theta_0)\| \leq b} \mathbf{E}_\theta^n W\left(\varphi^{-1}(n, \theta_0)\left(\widehat{\theta}^n - \theta\right)\right) = \mathcal{R}(\mathcal{E}(\theta_0), W). \tag{4.7}$$

For constructing estimators T^n of the parameter θ which attain the asymptotic lower bound of risks from (4.6) or (4.7) we follow Remark 2.9 in Section 2.7 of Chapter 2. Let $\varkappa^* = \varkappa^*_{\theta_0}(\lambda)$, be a minimax estimator for the experiment $\mathcal{E}(\theta_0)$, i.e. an estimator for which

$$\sup_{\alpha \in \mathcal{A}} E_{\alpha, \theta_0} W\left(\varkappa^*_{\theta_0}(\lambda) - \alpha\right) = \mathcal{R}(\mathcal{E}(\theta_0), W).$$

Then the estimators

$$\widehat{T}^n_{\theta_0} = \theta_0 + \varphi(n, \theta_0) \varkappa^*_{\theta_0}(\widehat{\lambda}^n_{\theta_0}), \tag{4.8}$$

with the random elements $\widehat{\lambda}^n_{\theta_0}$ constructed due to Theorem 2.8 are (locally) efficient in the asymptotic minimax sense:

$$\sup_{K \in \mathcal{K}} \liminf_n \inf_{\widehat{\theta}^n} \sup_{\theta \in U^n(\theta_0, K)} \mathbf{E}_\theta^n W\left(\varphi^{-1}(n, \theta_0)\left(\widehat{T}^n_{\theta_0} - \theta\right)\right) = \mathcal{R}(\mathcal{E}(\theta_0), W).$$

In Section 3.5 we explained that in many cases for the construction of asymptotically minimax estimators one can use the random elements $\lambda^n_{\theta_0}$ instead of their modifications $\widehat{\lambda}^n_{\theta_0}$. Therefore, the estimators

$$T^n_{\theta_0} = \theta_0 + \varphi(n, \theta_0)\varkappa^*_{\theta_0}(\lambda^n_{\theta_0}), \tag{4.9}$$

are often also (locally) asymptotically efficient. For more details see Sections 4.5 and 4.7 below.

Example 4.1 Let us consider the first-order autoregressive model with Gaussian errors and the autoregression parameter $|\theta| < 1$ (see also Section 5.1 of Chapter 5). The corresponding experiments $\mathcal{E}^n(\theta_0)$ localized at a point θ_0 are $(\gamma^n, \Gamma^n; \varphi)$-models (see Chapter 3) with

$$\begin{aligned}
(\Gamma^n)^2 &= \varphi^2(n, \theta_0)\,\langle M \rangle_n = \varphi^2(n, \theta_0) \sum_{k=1}^n x^2_{k-1}\,, \\
\Gamma^n \gamma^n &= \varphi(n, \theta_0) M_n = \varphi(n, \theta_0) \sum_{k=1}^n x_{k-1}\,(x_k - \theta_0 x_{k-1})\,.
\end{aligned}$$

If the localizing sequence is taken in the form

$$\varphi^2(n, \theta_0) = \frac{1 - \theta_0^2}{n},$$

then the experiments $\mathcal{E}^n(\theta_0)$ λ-converge to a $(\gamma, \Gamma; \varphi)$-model with $\Gamma \equiv 1$ (see Example 3.1), i.e. to the standard Gaussian shift experiment. But for this experiment the minimax estimator is $\varkappa^*(\lambda) = \lambda$ with $\lambda = \gamma/\Gamma$ (see Example 3.12). Hence, the estimator $T^n_{\theta_0}$ from (4.9) is of the form

$$\begin{aligned}
T^n_{\theta_0} &= \theta_0 + \varphi(n, \theta_0) \frac{\gamma^n}{\Gamma^n} \\
&= \theta_0 + \varphi(n, \theta_0) \frac{\varphi(n, \theta_0) M_n}{\varphi^2(n, \theta_0)\,\langle M \rangle_n} \\
&= \theta_0 + \frac{M_n}{\langle M \rangle_n} \\
&= \frac{\sum_{k=1}^n x_{k-1} x_k}{\sum_{k=1}^n x^2_{k-1}} \qquad \left(= \widehat{\theta}_n\right).
\end{aligned}$$

Therefore, $T^n_{\theta_0}$ coincides with the maximum likelihood estimator $\widehat{\theta}_n$ (see also Chapter 5) and hence, the *maximum likelihood estimators* $\widehat{\theta}_n$ are locally asymptotically efficient in the sense that for each $|\theta_0| < 1$

$$\lim_{b\to\infty} \lim_{n} \sup_{\|\varphi^{-1}(n,\theta_0)(\theta-\theta_0)\|\le b} \mathbf{E}_\theta^n W\left(\varphi^{-1}(n,\theta_0)\left(\widehat{\theta}^n - \theta\right)\right)$$
$$= \lim_{b\to\infty} \lim_{n} \inf_{\theta^n} \sup_{\|\varphi^{-1}(n,\theta_0)(\theta-\theta_0)\|\le b} \mathbf{E}_\theta^n W\left(\varphi^{-1}(n,\theta_0)\left(\theta^n - \theta\right)\right)$$
$$= \mathcal{R}\left(\mathcal{E}(\theta_0), W\right). \tag{4.10}$$

Moreover, under the standard condition on the loss function W one has

$$\mathcal{R}\left(\mathcal{E}(\theta_0), W\right) = \int W(x)\,\varphi(x)\,dx, \tag{4.11}$$

where $\varphi(x) = (2\pi)^{-1/2}\exp\left\{-x^2/2\right\}$ the standard normal density.

Taking in (4.10), for example, $\theta_0 = 0$ we get the following result:

$$\lim_{b\to\infty} \lim_{n} \sup_{|\theta|\le b/\sqrt{n}} \mathbf{E}_\theta^n W\left(\sqrt{n}\left(\widehat{\theta}^n - \theta\right)\right) = \int W(x)\,\varphi(x)\,dx \tag{4.12}$$

that can be easily recognized from the beginning since for $|\theta| < 1$

$$\mathcal{L}\left(\sqrt{n}\left(\widehat{\theta}^n - \theta\right) \mid \mathbf{P}_\theta^n\right) \to \mathcal{N}\left(0, \frac{1}{1-\theta^2}\right) \tag{4.13}$$

(see Section 5.1 of Chapter 5).

It is important that the estimators $T_{\theta_0}^n$ are apparently independent of θ_0 and coinciding with the maximum likelihood estimator $\widehat{\theta}^n$. This immediately yields that the estimator $\widehat{\theta}^n$ is locally efficient in the asymptotically small neighborhood of any point $|\theta| < 1$. Moreover, these estimators are asymptotically minimax, not only locally but also uniformly over any interval $[-1+\delta, 1-\delta]$, $\delta > 0$ (see Section 5.3 below).

The assertion (4.6) gives the exact lower bound for the local asymptotic minimax risk of arbitrary estimators. But usually the goal of the investigation is the value of "global" asymptotic minimax risk over the whole parametric set Θ (or over some fixed subset), defined as follows

$$\liminf_{n} \inf_{T^n} \sup_{\theta\in\Theta} \mathbf{E}_\theta^n\, W\left(\varphi^{-1}(n,\theta)\left(T^n - \theta\right)\right). \tag{4.14}$$

This expression differs from the left-hand side of (4.6), not only by the the class of all considered θ-values but also by the normalizing factor for the deviation of the estimator $T^n - \theta$. In (4.6) this factor is fixed and equal to $\varphi^{-1}(n,\theta_0)$. Studying the global risk one cannot, generally speaking, take this factor as fixed. For instance, in the models of explosive autoregression the values $\varphi(n,\theta)$ are of the order $|\theta|^{-n}$ and the fraction $|\theta|^{-n} / |\theta'|^{-n}$ tends to infinity for $|\theta'| > |\theta|$. Nevertheless, the result (4.6) can be applied to the investigation of the risk (4.14) under some auxiliary restrictions on the choice of the localizing sequences $\varphi(n,\theta)$, $n \ge 1$.

Before formulating the corresponding condition we make one remark. Although the rate with which the localizing sequences $\varphi(n,\theta)$ tend to zero is determined by contiguity considerations, there is some deliberation in their choice. In fact, if we multiply all members of a localizing sequence $\varphi(n,\theta)$ by a common constant $c > 0$, we simply make a reparametrization of the limit experiment $\mathcal{E}(\theta)$ with α substituted by $c^{-1}\alpha$. Note meanwhile that it is of great importance to have a consistent choice of sequences $\varphi(n,\theta)$ and, respectively, limit experiments $\mathcal{E}(\theta)$ at least in an asymptotically small neighborhood of each point θ_0. This motivates to introduce the following condition:

(Φ) For any point $\theta_0 \in \Theta$ and any $K \in \mathcal{K}$

$$\sup_{\theta \in U^n(\theta_0,K)} \left|\varphi^{-1}(n,\theta_0)\varphi(n,\theta) - 1\right| \to 0, \quad n \to \infty,$$

with $U^n(\theta_0,K) = \left\{\theta : \varphi^{-1}(n,\theta_0)(\theta - \theta_0) \in K\right\}$.

Theorem 4.1 *Let the conditions* (Θ), (Φ) *hold and let experiments* $\mathbf{E}^n$ *λ-converge locally to the family of regular experiments* $\mathcal{E} = (\mathcal{E}(\theta),\, \theta \in \Theta)$. *Let the loss function* W *satisfy the following condition* (W):

(W) (i) $W(\alpha) \geq 0,\ \alpha \in \mathcal{A}, \qquad W(0) = 0;$

(ii) *For any number* $\varepsilon > 0$ *there exists a number* $\delta_0 > 0$ *such that for all* $|\delta| \leq \delta_0$

$$W\left((1+\delta)\alpha\right) \leq (1+\varepsilon)W(\alpha), \qquad \forall \alpha \in \mathcal{A}.$$

Then it holds for any sequence of estimators $(T^n)_{n\geq 1}$ *one obtains*

$$\liminf_n \sup_{\theta\in\Theta} \mathbf{E}_\theta^n W\left(\varphi^{-1}(n,\theta)\left(T^n - \theta\right)\right) \geq \sup_{\theta\in\Theta} \mathcal{R}\left(\mathcal{E}(\theta), W\right). \tag{4.15}$$

Proof. Fix an arbitrary sequence of estimators T^n, $n \geq 1$. It suffices to prove that for any $\theta_0 \in \Theta$, the following inequality does hold:

$$\liminf_n \sup_{\theta\in\Theta} \mathbf{E}_\theta^n W\left(\varphi^{-1}(n,\theta)\left(T^n - \theta\right)\right) \geq \mathcal{R}\left(\mathcal{E}(\theta_0), W\right). \tag{4.16}$$

By (4.4)

$$\mathcal{R}\left(\mathcal{E}(\theta_0), W\right) = \sup_{K\in\mathcal{K}} \mathcal{R}\left(\mathcal{E}(\theta_0), W; K\right),$$

and it suffices to prove that for each $K \in \mathcal{K}$

$$\liminf_n \sup_{\theta\in U^n(\theta_0,K)} \mathbf{E}_\theta^n W\left(\varphi^{-1}(n,\theta)\left(T^n - \theta\right)\right) \geq \mathcal{R}\left(\mathcal{E}(\theta_0), W; K\right). \tag{4.17}$$

Let an arbitrary $\varepsilon > 0$ be fixed and let $\delta_0 = \delta_0(\varepsilon)$ be taken according to (W.ii). Let now the index $n(\varepsilon)$ be chosen in such a way that for $n \geq n(\varepsilon)$ we have

$$\left|\varphi^{-1}(n,\theta_0)\varphi(n,\theta)\right| \leq 1 + \delta_0, \qquad \theta \in U^n(\theta_0,K)$$

(it is possible by the condition (Φ)). Then for $n \geq n(\varepsilon)$

$$W\left(\varphi^{-1}(n,\theta_0)\left(T^n-\theta\right)\right) \leq (1+\varepsilon)W\left(\varphi^{-1}(n,\theta)\left(T^n-\theta\right)\right)$$

and (4.17) follows from (4.5) since ε was taken arbitrarily positive. □

4.3 Global Estimation. Preliminary Considerations

We have already noted that asymptotically efficient (in the local minimax sense) estimators $T_{\theta_0}^n = \theta_0 + \varphi(n,\theta_0)\varkappa_{\theta_0}^*(\lambda_{\theta_0}^n)$ (see (4.8)) depend on the localization point θ_0 which is, generally speaking, unknown a priori. It is natural to try to avoid this difficulty taking instead of θ_0 some *preliminary* estimator ρ^n of the unknown parameter θ possessing some consistency property like follows: that for each $\theta \in \Theta$

$$\lim_{b\to\infty} \limsup_n \mathbf{P}_\theta^n\left(\left\|\varphi^{-1}(n,\theta)(\rho^n-\theta)\right\| \geq b\right) = 0.$$

If such *pilot* estimators ρ^n, $n \geq 1$, are available, then, in order to construct a "good" estimator of the parameter θ, one can proceed in the following way. Since ρ^n belongs with a high probability to a $\varphi(n,\theta)$-neighborhood of the "true" point θ, this pilot estimator ρ^n can be considered as an "initial approximation" of θ and hence, analogously to the estimator $T_{\theta_0}^n$, one can consider the following expression:

$$\widehat{T}^n = \rho^n + \varphi(n,\rho^n)\varkappa_{\rho^n}^*(\widehat{\lambda}_{\rho^n}^n). \tag{4.18}$$

Estimators constructed in this way can be called *global estimators* and below we show that they are in many cases asymptotically minimax in global sense, i.e. independently of a priori information like knowing that the unknown value of θ lies near θ_0. Before starting with exploring the estimators (4.18), some questions should be addressed:

(1) Since the preliminary estimator ρ^n is random, it can take, generally speaking, any value in the parametric set Θ, the conditions like local λ-convergence should be satisfied not only for each point $\theta \in \Theta$ but even uniformly in θ;

(2) For the same reason we should require continuity of localizing sequences $\varphi(n,\theta)$ and limit experiments $\mathcal{E}(\theta)$ as functions of θ (see the condition (Φ) in Section 4.2);

(3) The pilot estimators ρ^n are constructed from the same observations as the global estimators $\widehat{T}^n$. This fact makes all the consideration essentially more complicate;

(4) Our exposition what follows is rather similar to the Le Cam's global estimator under the condition of "local asymptotic normality, see e.g. Le Cam (1960, 1986)). However, his construction essentially utilizes the linearity of the limit experiment (which is a Gaussian shift) and we do not assume

in general any special structure of the limit experiment. Instead, we require that the estimators $\varkappa_\theta^*$ entering in the construction (4.18) satisfy the equivariance property (see the conditions of Group 3 in Section 4.5).

Under the LAN condition, the limit experiment $\mathcal{E}(\theta)$ is for each θ a Gaussian shift experiment and the minimax estimator $\varkappa_\theta^*$ is of the form (see Example 3.12):

$$\varkappa_\theta^*(\lambda) = I_\theta^{-1}\lambda$$

where I_θ is the Fisher information matrix and $\lambda \in R^d$. (We assume here that $X = \mathcal{A} = R^d$.) Therefore, in the LAN case the proposed estimators $\widehat{T}^n$ are of the following form

$$\widehat{T}^n = \rho^n + \varphi(n, \rho^n) I_{\rho^n}^{-1} \widehat{\lambda}_{\rho^n}^n . \tag{4.19}$$

In order to state the asymptotic minimax properties of these estimators, i.e. to verify the (global) condition (see Theorem 4.1)

$$\lim_n \sup_{\theta \in \Theta} \mathbf{E}_\theta^n W\left(\varphi^{-1}(n,\theta)(\widehat{T}^n - \theta)\right) = \sup_{\theta \in \Theta} \mathcal{R}\left(\mathcal{E}(\theta), W\right), \tag{4.20}$$

it is sufficient to show that for each $\theta_0 \in \Theta$ the estimator $\widehat{T}^n$ differs only "a little" (under the measure $\mathbf{P}_{\theta_0}^n$) from the value $\theta_0 + \varphi(n,\theta_0) I_{\theta_0}^{-1} \widehat{\lambda}_{\theta_0}^n$, i.e.

$$\widehat{T}^n = \theta_0 + \varphi(n, \theta_0) I_{\theta_0}^{-1} \widehat{\lambda}_{\theta_0}^n + o(1). \tag{4.21}$$

The proof of (4.21) is based on some clear geometrical consideration (see below) which allows (in the case of a uniform version of LAN condition) to state the following relation between the random elements $\widehat{\lambda}_\theta^n$ for different, but close, values of θ:

$$\widehat{\lambda}_{\theta_0 + \varphi(n,\theta_0)\alpha}^n - \widehat{\lambda}_{\theta_0}^n + I_{\theta_0}\alpha \to 0, \qquad \alpha \in \mathcal{A}, \tag{4.22}$$

under the measure $\mathbf{P}_{\theta_0}^n$.

Assume the following conditions:

(i) the function I_θ^{-1} is continuous in θ;

(ii) for each $b > 0$

$$\sup_{\{\theta : |\varphi^{-1}(n,\theta_0)(\theta - \theta_0)| \le b\}} \left| \frac{\varphi(n,\theta)}{\varphi(n,\theta_0)} - 1 \right| \to 0, \quad n \to \infty;$$

(iii) for each $\varepsilon > 0$ there exists $b > 0$ such that

$$\limsup_n \mathbf{P}_\theta^n \left(\|\varphi^{-1}(n,\theta)(\rho^n - \theta)\| \ge b \right) \le \varepsilon.$$

Under these conditions (4.22) implies (omitting asymptotically small terms)

$$\begin{aligned}\widehat{T}^n &= \rho^n + \varphi(n,\rho^n)I_{\rho^n}^{-1}\widehat{\lambda}_{\rho^n}^n \\ &\approx \rho^n + \varphi(n,\theta_0)I_{\theta_0}^{-1}\left[\widehat{\lambda}_{\theta_0}^n - \varphi^{-1}(n,\theta_0)I_{\theta_0}(\rho^n - \theta_0)\right] \\ &= \rho^n + \varphi(n,\theta_0)I_{\theta_0}^{-1}\widehat{\lambda}_{\theta_0}^n - \rho^n + \theta_0 \\ &= \theta_0 + \varphi(n,\theta_0)I_{\theta_0}^{-1}\widehat{\lambda}_{\theta_0}^n,\end{aligned}$$

which is just the required property (4.21).

Note that the accurate derivation here is a rather delicate procedure since the estimators ρ^n are random and some more conditions are required. In particular, one should provide the uniform version of the property (4.22) and, moreover, the estimators ρ^n should be discretized. This means that these estimators take values in some discrete subset Θ^n in Θ and the number of points from Θ entering in each neighborhood $U^n(\theta_0, K)$ should be finite for all $\theta_0 \in \Theta$ and $K \in \mathcal{K}$, and bounded by some value $N(K)$ depending only on K but not on n. Actually, for any estimator ρ^n one can construct its discretized version (satisfying (iii)) under some uniform versions of the conditions (ii)–(iii).

The relation (4.22) plays the central role both in the LAN case and in the general situation. It was already noted that under the LAN condition the proof of (4.22) is based essentially on the linearity of the limit experiments $\mathcal{E}(\theta)$ (see Section 4.6 below for a generalization of this property). In the general situation, the property (4.22) can be stated in a different form which connects the random field

$$\widehat{\mathcal{Z}}^n(\theta) = \left(\widehat{Z}_\alpha^n(\theta),\, \alpha \in \mathcal{A}\right),$$

and

$$\widehat{Z}_\alpha^n(\theta) = Z_\alpha(\theta, \widehat{\lambda}_\theta^n), \qquad \alpha \in \mathcal{A}.$$

For a precise formulation we need some more notation. Let $\mathcal{E} = (\Omega, \mathcal{F}; P_\alpha\,,\, \alpha \in \mathcal{A})$ be a ***homogeneous*** experiment, i.e. $P_\alpha \sim P_\beta$ for all $\alpha, \beta \in \mathcal{A}$. Denote for $\beta \in \mathcal{A}$ by $T_\beta\mathcal{E}$ the experiment obtained from $\mathcal{E}$ by the shift of the parameter α by β:

$$T_\beta\mathcal{E} = (\Omega, \mathcal{F}; T_\beta P_\alpha\,,\, \alpha \in \mathcal{A})$$

with $T_\beta P_\alpha = P_{\alpha+\beta}$. If $Z_\alpha = dP_\alpha/dP_0$, then it holds for the experiment $T_\beta\mathcal{E}$

$$\frac{d(T_\beta P_\alpha)}{d(T_\beta P_0)} = \frac{dP_{\alpha+\beta}}{dP_\beta} = \frac{Z_{\alpha+\beta}}{Z_\beta}, \qquad \alpha \in \mathcal{A}.$$

Define

$$T_\beta Z_\alpha = \frac{Z_{\alpha+\beta}}{Z_\beta}, \qquad \alpha \in \mathcal{A}, \tag{4.23}$$

and let $T_\beta\mathcal{Z} = (T_\beta Z_\alpha\,,\, \alpha \in \mathcal{A})$ be the likelihood process for the experiment $T_\beta\mathcal{E}$.

Now we turn to the local convergence $\mathbb{E}^n \xrightarrow{(\lambda,\varphi)} \mathcal{E}$. Given a pair θ_0, $\theta \in \Theta$, define

$$\beta^n(\theta,\theta_0) = \varphi^{-1}(n,\theta_0)(\theta-\theta_0).$$

We claim that the process

$$\widehat{\mathcal{Z}}^n(\theta) = \left(\widehat{Z}^n_\alpha(\theta),\ \alpha \in \mathcal{A}\right) = \left(Z_\alpha(\theta,\widehat{\lambda}^n_\theta),\ \alpha \in \mathcal{A}\right),$$

for the values of θ close to θ_0, can be approximated by the process $T_{\beta^n(\theta,\theta_0)}\widehat{\mathcal{Z}}^n(\theta_0)$ which is obtained from $\widehat{\mathcal{Z}}^n(\theta_0)$ by the action of the operator T_{β^n}, according to (4.23):

$$\widehat{\mathcal{Z}}^n(\theta) \approx T_{\beta^n(\theta,\theta_0)}\widehat{\mathcal{Z}}^n(\theta_0) \tag{4.24}$$

(up to a small term under the measure $\mathbf{P}^n_{\theta_0}$).

Relation like in (4.24) will be later called the *connectedness equation*. Accurate formulations and proofs will be given in Section 4.4. Now we discuss some heuristics related to the connectedness equation and its application to the investigation of the asymptotic properties of the global estimators (4.18). For this purpose let us fix for a moment the index n, a point $\theta_0 \in \Theta$ and a set $K \in \mathcal{K}$, and let $\theta \in U^n(\theta_0, K)$. For simplicity we suppose that $\varphi(n,\theta)$ does not depend on θ ($\varphi(n,\theta) \equiv \varphi(n)$). (Recall that due to ($A$.ii) the value $\varphi(n,\theta)$ is "close" to $\varphi(n,\theta_0)$.)

Consider the experiment $\mathcal{E}^n(\theta) = (\Omega^n, \mathcal{F}^n;\ P^n_{\alpha,\theta},\ \alpha \in \mathcal{A})$ with $P^n_{\alpha,\theta} = \mathbf{P}^n_{\theta+\varphi(n)\alpha}$, $\alpha \in \mathcal{A}$, which can be considered as the shift of the experiment $\mathcal{E}^n(\theta_0) = (\Omega^n, \mathcal{F}^n;\ P^n_{\alpha,\theta_0},\ \alpha \in \mathcal{A})$ by the vector $\beta^n(\theta,\theta_0) = \varphi^{-1}(n)(\theta-\theta_0)$. In fact, $\theta = \theta_0 + \varphi(n)\beta^n(\theta,\theta_0)$ and defining $\beta^n = \beta^n(\theta,\theta_0)$ we obtain

$$P^n_{\alpha,\theta} = \mathbf{P}^n_{\theta+\varphi(n)\alpha} = \mathbf{P}^n_{\theta_0+\varphi(n)(\alpha+\beta^n)} = P^n_{\alpha+\beta^n,\theta} = T_{\beta^n}P^n_{\alpha,\theta_0}.$$

Thus

$$Z^n_\alpha(\theta) = \frac{dP^n_{\alpha,\theta}}{dP^n_{0,\theta}} = \frac{d\left(T_{\beta^n}P^n_{\alpha,\theta_0}\right)}{d\left(T_{\beta^n}P^n_{0,\theta_0}\right)} = \frac{Z^n_{\alpha+\beta^n}(\theta_0)}{Z^n_{\beta^n}(\theta_0)} = T_{\beta^n}Z^n_\alpha(\theta_0).$$

The assumed condition of local λ-convergence yields that the likelihood process $\mathcal{Z}^n(\theta) = (Z^n_\alpha(\theta),\ \alpha \in \mathcal{A})$ for the experiment $\mathcal{E}^n(\theta)$ is approximated by the process $\widehat{\mathcal{Z}}^n(\theta) = \left(Z_\alpha(\theta,\widehat{\lambda}^n_\theta),\ \alpha \in \mathcal{A}\right)$. The same is true for the likelihood process of the experiment $\mathcal{E}^n(\theta_0)$ as well as for the experiment shifted by β^n. Therefore, the exact equality $\mathcal{Z}^n(\theta) = T_{\beta^n}\mathcal{Z}^n(\theta_0)$, together with the asymptotic equalities $\mathcal{Z}^n(\theta) \approx \widehat{\mathcal{Z}}^n(\theta)$ and $T_{\beta^n}\mathcal{Z}^n(\theta_0) \approx T_{\beta^n}\widehat{\mathcal{Z}}^n(\theta_0)$, imply the desired property (4.24).

It immediately leads to the natural question of whether the connectedness equation (4.24) between the random fields $\widehat{\mathcal{Z}}^n(\theta)$ and $\widehat{\mathcal{Z}}^n(\theta_0)$ implies some relation between the elements $\widehat{\lambda}^n_\theta$ and $\widehat{\lambda}^n_{\theta_0}$ which determine these random fields. Usually the answer to this question is positive, although we can obtain this relation in an explicit form only under some additional assumptions upon the structure of the limit experiments $\mathcal{E}(\theta)$. The most typical case is considered below in Section 4.6 where we assume that the limit experiments $\mathcal{E}(\theta)$ are *transitive*. We also assume a continuous dependence of the fields $\mathcal{Z}(\theta) = (Z_\alpha(\theta), \alpha \in \mathcal{A})$ on the localization point $\theta \in \Theta$. Under these assumptions we state a connectedness relation for the random elements $\widehat{\lambda}_\theta^n$ which coincides with (4.24) in the LAN case.

Now we concentrate ourselves on the question of how the connectedness equation can be used for studying of the asymptotic properties of the global estimators of the type of (4.18). For this purpose we modify slightly the expression for these estimators. The problem is that the connectedness equation links not the elements $\widehat{\lambda}_\theta^n$ and $\widehat{\lambda}_{\theta_0}^n$ but the corresponding random fields $\widehat{\mathcal{Z}}^n(\theta)$ and $\widehat{\mathcal{Z}}^n(\theta_0)$. That is why it is desirable to have the expression for the global estimators in a form depending on these random fields. We shall assume that minimax decisions $\varkappa_\theta^*$ for the experiments $\mathcal{E}(\theta)$ are functions not of observations $x \in X$, but of the statistics $\mathcal{Z}(\theta, x) = (Z_\alpha(\theta, x), \alpha \in \mathcal{A})$, $x \in X$. It is obvious that this assumption is not restrictive since $\mathcal{Z}(\theta)$ is the sufficient statistic for the experiment $\mathcal{E}(\theta)$. Moreover, we will suppose that the decisions (estimators) $\varkappa_\theta^*$, as functions of $\mathcal{Z}(\theta)$, satisfy the following conditions:

(i) for each $\theta \in \Theta$ the decision $\varkappa_\theta^*$ depends continuously on the trajectories $\mathcal{Z}(\theta, x) = (Z_\alpha(\theta, x), \alpha \in \mathcal{A})$;
(ii) the decisions $\varkappa_\theta^*$ depend continuously on θ;
(iii) the decisions $\varkappa_\theta^*$ satisfy, for each $\theta \in \Theta$, the equivariance condition

$$\varkappa_\theta^* \left(T_\beta \mathcal{Z}(\theta)\right) = \varkappa_\theta^* \left(\mathcal{Z}(\theta)\right) - \beta, \qquad \beta \in \mathcal{A},$$

where $T_\beta \mathcal{Z}(\theta) \doteq \left(\frac{Z_{\alpha+\beta}(\theta)}{Z_\beta(\theta)}, \alpha \in \mathcal{A}\right)$.

Remark 4.3 The last condition is, of course, the most crucial one. If the experiment $\mathcal{E}(\theta)$ is transitive, then this property is satisfied under rather general assumptions. But in the general situation we have to assume that it holds. It is important to note that the most popular maximum likelihood and Pitman estimators (i.e. generalized Bayes estimator with respect to the uniform prior) are equivariant independently of the structure of the experiment.

Now it is natural to construct the global estimators $\widehat{T}^n$, in the following form:

$$\widehat{T}^n = \rho^n + \varphi(n, \rho^n) \varkappa_{\rho^n}^* \left(\widehat{\mathcal{Z}}^n(\rho^n)\right).$$

Taking into account the second property of the estimators $\varkappa_\theta^*$ from above we obtain for $\widehat{T}^n$ the following asymptotic expansion

$$\widehat{T}^n \approx \rho^n + \varphi(n) \varkappa_\theta^* \left(\widehat{\mathcal{Z}}^n(\rho^n)\right). \tag{4.25}$$

Furthermore, using the connectedness equation (4.24) and the properties (i), (iii) of $\varkappa_\theta^*$ we find that the following relation holds with $\beta^n = \beta^n(\rho^n, \theta) = \varphi^{-1}(n)(\rho^n - \theta)$:

$$\begin{aligned}\widehat{T}^n &\approx \rho^n + \varphi(n)\varkappa_\theta^*\left(T_{\beta^n}\widehat{\mathcal{Z}}^n(\theta)\right)\\ &= \rho^n + \varphi(n)\left[\varkappa_\theta^*\left(\widehat{\mathcal{Z}}^n(\theta)\right) - \beta^n\right]\\ &= \theta_0 + \varphi(n)\varkappa_\theta^*\left(\widehat{\mathcal{Z}}^n(\theta)\right).\end{aligned}$$

Hence

$$\varphi^{-1}(n)\left(\widehat{T}^n - \theta\right) \approx \varkappa_\theta^*\left(\widehat{\mathcal{Z}}^n(\theta)\right)$$

and

$$\mathcal{L}\left(\varphi^{-1}(n)\left(\widehat{T}^n - \theta\right) \mid \mathbf{P}_\theta^n\right) \to \mathcal{L}\left(\varkappa_\theta^*\left(\mathcal{Z}(\theta)\right) \mid P_{0,\theta}\right).$$

This relation gives us some reason to say that the estimators $\widehat{T}^n$ possess some asymptotic minimax properties. The exact formulations and proofs of these properties of global estimators are given in Section 4.5 (for the general case of local λ-convergence) and in Section 4.7 (for the case of local λ-convergence to transitive and weakly transitive experiments).

4.4 Connectedness Equation for Local λ-Convergence

The goal of the present section is to state the connectedness equation under the condition of local λ-convergence of experiments

$$\mathbb{E}^n = (\Omega^n, \mathcal{F}^n; \mathbf{P}_\theta^n, \theta \in \Theta), \quad n \geq 1,$$

to the family $\boldsymbol{\mathcal{E}} = (\mathcal{E}(\theta), \theta \in \Theta)$ with

$$\mathcal{E}(\theta) = (X, \mathcal{X}; P_{\alpha,\theta}, \alpha \in \mathcal{A}).$$

First we formulate our assumptions. The parameter set Θ is supposed to be a subset of some Banach space $\mathcal{A}$. Further, let a class $\mathcal{K}$ of subsets be fixed in the space $\mathcal{A}$, $\mathcal{K} = \{K, K \subseteq \mathcal{A}\}$, (cf. Section 4.2), and the following condition holds:

(K) (i) All subsets K from $\mathcal{K}$ are compact convex and contain a "zero" element $\alpha = 0$;
(ii) $\bigcup_{K\in\mathcal{K}} K = \mathcal{A}$, i.e. any point $\alpha \in \mathcal{A}$ belongs to some $K \in \mathcal{K}$;
(iii) $cK \in \mathcal{K}$ and $K + K' \in \mathcal{K}$ for any number $c > 0$ and any $K, K' \in \mathcal{K}$, where $cK = \{c\alpha : \alpha \in K\}$, $K + K' = \{\alpha + \alpha' : \alpha \in K, \alpha' \in K'\}$.

In the case of a finite-dimensional space $\mathcal{A}$ the class $\mathcal{K}$ is usually taken as the set of all balls $K_b = \{\alpha \in \mathcal{A} : \|\alpha\| \leq b\}$ with the center at zero, $b > 0$.

With respect to the parameter set Θ we suppose that it satisfies the condition (Θ) from Section 4.2, i.e. each point θ_0 is contained in Θ together with its neighborhoods $U^n(\theta_0, K)$ of the form

$$U^n(\theta_0, K) = \{\theta : \varphi^{-1}(n, \theta_0) \in K\}, \quad K \in \mathcal{K}, \tag{4.26}$$

for n large enough, $n \geq n(\theta_0, K)$. In the case of a finite-dimensional $\mathcal{A}$ this condition means that the set Θ is open.

Let now T_0 be some subset of Θ. We shall prove the connectedness equation uniformly in point θ_0 from T_0. For this we formulate all the conditions, including the condition of local λ-convergence, also uniformly in θ from a set T which is some neighborhood of the set T_0. In its turn, for this we need the set T to lie in Θ together with some its neighborhood. We introduce the corresponding condition on the set T_0 :

(T_0) The subset T_0 from Θ is such that there exists a larger set T in Θ (i.e. $T_0 \subseteq T, T \subseteq \Theta$) and for any $K \in \mathcal{K}$ and sufficiently large n, $n \geq n(K)$ the following inclusions hold:

$$U^n(T_0, K) \subseteq T,$$
$$U^n(T, K) \subseteq \Theta,$$

where for each subset A from Θ we denote

$$U^n(A, K) = \bigcup_{\theta \in A} U^n(\theta, K),$$

and $U^n(\theta, K)$ is defined by (4.26).

Now we give the uniform version (in the localization point θ from T) of the conditions of local λ-convergence and of regularity of limit experiments $\mathcal{E}(\theta)$.

Definition 4.4 We say that the experiments $\mathbb{E}^n$, $n \geq 1$ obey the condition of $(\mathcal{K}, T)$-*uniform local λ-convergence* to the family $\mathcal{E} = (\mathcal{E}(\theta), \theta \in \Theta)$, if there exist $\varphi(n, \theta) \to 0$, $\lambda_\theta^n : \Omega^n \to X$, $n \geq 1$, $\theta \in \Theta$, such that for each $K \in \mathcal{K}$ the following conditions are fulfilled:

(i) for any $\varepsilon > 0$

$$\sup_{\theta \in T} \sup_{\alpha \in K} \mathbf{P}_\theta^n \left(|Z_\alpha^n(\theta) - Z_\alpha(\theta, \lambda_\theta^n)| > \varepsilon\right) \to 0, \quad n \to \infty;$$

(ii) the distributions of the random elements λ_θ^n under the measures $\mathbf{P}_\theta^n$ $(= P_{0,\theta}^n)$ weakly converge to the measures $P_{0,\theta}$ $(= P_{\alpha,\theta}$ for $\alpha = 0)$, uniformly in $\theta \in \Theta$, i.e. for any continuous bounded function $g = g(x)$ on X

$$\sup_{\theta \in T} |\mathbf{E}_\theta^n g(\lambda_\theta^n) - E_{0,\theta} g(\lambda)| \to 0, \quad n \to \infty,$$

where $\lambda = \lambda(x)$ is the identity operator in X, $\lambda(x) = x$.

Definition 4.5 A family of experiments $\mathcal{E} = (\mathcal{E}(\theta),\ \theta \in \Theta)$ is called $(\mathcal{K}, T)$*-regular* if the following conditions hold:

$(\mathcal{K}T)$ (i) for any $K \in \mathcal{K}$ and $\varepsilon > 0$

$$\sup_{\theta \in T} \sup_{\alpha \in K} P_{0,\theta}\left(D_\delta Z_\alpha(\theta, \lambda) > \varepsilon\right) \to 0, \quad \delta \to 0,$$

where for an arbitrary function $f = f(x)$ we denote

$$D_\delta f(x) = \sup_{\{x' \in X : \rho(x,x') \le \delta\}} |f(x) - f(x')|,$$

and $\rho(x, x')$ is a metric in X;

(ii) for each $K \in \mathcal{K}$

$$\sup_{\theta \in T} \sup_{\alpha \in K} E_{0,\theta} Z_\alpha(\theta, \lambda) I_{\{\lambda : Z_\alpha(\theta,\lambda) > h\}} \to 0, \quad h \to \infty,$$

$E_{0,\theta}$ being the expectation with respect to the measure $P_{0,\theta}$;

(iii) for each $K \in \mathcal{K}$

$$\sup_{\theta \in T} \sup_{\alpha \in K} P_{0,\theta}\left(Z_\alpha(\theta, \lambda) < \delta\right) \to 0, \quad \delta \to 0;$$

(iv) for each $K \in \mathcal{K}$ and any $\varepsilon > 0$

$$\sup_{\theta \in T} \sup_{\alpha \in K} P_{0,\theta}\left(D_\delta^{(\alpha)} Z_\alpha(\theta, \lambda) > \varepsilon\right) \to 0, \quad \delta \to 0,$$

where

$$D_\delta^{(\alpha)} f_\alpha(x) = \sup_{\{\alpha' \in \mathcal{A} : \|\alpha' - \alpha\| \le \delta\}} |f_{\alpha'}(x) - f_\alpha(x)|, \quad \alpha \in \mathcal{A},\ x \in X.$$

Remark 4.4 This definition can be treated as the uniform version (in $\theta \in T$) of the definition of a $\mathcal{K}$-regular experiment (see Definition 2.9 in Section 2.6 of Chapter 2).

Remark 4.5 The second condition in the definition of $(\mathcal{K}, T)$-regularity means that if the $P_{0,\theta}$-measure of a set A is small then the $P_{\alpha,\theta}$-measure of this set is also small, and moreover, that this is true uniformly in $\alpha \in K$. The third condition, to the contrary, says that the smallness of $P_{\alpha,\theta}(A)$ implies the same for $P_{0,\theta}(A)$.

Remark 4.6 Although the condition of $(\mathcal{K}, T)$-regularity looks rather cumbersome, it is not very restrictive. The following set of conditions is sufficient:

(i) the function $Z_\alpha(\theta, x)$ is *continuous* in the set of arguments $\alpha \in \mathcal{A}$, $\theta \in \Theta$ and $x \in X$;

(ii) for each $K \in \mathcal{K}$ the family of measures $(P_{0,\theta}\,,\ \theta \in T)$ on $(X, \mathcal{X})$ is *tight*, i.e. for each $\varepsilon > 0$ there is a compact C in X such that

$$\sup_{\theta \in T} P_{0,\theta}(X \setminus C) \le \varepsilon; \tag{4.27}$$

(iii) The set T is *subcompact*, i.e. its closure is compact.

The proof of the fact that these conditions imply the property of $(\mathcal{K},T)$-regularity does not differ essentially from the proof of Lemma 2.3 in Section 2.6. As an example we explain why the first condition in Definition 4.5holds. Let C be a compact in X satisfying (4.27), let $\overline{T}$ be the compact in $\mathcal{A}$ which is the closure of T, and let $K \in \mathcal{K}$. Since the function $Z_\alpha(\theta,x)$ is continuous in the set of arguments it is uniformly continuous on the compact $\{(\alpha,x,\theta) : \alpha \in K,\ x \in C,\ \theta \in \overline{T}\}$. Therefore, for any $\varepsilon > 0$ there exists a $\delta > 0$ such that

$$\sup_{\theta\in\overline{T}}\sup_{\alpha\in K}\sup_{x\in C} D_\delta Z_\alpha(\theta,x) \le \varepsilon.$$

This implies the assertion (i) in Definition 4.5 in view of (4.27).

In order to formulate the main result we need also some strengthening of the condition of $(\mathcal{K},T)$-regularity.

Definition 4.6 A family of experiments $\mathcal{E} = (\mathcal{E}(\theta),\ \theta \in \Theta)$ is *strictly* $(\mathcal{K}T)$*-regular* if it is $(\mathcal{K},T)$-regular and instead of the condition $(\mathcal{K},T.i)$ the following stronger condition holds:

(i') for each $K \in \mathcal{K}$ and $\varepsilon > 0$

$$\sup_{\theta\in T} P_{0,\theta}\left(\sup_{\alpha\in K} D_\delta Z_\alpha(\theta,\lambda) > \varepsilon\right) \to 0, \quad \delta \to 0.$$

Remark 4.7 In the condition $((\mathcal{K},T.\text{i})$ of Definition 4.5 $\sup_{\alpha\in K}$ is taken for the probability of the event $\{D_\delta Z_\alpha(\theta,\lambda) > \varepsilon\}$. Here this sup is related to each trajectory of the random field $\mathcal{Z}(\theta,\lambda) = \{Z_\alpha(\theta,\lambda),\ \alpha \in \mathcal{A}\}$. That is why this condition is much stronger than the condition (i). We illustrate this fact on the example with shift experiments.

Example 4.2 Let P be a measure on the real line with a density $p(x)$ with respect to the Lebesgue measure, and let $\mathcal{E} = (R,\mathcal{B}(R);\ P_\alpha,\ \alpha \in R)$ where $P_\alpha(dx) = p(x-\alpha)\,dx$. Then the condition (i) in Definition 4.5 is satisfied for densities with a finite number of discontinuities of the first order but the condition (i') requires continuity of the density $p(x)$.

Remark 4.8 It is worth noting that the conditions (i)-(iii) from the above are sufficient not only for $(\mathcal{K},T)$-regularity (see Remark 3) but also for strict $(\mathcal{K},T)$-regularity, see Definition 4.6.

As a conclusion we give the T-uniform version of the condition (Φ) from Section 4.2:

(Φ') For each $K \in \mathcal{K}$

$$\sup_{\theta_0\in T}\ \sup_{\theta\in U^n(\theta_0,K)} \left|\varphi(n,\theta_0)\varphi^{-1}(n,\theta) - 1\right| \to 0, \quad n \to \infty,$$

and

$$\sup_{\theta\in T}\varphi(n,\theta)\to 0,\quad n\to\infty.$$

Now we are in a position to formulate the main result of the section. The condition of local λ-convergence $\mathbb{E}^n \overset{(\lambda,\varphi)}{\longrightarrow} \mathcal{E}$ means that for each point $\theta\in\Theta$ the localized experiments $\mathcal{E}^n(\theta)$ λ-converge to $\mathcal{E}(\theta)$ with the corresponding random elements $\lambda_\theta^n : \Omega^n \to X$. Let $\widehat{\lambda}_\theta^n$ be modifications of these random elements constructed according to Theorem 2.5 from Section 2.5 of Chapter 2 such that $\mathcal{L}\left(\widehat{\lambda}_\theta^n \mid \mathbf{P}_\theta^n\right) = \mathcal{L}(\lambda \mid P_{0,\theta}) = P_{0,\theta}$. We introduce the random fields

$$\widehat{\mathcal{Z}}^n(\theta) = \left(Z_\alpha(\theta,\widehat{\lambda}_\theta^n),\ \alpha\in\mathcal{A}\right),\quad \theta\in\Theta,$$

which we denote also in the form

$$\widehat{\mathcal{Z}}^n(\theta) = \mathcal{Z}(\theta)\circ\widehat{\lambda}_\theta^n,$$

where $\mathcal{Z}(\theta) = (Z_\alpha(\theta,\lambda),\ \alpha\in\mathcal{A})$ is the likelihood process for the limit experiment $\mathcal{E}(\theta)$ with

$$Z_\alpha(\theta,x) = \frac{dP_{\alpha,\theta}}{dP_{0,\theta}}(x),\quad x\in X,\ \alpha\in\mathcal{A},\ \theta\in\Theta.$$

Fix now a point $\theta_0\in T_0$, a set $K\in\mathcal{K}$ and an index n, and let a point θ lie in the neighborhood $U^n(\theta_0,K)$ of the point θ_0, i.e.

$$\beta^n(\theta,\theta_0) = \varphi^{-1}(n,\theta_0)(\theta-\theta_0)\in K.$$

Our goal is to prove the following asymptotic relation, uniformly in $\theta_0\in T_0$ and $\theta\in U^n(\theta_0,K)$ (cf. (4.24)):

$$\widehat{\mathcal{Z}}^n(\theta)\approx T_{\beta^n(\theta,\theta_0)}\widehat{\mathcal{Z}}^n(\theta_0) \tag{4.28}$$

where, for $\beta\in\mathcal{A}$,

$$T_\beta\widehat{\mathcal{Z}}^n(\theta_0) = T_\beta\left(\mathcal{Z}(\theta_0)\circ\widehat{\lambda}_{\theta_0}^n\right) = \left(T_\beta Z_\alpha(\theta_0,\widehat{\lambda}_{\theta_0}^n),\ \alpha\in\mathcal{A}\right) \tag{4.29}$$

and

$$T_\beta Z_\alpha(\theta,x) = \frac{Z_{\alpha+\beta}(\theta,x)}{Z_\beta(\theta,x)},\quad \alpha,\beta\in\mathcal{A},\ x\in X. \tag{4.30}$$

Theorem 4.2 *Let conditions $(\mathcal{K})$, (T_0) and (Φ') hold and let experiments $\mathbb{E}^n$, $n\geq 1$, satisfy the condition of $(\mathcal{K},T)$-uniform local λ-convergence to strictly regular experiments $\mathcal{E} = (\mathcal{E}(\theta),\ \theta\in\Theta)$. Then for any $K\in\mathcal{K}$ and any $\varepsilon>0$*

$$\sup_{\theta_0\in T_0}\ \sup_{\theta\in U^n(\theta_0,K)} \mathbf{P}_{\theta_0}^n\left(\left\|\widehat{\mathcal{Z}}^n(\theta) - T_{\beta^n(\theta,\theta_0)}\widehat{\mathcal{Z}}^n(\theta_0)\right\|_K > \varepsilon\right)\to 0,\qquad n\to\infty \tag{4.31}$$

where $\beta^n(\theta,\theta_0) = \varphi^{-1}(n,\theta_0)(\theta-\theta_0)$, $\widehat{\mathcal{Z}}^n(\theta)$ *and* $T_{\beta^n(\theta,\theta_0)}\widehat{\mathcal{Z}}^n(\theta_0)$ *are defined by (4.28)-(4.30), and for any fields* $\mathcal{Z}^{(1)} = \left(\mathcal{Z}^{(1)}_\alpha,\ \alpha\in\mathcal{A}\right)$, $\mathcal{Z}^{(2)} = \left(\mathcal{Z}^{(2)}_\alpha,\ \alpha\in\mathcal{A}\right)$ *we denote*

$$\left\|\mathcal{Z}^{(1)} - \mathcal{Z}^{(2)}\right\|_K = \sup_{\alpha\in K}\left|Z^{(1)}_\alpha - Z^{(2)}_\alpha\right|, \quad K\in\mathcal{K}.$$

Remark 4.9 The assertion of the theorem makes sense even if the set T_0 consists of only one point θ_0. Note that even in this case we need that the conditions of local λ-convergence and regularity will be satisfied uniformly in some neighborhood of this point θ_0.

Remark 4.10 The condition of strict $(\mathcal{K},T)$-regularity can be weakened to the condition of $(\mathcal{K},T)$-regularity if we consider instead of deviations of the random fields $\widehat{\mathcal{Z}}^n(\theta)$ and $T_{\beta^n(\theta,\theta_0)}\widehat{\mathcal{Z}}^n(\theta_0)$ on compacts $K\in\mathcal{K}$ deviations over possible finite sets S from $\mathcal{A}$. We give the corresponding assertion as a separate result.

Theorem 4.3 *Let the conditions* (K), (T_0) *and* (Φ') *hold and let experiments* $\mathbb{E}^n$, $n\geq 1$, *satisfy the condition of* $(\mathcal{K},T)$*-uniform local* λ*-convergence to regular experiments* $\mathcal{E} = (\mathcal{E}(\theta),\ \theta\in\Theta)$. *Then for any* $K\in\mathcal{K}$, *any finite subset* S *from* K *and any* $\varepsilon>0$

$$\sup_{\theta_0\in T_0}\sup_{\theta\in U^n(\theta_0,K)} \mathbf{P}^n_{\theta_0}\left(\left\|\widehat{\mathcal{Z}}^n(\theta) - T_{\beta^n(\theta,\theta_0)}\widehat{\mathcal{Z}}^n(\theta_0)\right\|_S > \varepsilon\right)\to 0, \qquad n\to\infty. \tag{4.32}$$

Proof. This assertion is a particular case of Theorem 4.2. □

Remark 4.11 We formulate the results with the random elements $\widehat{\lambda}^n_\theta$ instead of λ^n_θ entering in the definition of local λ-convergence. However, it easily follows from of Theorem 4.6, (i) below and from the condition (i') of Definition 4.6 that the result of Theorem 4.2 remains valid if we substitute of $\widehat{\lambda}^n_\theta$ with λ^n_θ. We prefer to state the connectedness equation in the form (4.31) because it is technically more convenient due to the useful equality

$$\mathcal{L}\left(\widehat{\lambda}^n_\theta \mid \mathbf{P}^n_\theta\right) = P_{0,\theta}, \quad \theta\in\Theta,\ n\geq 1$$

which will be used for further statistical applications when studying the asymptotic properties of global estimators. For the sake of completeness we formulate the corresponding result using the elements λ^n_θ.

Theorem 4.4 *Let the conditions of Theorem 4.2 be satisfied. Then for any* $K\in\mathcal{K}$ *and each* $\varepsilon>0$

$$\sup_{\theta_0\in T_0}\sup_{\theta\in U^n(\theta_0,K)} \mathbf{P}^n_{\theta_0}\left(\left\|\mathcal{Z}(\theta,\lambda^n_\theta) - T_{\beta^n(\theta,\theta_0)}\mathcal{Z}(\theta_0,\lambda^n_{\theta_0})\right\|_K > \varepsilon\right)\to 0, \qquad n\to\infty.$$

Similarly can be changed the assertion of Theorem 4.3.

Theorem 4.5 *Under the conditions of Theorem 4.3 for each $K \in \mathcal{K}$, $\varepsilon > 0$ and any finite set S from K*

$$\sup_{\theta_0 \in T_0} \sup_{\theta \in U^n(\theta_0, K)} \mathbf{P}^n_{\theta_0} \left(\left\| \mathcal{Z}(\theta, \lambda^n_\theta) - T_{\beta^n(\theta,\theta_0)} \mathcal{Z}(\theta_0, \lambda^n_{\theta_0}) \right\|_S > \varepsilon \right) \to 0, \qquad n \to \infty.$$

Before starting with proofs of Theorems 4.2-4.3 we present a uniform version of Theorem 2.5, Section 2.5 of Chapter 2 and some of its corollaries.

Theorem 4.6 *Let $(\mathcal{K}, T)$-uniform local λ-convergence of experiments $\mathbb{E}^n$ to $(\mathcal{K}, T)$-regular experiments $\mathcal{E} = (\mathcal{E}(\theta),\ \theta \in \Theta)$ hold. Then there exist random elements $\widehat{\lambda}^n_\theta$, $\theta \in \Theta$, $n \geq 1$, with values in X defined possibly on some extensions $\left(\widehat{\Omega}^n, \widehat{\mathcal{F}}^n\right)$ of the original spaces $(\Omega^n, \mathcal{F}^n)$ such that the following conditions are fulfilled:*

(i) *for each $\varepsilon > 0$*

$$\sup_{\theta \in T} \mathbf{P}^n_\theta \left(\rho(\widehat{\lambda}^n_\theta, \lambda^n_\theta) > \varepsilon \right) \to 0, \quad n \to \infty;$$

(ii) *for each $\theta \in T$*

$$\mathcal{L}\left(\widehat{\lambda}^n_\theta \mid \mathbf{P}^n_\theta \right) = \mathcal{L}\left(\lambda \mid P_{0,\theta} \right) = P_{0,\theta};$$

(iii) *for each $K \in \mathcal{K}$*

$$\sup_{\theta \in T} \sup_{\alpha \in K} \mathbf{E}^n_\theta \left| Z^n_\alpha(\theta, \omega) - Z_\alpha(\theta, \widehat{\lambda}^n_\theta(\omega)) \right| \to 0, \quad n \to \infty.$$

Proof. Similarly to the case of Theorem 2.8, the proof is essentially the same as for Theorem 2.5 where all the assertions should be substituted by their uniform (in $\theta \in \Theta$ and $\alpha \in K$) versions. □

Now we present a corollary of this theorem which particularly yields contiguity of the family of measures $(\mathbf{P}^n_\theta)$ to the family of measures $\left(\mathbf{P}^n_{\theta_0}\right)$ if θ is "close" to θ_0. This corollary (as well as its proof) is the uniform version (in $\theta_0 \in T_0$ and $\alpha \in K$) of one of the assertions of Theorem 2.6.

Corollary 4.1 *Let for experiments $\mathbb{E}^n$ the condition of $(\mathcal{K}, T)$-uniform local λ-convergence to $(\mathcal{K}, T)$-regular experiments $\mathcal{E} = (\mathcal{E}(\theta),\ \theta \in \Theta)$ hold. Let for each $n \geq 1$ random events $A^n_{\alpha,\theta} \in \mathcal{F}^n$, $\theta \in \Theta$, $\alpha \in \mathcal{A}$, be fixed. If for $K \in \mathcal{K}$*

$$\sup_{\theta_0 \in T_0} \sup_{\alpha \in K} \mathbf{P}^n_{\theta_0} \left(A^n_{\alpha,\theta_0} \right) \to 0, \qquad n \to \infty,$$

then

$$\sup_{\theta_0 \in T_0} \sup_{\alpha \in K} \mathbf{P}^n_{\theta_0 + \varphi(n,\theta_0)} \left(A^n_{\alpha,\theta_0} \right) \to 0, \qquad n \to \infty.$$

We will use this assertion in a slightly different form. Namely, let $\theta_0 \in T_0$, $\alpha \in K$ and

$$\begin{aligned} \theta &= \theta_0 + \varphi(n, \theta_0)\alpha, \\ B^n_{\theta,\theta_0} &= A^n_{\alpha,\theta_0}. \end{aligned}$$

Then the following implication holds:

$$\begin{aligned} &\sup_{\theta_0 \in T_0} \sup_{\theta \in U^n(\theta_0, K)} \mathbf{P}^n_{\theta_0}\left(B^n_{\theta,\theta_0}\right) \to 0,\ n \to \infty, \\ \Longrightarrow\quad &\sup_{\theta_0 \in T_0} \sup_{\theta \in U^n(\theta_0, K)} \mathbf{P}^n_{\theta}\left(B^n_{\theta,\theta_0}\right) \to 0,\ n \to \infty. \end{aligned}$$

Obviously, the inverse implication is also valid.

Corollary 4.2 *Let experiments $\mathbb{E}^n$ fulfill the condition of $(\mathcal{K}, T)$-uniform local λ-convergence to $(\mathcal{K}, T)$-regular experiments $\mathcal{E} = (\mathcal{E}(\theta),\ \theta \in \Theta)$ hold. Then for any events $B^n_{\theta,\theta_0} \in \mathcal{F}^n$, $n \geq 1$, all $\theta, \theta_0 \in \Theta$, and any $K \in \mathcal{K}$ the following conditions are equivalent:*

(i)

$$\sup_{\theta_0 \in T_0} \sup_{\theta \in U^n(\theta_0, K)} \mathbf{P}^n_{\theta_0}\left(B^n_{\theta,\theta_0}\right) \to 0,\ n \to \infty;$$

(ii)

$$\sup_{\theta_0 \in T_0} \sup_{\theta \in U^n(\theta_0, K)} \mathbf{P}^n_{\theta}\left(B^n_{\theta,\theta_0}\right) \to 0,\ n \to \infty.$$

Proof. It is an easy combination of assertions (ii)–(iii) of Theorem 4.6 and conditions (ii)–(iii) of Definition 4.5. □

Proof of Theorem 4.3. The idea of the proof was discussed in the previous section. The proof can be split in two steps: at the first step we check coordinatewise closeness of the random fields $\widehat{\mathcal{Z}}^n(\theta)$ and $T_{\beta^n(\theta,\theta_0)}\widehat{\mathcal{Z}}^n(\theta_0)$, i.e. we actually prove the statement of Theorem 4.3. More precisely, we show that for any $K \in \mathcal{K}$, $\varepsilon > 0$

$$\sup_{\theta_0 \in T_0} \sup_{\theta \in U^n(\theta_0, K)} \sup_{\alpha \in K} \mathbf{P}^n_{\theta_0}\left(\left|\widehat{Z}^n_\alpha(\theta) - T_{\beta^n(\theta,\theta_0)}\widehat{Z}^n_\alpha(\theta_0)\right| > \varepsilon\right) \to 0, \qquad n \to \infty. \tag{4.33}$$

In the next step we show that (4.33) yields the required assertion (4.31) in view of the condition (i') from Definition 4.6.

Let us examine the first step. We fix $K \in \mathcal{K}$ and $\varepsilon > 0$ and choose any $\theta_0 \in T_0$, $\alpha, \beta \in K$. In order to prove (4.33) it suffices to check that

$$\mathbf{P}^n_{\theta_0}\left(\left|\widehat{Z}^n_\alpha(\theta_0 + \varphi(n, \theta_0)\beta) - T_\beta\widehat{Z}^n_\alpha(\theta_0)\right| > \varepsilon\right) \leq \varepsilon(n), \tag{4.34}$$

where the sequence $\varepsilon(n)$ depends only on $K \in \mathcal{K}$ and $\varepsilon(n) \to 0$ as $n \to \infty$.

We start the proof of (4.34) with the case when

$$\varphi(n,\theta_0+\varphi(n,\theta_0)\beta)=\varphi(n,\theta_0),\quad n\geq 1. \tag{4.35}$$

(Recall that by the condition (Φ') it holds $\varphi(n,\theta_0+\varphi(n,\theta_0)\beta)\approx\varphi(n,\theta_0)$.)

Define $\varphi(n)=\varphi(n,\theta_0)$, $\theta^n=\theta_0+\varphi(n)\beta$. Then

$$I_{\{\omega:Z^n_\beta(\theta_0,\omega)>0\}}\left[Z^n_\alpha(\theta^n,\omega)-T_\beta Z^n_\alpha(\theta_0,\omega)\right]=0\qquad \mathbf{P}^n_{\theta_0}\text{-a.s..} \tag{4.36}$$

In fact, if $Z^n_\beta(\theta_0,\omega)>0$, then

$$\begin{aligned} Z^n_\alpha(\theta^n,\omega) &= \frac{d\mathbf{P}^n_{\theta^n+\varphi(n)\alpha}}{d\mathbf{P}^n_{\theta^n}}(\omega)=\frac{d\mathbf{P}^n_{\theta_0+\varphi(n)(\alpha+\beta)}}{d\mathbf{P}^n_{\theta_0+\varphi(n)\beta}}(\omega)\\ &= \frac{Z^n_{\alpha+\beta}(\theta_0,\omega)}{Z^n_\beta(\theta_0,\omega)}=T_\beta Z^n_\alpha(\theta_0,\omega),\quad \omega\in\Omega^n. \end{aligned}$$

Further, since $\mathbf{P}^n_{\theta_0}\left(Z^n_\beta(\theta_0,\omega)=0\right)$ is the mass of the singular component of the measure $\mathbf{P}^n_{\theta_0}$ with respect to the measure $\mathbf{P}^n_{\theta^n}$, due to Corollary 4.2 from Theorem 4.6

$$\mathbf{P}^n_{\theta_0}\left(Z^n_\beta(\theta_0,\omega)=0\right)<\varepsilon_1(n)\to 0,\qquad n\to\infty, \tag{4.37}$$

where $\varepsilon_1(n)$ does not depend on θ_0,α,β.

By (4.36) and (4.37) it suffices for (4.34) to verify the conditions

$$\mathbf{P}^n_{\theta_0}\left(\left|\widehat{Z}^n_\alpha(\theta^n)-Z^n_\alpha(\theta^n)\right|>\varepsilon/2\right)\leq\varepsilon_2(n)\to 0, \tag{4.38}$$

$$\mathbf{P}^n_{\theta_0}\left(\left|T_\beta\widehat{Z}^n_\alpha(\theta_0)-T_\beta Z^n_\alpha(\theta_0)\right|>\varepsilon/2\right)\leq\varepsilon_3(n)\to 0, \tag{4.39}$$

as $n\to\infty$, where $\varepsilon_2(n)$, $\varepsilon_3(n)$ are numeric sequences independent of θ_0,α,β.

The result of Corollary 4.2 allows us to replace $\mathbf{P}^n_{\theta_0}$ in (4.38) by $\mathbf{P}^n_{\theta^n}$. Now, due to condition (T_0), for sufficiently large n, $n\geq n(K)$, it holds $\theta^n\in T$, and the assertion follows directly from (iii) of Theorem 4.6.

Now we turn to the proof of (4.39). It suffices to check that for each $\varepsilon_3>0$ for n sufficiently large, $n\geq n(\varepsilon_3)$, the following inequality holds:

$$\mathbf{P}^n_{\theta_0}\left(\left|T_\beta\widehat{Z}^n_\alpha(\theta_0)-T_\beta Z^n_\alpha(\theta_0)\right|>\varepsilon/2\right)\leq\varepsilon_3. \tag{4.40}$$

Define for each $\delta\in(0,1)$ the event

$$\begin{aligned} C^n_\delta = &\Big\{\widehat{Z}^n_{\alpha+\beta}(\theta_0)\in(\delta,\delta^{-1}),\ Z^n_{\alpha+\beta}(\theta_0)\in(\delta,\delta^{-1}),\\ &\left|\widehat{Z}^n_\beta(\theta_0)-Z^n_\beta(\theta_0)\right|+\left|\widehat{Z}^n_{\alpha+\beta}(\theta_0)-Z^n_{\alpha+\beta}(\theta_0)\right|<\varepsilon\delta^3/4\Big\}. \end{aligned}$$

Below we shall show that there are some $\delta=\delta(\varepsilon_3)$ and $n(\varepsilon_3)$ such that for $n\geq n(\varepsilon_3)$

$$\mathbf{P}^n_{\theta_0}\left(C^n_\delta\right)\geq 1-\varepsilon_3. \tag{4.41}$$

This would imply that on the set C_δ^n

$$\begin{aligned}\left|T_\beta \widehat{Z}_\alpha^n(\theta_0) - T_\beta Z_\alpha^n(\theta_0)\right| &= \left|\frac{Z_{\alpha+\beta}(\theta_0)}{Z_\beta(\theta_0)}\left(\widehat{\lambda}_{\theta_0}^n\right) - \frac{Z_{\alpha+\beta}^n(\theta_0)}{Z_\beta^n(\theta_0)}\right| \\ &\le \delta^{-2}\left|\widehat{Z}_{\alpha+\beta}^n(\theta_0) Z_\beta^n(\theta_0) - Z_{\alpha+\beta}^n(\theta_0)\widehat{Z}_\beta^n(\theta_0)\right| \\ &\le \delta^{-3}\left|\widehat{Z}_{\alpha+\beta}^n(\theta_0) - Z_{\alpha+\beta}^n(\theta_0)\right| + \delta^{-3}\left|Z_\beta^n(\theta_0) - \widehat{Z}_\beta^n(\theta_0)\right| \\ &\le \varepsilon/2\end{aligned}$$

and (4.40) follows.

It remains to prove (4.41). By the conditions (ii)-(iii) of Definition 4.5 there is a number δ depending only on the compact K and ε_3 such that

$$P_{0,\theta_0}\left(Z_{\alpha+\beta}(\theta_0) \in (2\delta, \delta^{-1} - \delta)\right) \ge 1 - \varepsilon_3/2. \tag{4.42}$$

Since $\mathcal{L}\left(\widehat{\lambda}_{\theta_0}^n \mid \mathbf{P}_{\theta_0}^n\right) = P_{0,\theta_0}$, the inequality (4.42) implies

$$\mathbf{P}_{\theta_0}^n\left(\widehat{Z}_{\alpha+\beta}^n(\theta_0) \in (2\delta, \delta^{-1} - \delta)\right) \ge 1 - \varepsilon_3/2. \tag{4.43}$$

Let now a number $n(\varepsilon_3)$ be chosen in such a way that for $n \ge n(\varepsilon_3)$

$$\mathbf{P}_{\theta_0}^n\left(\left|\widehat{Z}_{\alpha+\beta}^n(\theta_0) - Z_{\alpha+\beta}^n(\theta_0)\right| > \varepsilon\delta^3/2\right) < \varepsilon_3/2. \tag{4.44}$$

(Existence of such $n(\varepsilon_3)$ follows from Theorem 4.6.) Then the required relation (4.41) (and hence (4.40)) follows directly from (4.43) and (4.44). Thus, (4.34) is proved under the assumption (4.35).

To prove (4.34) in the general case we note that the assumption (4.35) has been used only once to state (4.36). In general, one can state an assertion similar to (4.36) substituting $Z_\alpha^n(\theta^n)$ by $Z_{a^n\alpha}^n(\theta^n)$, where

$$a^n = \frac{\varphi(n,\theta_0)}{\varphi(n,\theta^n)}.$$

In fact, for $Z_\beta^n(\theta_0) > 0$ we easily obtain

$$Z_{a^n\alpha}^n(\theta^n) = \frac{d\mathbf{P}_{\theta_0+\varphi(n,\theta_0)\beta+\varphi(n,\theta^n)\alpha}^n}{d\mathbf{P}_{\theta_0+\varphi(n,\theta_0)\beta}^n} = \frac{Z_{\alpha+\beta}^n(\theta_0)}{Z_\beta^n(\theta_0)} = T_\beta Z_\alpha^n(\theta_0).$$

Therefore, (4.38) follows from the inequality

$$\mathbf{P}_{\theta_0}^n\left(\left|\widehat{Z}_\alpha^n(\theta^n) - Z_{a^n\alpha}^n(\theta^n)\right| > \varepsilon/2\right) \le \varepsilon_2(n) \to 0. \tag{4.45}$$

The condition (Φ') provides for n large enough the inequality $a^n \le 2$ and hence $a^n\alpha \in 2K \in \mathcal{K}$ and similarly to (4.38) we obtain

$$\mathbf{P}_{\theta_0}^n\left(\left|\widehat{Z}_{a^n\alpha}^n(\theta^n) - Z_{a^n\alpha}^n(\theta^n)\right| > \varepsilon/4\right) \le \varepsilon_2(n) \to 0.$$

Now (4.45) follows from

$$\mathbf{P}^n_{\theta_0}\left(\left|\widehat{Z}^n_\alpha(\theta^n)-\widehat{Z}^n_{a^n\alpha}(\theta^n)\right|>\varepsilon/4\right)\le\varepsilon_4(n)\to 0. \tag{4.46}$$

Corollary 4.2 again allows us to replace $\mathbf{P}^n_{\theta_0}$ by $\mathbf{P}^n_{\theta^n}$. Moreover, the conditions $\theta_0\in T_0$ and (T_0) imply that $\theta^n\in T$. Now the equality $\mathcal{L}\left(\widehat{Z}^n_\alpha(\theta)\mid\mathbf{P}^n_\theta\right)=\mathcal{L}\left(Z_\alpha(\theta)\mid P_{0,\theta}\right)$ reduces the assertion (4.46) to the following one:

$$P_{0,\theta^n}\left(|Z_\alpha(\theta^n)-Z_{a^n\alpha}(\theta^n)|>\varepsilon/4\right)\le\varepsilon_4(n)\to 0. \tag{4.47}$$

The condition (Φ') and the compactness of K guarantee the existence of a sequence $\delta(n)\to 0$ (depending only on K) such that

$$|\alpha-a^n\alpha|\le\|K\|\,|a^n-1|\le\delta(n),$$

where $\|K\|=\sup_{\alpha\in K}\|\alpha\|<\infty$. Thus

$$|Z_\alpha(\theta^n)-Z_{a^n\alpha}(\theta^n)|\le D^{(\alpha)}_{\delta(n)}Z_\alpha(\theta^n)$$

and (4.47) follows from Definition 4.5, (iv). The first step of the proof related to (4.33) is complete.

Now we turn to the second step, i.e. we shall prove that (4.33) implies (4.31). According to (4.33) for each $\varepsilon>0$ and any $K\in\mathcal{K}$ there is a number $n(\varepsilon)$ such that for $n\ge n(\varepsilon)$

$$\mathbf{P}^n_{\theta_0}\left(\left|\widehat{Z}^n_\alpha(\theta)-T_{\beta^n(\theta,\theta_0)}\widehat{Z}^n_\alpha(\theta_0)\right|>\varepsilon\right)<\varepsilon$$

with arbitrary $\theta_0\in T_0$, $\theta\in U^n(\theta_0,K)$ and $\alpha\in K$. This obviously implies that for any fixed finite set S from K there is a number $n_1(\varepsilon)$ such that

$$\mathbf{P}^n_{\theta_0}\left(\sup_{\alpha\in S}\left|\widehat{Z}^n_\alpha(\theta)-T_{\beta^n(\theta,\theta_0)}\widehat{Z}^n_\alpha(\theta_0)\right|>\varepsilon\right)<\varepsilon \tag{4.48}$$

with arbitrary $\theta_0\in T_0$, $\theta\in U^n(\theta_0,K)$.

Notice that (4.48) is exactly the assertion of Theorem 4.3 and hence this theorem is proved since the condition of strict $(\mathcal{K},T)$-regularity has not yet been used.

In order to extend the property (4.48) to the case when sup is taken over the whole set K we use the condition of strict $(\mathcal{K},T)$-regularity of the experiments $\mathcal{E}=(\mathcal{E}(\theta),\,\theta\in\Theta)$. This condition implies for $\delta>0$ small enough, say $\delta\le\delta_1(\varepsilon)$, the following inequality: for any $\theta\in T$

$$P_{0,\theta}\left(\sup_{\alpha\in K}D_\delta Z_\alpha(\theta)>\varepsilon\right)<\varepsilon. \tag{4.49}$$

(Recall that

$$D_\delta f(x)=\sup_{\{x'\in X:\rho(x,x')\le\delta\}}|f(x)-f(x')|,\quad x\in X.)$$

The condition (iii) of Definition 4.5 and the equality

$$T_\beta Z_\alpha(\theta_0) = \frac{Z_{\alpha+\beta}(\theta_0)}{Z_\beta(\theta_0)}$$

imply that for $\delta > 0$ small enough, say $\delta < \delta_2(\varepsilon)$,

$$P_{0,\theta}\left(\sup_{\alpha\in K} D_\delta\left(T_\beta Z_\alpha(\theta)\right) > \varepsilon\right) < \varepsilon \tag{4.50}$$

with arbitrary $\theta_0 \in T_0$, $\theta \in U^n(\theta_0, K)$.

Let $\delta = \min\{\delta_1(\varepsilon), \delta_2(\varepsilon)\}$ and K_δ be an arbitrary finite δ-net in the compact K. Then it follows from (4.49) that for any $\theta \in T$

$$P_{0,\theta}\left(\sup_{\alpha\in K} \left|Z_\alpha(\theta) - Z_{\alpha(\delta)}(\theta)\right| > \varepsilon\right) < \varepsilon \tag{4.51}$$

where $\alpha(\delta)$ means the point from K_δ closest to α and such that $\|\alpha - \alpha(\delta)\| \le \delta$. Similarly (4.50) implies that for any $\theta_0 \in T_0$, $\beta \in K$

$$P_{0,\theta_0}\left(\sup_{\alpha\in K} \left|T_\beta Z_\alpha(\theta_0) - T_\beta Z_{\alpha(\delta)}(\theta_0)\right| > \varepsilon\right) < \varepsilon. \tag{4.52}$$

Using again the equality $\mathcal{L}\left(\widehat{\lambda}_\theta^n \mid \mathbf{P}_\theta^n\right) = P_{0,\theta}$ we obtain by (4.51) that for each $\theta \in T$

$$\mathbf{P}_\theta^n\left(\sup_{\alpha\in K} \left|\widehat{Z}_\alpha^n(\theta) - \widehat{Z}_{\alpha(\delta)}^n(\theta)\right| > \varepsilon\right) < \varepsilon, \quad n \ge 1. \tag{4.53}$$

Similarly, (4.52) implies for each $\theta_0 \in T_0$, $\beta \in K$,

$$\mathbf{P}_{\theta_0}^n\left(\sup_{\alpha\in K} \left|T_\beta \widehat{Z}_\alpha^n(\theta_0) - T_\beta \widehat{Z}_{\alpha(\delta)}^n(\theta_0)\right| > \varepsilon\right) < \varepsilon, \quad n \ge 1. \tag{4.54}$$

If $\theta \in U^n(\theta_0, K)$, then by Corollary 4.2 $\mathbf{P}_\theta^n$ in (4.54) can be replaced by $\mathbf{P}_{\theta_0}^n$ for n large enough. The previously proved assertion yields that for sufficiently large n and for any $\theta_0 \in T_0$, $\theta \in U^n(\theta_0, K)$

$$\mathbf{P}_{\theta_0}^n\left(\sup_{\alpha\in K} \left|T_\beta \widehat{Z}_\alpha^n(\theta_0) - \widehat{Z}_\alpha^n(\theta)\right| > 3\varepsilon\right) \le 2\varepsilon + \mathbf{P}_{\theta_0}^n\left(\sup_{\alpha\in K_\delta} \left|T_\beta \widehat{Z}_\alpha^n(\theta_0) - \widehat{Z}_\alpha^n(\theta)\right| > \varepsilon\right)$$

with $\beta = \beta^n(\theta, \theta_0)$. By (4.48) the last term here is small as $n \to \infty$, uniformly in $\theta_0 \in T_0$ and $\theta \in U^n(\theta_0, K)$. This completes the proof of the second step because $\varepsilon > 0$ was taken arbitrarily. Theorems 4.2 and 4.3 are proved.

4.5 Asymptotic Efficiency of Global Estimators under Local λ-Convergence. General Case

In the present section we state the (global) asymptotic efficiency (in the minimax sense, see (4.59)) of the global estimators introduced in Section 4.5. We base our results on the connectedness equations obtained in Section 4.4 do not imposing auxiliary assumptions on the structure of limit experiments. Instead we assume some properties of the estimators $\varkappa_\theta^*$ (for limit experiments) entering in the definition of global estimators (see (4.18)).

As in the previous sections we assume to be given a sequence of experiments $\mathbb{E}^n = (\Omega^n, \mathcal{F}^n; \mathbf{P}_\theta^n, \theta \in \Theta)$, $n \geq 1$, satisfying the condition of local λ-convergence to experiments $\boldsymbol{\mathcal{E}} = (\mathcal{E}(\theta), \theta \in \Theta)$. Suppose also to be given for each limit experiment $\mathcal{E}(\theta) = (X, \mathcal{X}; P_{\alpha,\theta}, \alpha \in \mathcal{A})$ some estimator $\varkappa_\theta^*$ of the parameter α which is a function of the trajectory of the likelihood process $\mathcal{Z}(\theta, x) = (Z_\alpha(\theta, x), \alpha \in \mathcal{A})$, $x \in X$, where

$$Z_\alpha(\theta, x) = \frac{dP_{\alpha,\theta}}{dP_{0,\theta}}(x), \quad x \in X.$$

As usual, non-randomized estimators are assumed to map the space of observations in in $\mathcal{A}$ and, in the general case, randomized estimators are valued in the set $\mathcal{M}(\mathcal{A})$ of probability measures on $\mathcal{A}$.

The goal of the following exposition is to study the asymptotic behavior of the global estimators

$$\widehat{T}^n = \rho^n + \varphi(n, \rho^n)\varkappa_{\rho^n}^*\left(\widehat{\mathcal{Z}}^n(\rho^n)\right) \tag{4.55}$$

where ρ^n is a preliminary estimator of the parameter θ which should be $\varphi(n,\theta)$-consistent for each $\theta \in \Theta$ and the random fields $\widehat{\mathcal{Z}}^n(\theta)$ are constructed in accordance with Theorem 4.2.

One of the assertions proved below is of the following form

$$\lim_n \sup_{\theta \in T_0} \left| \mathbf{E}_\theta^n W\left(\varphi^{-1}(n,\theta)(\widehat{T}^n - \theta)\right) - E_{0,\theta} W\left(\varkappa_\theta^*\left(\mathcal{Z}(\theta)\right)\right) \right| = 0 \tag{4.56}$$

where T_0 is a subset in Θ satisfying the condition (T_0) from Section 4.4 and W is a continuous bounded loss function on $\mathcal{A}$. In particular, (4.56) implies

$$\lim_n \sup_{\theta \in T_0} \mathbf{E}_\theta^n W\left(\varphi^{-1}(n,\theta)(\widehat{T}^n - \theta)\right) \leq \sup_{\theta \in T_0} E_{0,\theta} W\left(\varkappa_\theta^*\left(\mathcal{Z}(\theta)\right)\right). \tag{4.57}$$

If the decisions (estimators) $\varkappa_\theta^*$ are minimax for the experiments $\mathcal{E}(\theta)$ or, at least, the following condition holds

$$E_{0,\theta} W\left(\varkappa_\theta^*\left(\mathcal{Z}(\theta)\right)\right) \leq \mathcal{R}\left(\mathcal{E}(\theta), W\right) = \inf_{\varkappa} \sup_{\alpha \in \mathcal{A}} E_{\alpha,\theta} W(\varkappa - \alpha),$$

then we obtain from (4.57) the following inequality:

$$\lim_n \sup_{\theta \in T_0} \mathbf{E}_\theta^n W\left(\varphi^{-1}(n,\theta)(\widehat{T}^n - \theta)\right) \leq \sup_{\theta \in T_0} \mathcal{R}\left(\mathcal{E}(\theta), W\right).$$

Theorem 4.1 claims that under condition (T_0) any estimator T^n of the parameter θ satisfies the property:

$$\liminf_n \sup_{\theta \in T_0} \mathbf{E}_\theta^n W\left(\varphi^{-1}(n,\theta)(T^n - \theta)\right) \geq \sup_{\theta \in T_0} \mathcal{R}(\mathcal{E}(\theta), W). \tag{4.58}$$

Therefore, the combination of two asymptotic minimax results (in the present case (4.58) and (4.56)) yields the following (global) property of asymptotic efficiency of the global estimator $\widehat{T}^n$, $n \geq 1$:

$$\begin{aligned} &\liminf_n \sup_{\theta \in T_0} \inf_{T^n} \mathbf{E}_\theta^n W\left(\varphi^{-1}(n,\theta)(T^n - \theta)\right) \\ &= \lim_n \sup_{\theta \in T_0} \mathbf{E}_\theta^n W\left(\varphi^{-1}(n,\theta)(\widehat{T}^n - \theta)\right) = \sup_{\theta \in T_0} \mathcal{R}(\mathcal{E}(\theta), W). \end{aligned} \tag{4.59}$$

The proof of the property (4.56) given below is based on the following three groups of assumptions.

Group 1. The conditions (K), (T_0) and (Φ') from Section 4.4 hold as well as the condition of $(\mathcal{K}, T)$-uniform local λ-convergence and $(\mathcal{K}, T)$-regularity of the limit experiments $\boldsymbol{\mathcal{E}} = (\mathcal{E}(\theta), \theta \in \Theta)$.

Group 2. The preliminary estimators ρ^n are supposed to be consistent and discretized:

Consistency: For any $\varepsilon > 0$ there are a compact $K \in \mathcal{K}$ and a number $n(\varepsilon)$ such that for $n \geq n(\varepsilon)$

$$\mathbf{P}_\theta^n\left(\varphi^{-1}(n,\theta)(\rho^n - \theta) \notin K\right) < \varepsilon, \quad \forall \theta \in T_0.$$

In the case of a finite-dimensional space $\mathcal{A}$ this condition can be rewritten in the following form

$$\limsup_n \sup_{\theta \in T_0} \mathbf{P}_\theta^n\left(\left|\varphi^{-1}(n,\theta)(\rho^n - \theta)\right| \geq b\right) \to 0, \qquad b \to \infty.$$

Discreteness: For each $n \geq 1$ there is a discrete subset Θ^n in Θ such that the estimator ρ^n takes values only in Θ^n and for each $K \in \mathcal{K}$ the number of points in the intersection of Θ^n with each neighborhood of the form $U^n(\theta_0, K)$ is bounded by a universal constant $C(K)$ depending only on K, i.e.

$$\#\left(\Theta^n \cap U^n(\theta_0, K)\right) \leq C(K), \quad \forall \theta_0 \in T_0,$$

with $U^n(\theta_0, K) = \left\{\theta : \varphi^{-1}(n,\theta)(\theta - \theta_0) \in K\right\}$.

The meaning of the condition of discreteness is that in each $\varphi(n, \theta_0)$-small neighborhood $U^n(\theta_0, K)$ of a point $\theta_0 \in T_0$ the estimator ρ^n has only a finite number of values. The combination of the conditions of consistency and discreteness guarantees that with $\mathbf{P}_{\theta_0}^n$-probability close to 1 the estimator ρ^n has values in a finite subset of the set Θ.

Remark 4.12 The condition of discreteness is not very restrictive. Moreover, for any estimator ρ^n satisfying the consistency condition one can construct its modification $\overline{\rho}^n$ which satisfies both the conditions of consistency and discreteness. We explain this fact in the obvious case when the space $\mathcal{A}$ is finite-dimensional and besides the condition (Φ') the following one is also fulfilled:

$$\sup_{\theta,\theta'\in T}\left|\frac{\varphi(n,\theta)}{\varphi(n,\theta')}\right| \leq D < \infty \tag{4.60}$$

with some fixed D and each $n \geq 1$.

In this case one can take Θ^n as an arbitrary $\varphi(n,\theta_0)$-net in Θ where θ_0 is some point in T_0. The estimator $\overline{\rho}^n$ can be taken as the point of Θ^n closest to ρ^n. The property of discreteness is proved by (4.60). The consistency property remains valid.

The conditions of the third group are related to the decisions $\varkappa_\theta^*$, $\theta \in \Theta$, for the limit experiments.

Let $\mathbf{Z}(\mathcal{A})$ be the space of functions $\mathcal{Z} = (Z_\alpha\,,\, \alpha \in \mathcal{A})$ measurable with respect to the Borel σ-field $\mathcal{B}(\mathcal{A})$. Define on $\mathbf{Z}(\mathcal{A})$ the topology induced by the family of seminorms $\|\mathcal{Z}\|_K\,,\, K \in \mathcal{K}$, where

$$\|\mathcal{Z}\|_K = \sup_{\alpha\in K} |Z_\alpha|\,.$$

If the class $\mathcal{K}$ is rich enough, then this topology coincides with the topology of uniform convergence on compacts.

Let now $\mathbf{Z}^+(\mathcal{A})$ be the subspace in $\mathbf{Z}(\mathcal{A})$ consisting of non-negative functions and let $\mathcal{B}(\mathbf{Z}^+(\mathcal{A}))$ be the Borel σ-field in $\mathbf{Z}^+(\mathcal{A})$. The estimators $\varkappa_\theta$ considered below for the experiment $\mathcal{E}(\theta) = (X, \mathcal{X}; P_{\alpha,\theta}\,,\, \alpha \in \mathcal{A})$, $\theta \in \Theta$, are supposed to be Markov kernels from $(\mathbf{Z}^+(\mathcal{A}), \mathcal{B}(\mathbf{Z}^+(\mathcal{A})))$ to $(\mathcal{A}, \mathcal{B}(\mathcal{A}))$. The set of such estimators will be denoted by $\mathcal{M}(\mathbf{Z}^+(\mathcal{A}), \mathcal{A})$. In the general case estimators $\varkappa_\theta$ are treated as Markov kernels from $(X, \mathcal{X})$ to $(\mathcal{A}, \mathcal{B}(\mathcal{A}))$. Since the statistic $\mathcal{Z}(\theta, x) = (Z_\alpha(\theta, x),\, \alpha \in \mathcal{A})$ with values in $\mathbf{Z}^+(\mathcal{A})$ is sufficient for the experiment $\mathcal{E}(\theta)$, for any estimator $\varkappa_\theta$ from $\mathcal{M}(X, \mathcal{A})$ one can construct the estimator $\varkappa_\theta'$ from $\mathcal{M}(\mathbf{Z}^+(\mathcal{A}), \mathcal{A})$ (with the same probability properties) as the conditional expectation with respect to the σ-field generated by $\mathcal{Z}(\theta, x)$.

It is convenient to treat estimators $\varkappa_\theta \in \mathcal{M}(\mathbf{Z}^+(\mathcal{A}), \mathcal{A})$ as measurable mappings from $\mathbf{Z}^+(\mathcal{A})$ to the space $\mathcal{M}(\mathcal{A})$ of measures on $(\mathcal{A}, \mathcal{B}(\mathcal{A}))$ which is the metric space with the Lévy-Prokhorov metric. Of course, in non-randomized case these estimators are treated in the standard sense as measurable mappings from $\mathbf{Z}^+(\mathcal{A})$ to $\mathcal{A}$.

Group 3. (i) For each $\varepsilon > 0$ there is a compact $K \in \mathcal{K}$ and a number $\delta > 0$ such that the condition

$$\|\mathcal{Z} - \mathcal{Z}'\|_K < \delta, \qquad \mathcal{Z}, \mathcal{Z}' \in \mathbf{Z}^+(\mathcal{A}),$$

implies the condition

$$\sup_{\theta \in T} L\left(\varkappa_\theta^*(\mathcal{Z}), \varkappa_\theta^*(\mathcal{Z}')\right) \leq \varepsilon$$

where L is the Lévy-Prokhorov metric.
In the non-randomized case the latter condition can be rewritten in the form

$$\sup_{\theta \in T} \|\varkappa_\theta^*(\mathcal{Z}) - \varkappa_\theta^*(\mathcal{Z}')\| \leq \varepsilon$$

where $\|\cdot\|$ is the norm in the Banach space $\mathcal{A}$;

(ii) For each $\varepsilon > 0$ there is a number $\delta > 0$ such that the condition $\|\theta - \theta_0\| < \delta$ for $\theta_0 \in T_0$ and $\theta \in T$ implies the inequality

$$P_{0,\theta_0}\left(L\left(\varkappa_{\theta_0}^*(\mathcal{Z}(\theta_0)), \varkappa_\theta^*(\mathcal{Z}(\theta_0))\right) > \varepsilon\right) < \varepsilon.$$

In the non-randomized case this condition is of the form

$$P_{0,\theta_0}\left(\left\|\varkappa_{\theta_0}^*(\mathcal{Z}(\theta_0)) - \varkappa_\theta^*(\mathcal{Z}(\theta_0))\right\| > \varepsilon\right) < \varepsilon;$$

(iii)

$$\sup_{\theta \in T} P_{0,\theta}\left(\|\varkappa_\theta^*(\mathcal{Z}(\theta))\| > h\right) \to 0, \qquad h \to \infty;$$

(iv) The estimators $\varkappa_\theta^*$ are equivariant, i.e. for each $\theta \in \Theta$ and $\beta \in \mathcal{A}$

$$\varkappa_\theta^*(T_\beta \mathcal{Z}) = \varkappa_\theta^*(\mathcal{Z}) - \beta, \quad P_{0,\theta}\text{-a.s.},$$

where $\mathcal{Z} = (Z_\alpha, \alpha \in \mathcal{A})$, $T_\beta \mathcal{Z} = \left(\frac{Z_{\alpha+\beta}}{Z_\beta}, \alpha \in \mathcal{A}\right)$.

Remark 4.13 The condition (i) can be treated as the uniform continuity in $\theta \in T$ of the estimators $\varkappa_\theta^*(\mathcal{Z})$ as functions of the argument $\mathcal{Z} \in \mathbf{Z}^+(\mathcal{A})$. The condition (ii) means uniform continuity of $\varkappa_\theta^*$ in $\theta \in T$.

It is easy to show that the conditions (i) and (ii) are fulfilled if the estimators $\varkappa_\theta^*$ are continuous in the set of arguments $\mathcal{Z} \in \mathbf{Z}(\mathcal{A})$ and $\theta \in T$, the family of measures $\{P_{0,\theta}, \theta \in T\}$ is tight, and the set T is subcompact (cf. Remark 4.5 after Definition 4.5). The condition (iii) means uniform integrability of the estimators $\varkappa_\theta^*$ under the measures $P_{0,\theta}$, $\theta \in T$.

Remark 4.14 Of course, the condition (iv) is the most restrictive one. Nevertheless, it is surely satisfied for the two most widespread methods: *maximum likelihood* and *Pitman* (generalized Bayes with respect to uniform prior on $\mathcal{A}$) *estimators.*

Theorem 4.7 *Let the conditions of Group 1-3 hold. Then the estimators $\widehat{T}^n$ from (4.55) satisfy the property: for each $\varepsilon > 0$*

$$\lim_n \sup_{\theta \in T_0} \mathbf{P}_\theta^n\left(L\left(\varphi^{-1}(n,\theta)(\widehat{T}^n - \theta), \varkappa_\theta^*\left(\widehat{\mathcal{Z}}^n(\theta)\right)\right) > \varepsilon\right) = 0 \tag{4.61}$$

where $L(\cdot,\cdot)$ is the Lévy-Prokhorov metric.

Remark 4.15 The assertion (4.56) follows obviously from (4.61) if the function W is continuous and bounded and the set T_0 is subcompact.

Remark 4.16 In the non-randomized case the assertion of the theorem is of the form

$$\lim_n \sup_{\theta\in T_0} \mathbf{P}_\theta^n \left(\left| \varphi^{-1}(n,\theta)(\widehat{T}^n - \theta) - \varkappa_\theta^* \left(\widehat{Z}^n(\theta) \right) \right| > \varepsilon \right) = 0.$$

Proof of Theorem 4.7. To make the exposition clearer we consider only non-randomized estimators $\varkappa_\theta^*$. The general case can be treated similarly when changing the expressions of the form $\varkappa - \varkappa'$ by $L(\varkappa, \varkappa')$. The proof uses the following technical result.

Lemma 4.1 *Let for each $\theta, \theta_0 \in \Theta$ events $A^n(\theta,\theta_0) \in \mathcal{F}^n$ be fixed, $n \geq 1$, such that for any $K \in \mathcal{K}$*

$$\sup_{\theta_0\in T_0} \sup_{\theta\in U^n(\theta_0,K)} \mathbf{P}_{\theta_0}^n \left(A^n(\theta,\theta_0)\right) \to 0, \quad n \to \infty. \tag{4.62}$$

If estimators ρ^n satisfy the conditions of consistency and discreteness, then

$$\sup_{\theta_0\in T_0} \mathbf{P}_{\theta_0}^n \left(A^n(\rho^n,\theta_0)\right) \to 0, \quad n \to \infty. \tag{4.63}$$

Proof. Denote for $\theta_0 \in T_0$ and $K \in \mathcal{K}$

$$D^n(\theta_0,K) = \Theta^n \cap U^n(\theta_0,K), \quad n \geq 1,$$

where the set Θ^n was introduced in the discreteness condition. By this condition the set $D^n(\theta_0,K)$ is finite with cardinality $\#D^n(\theta_0,K) \leq C(K)$, $n \geq 1$, and the estimator ρ^n takes its values near the point θ_0 precisely in $D^n(\theta_0,K)$. Therefore, by (4.62)

$$\sup_{\theta_0\in T_0} \mathbf{P}_{\theta_0}^n \left(\bigcup_{\theta\in D^n(\theta_0,K)} A^n(\theta,\theta_0) \right) \to 0, \quad n \to \infty,$$

and

$$\mathbf{P}_{\theta_0}^n \left(A^n(\rho^n,\theta_0)\right) \leq \mathbf{P}_{\theta_0}^n \left(\rho^n \notin U^n(\theta_0,K)\right) + \mathbf{P}_{\theta_0}^n \left(\bigcup_{\theta\in D^n(\theta_0,K)} A^n(\theta,\theta_0) \right).$$

This inequality and the condition of consistency yield the assertion (4.63). □

Now we come back to the proof of the theorem. Define

$$\beta^n(\rho^n,\theta_0) = \varphi^{-1}(n,\theta_0)\left(\rho^n - \theta_0\right).$$

We have

$$\varphi^{-1}(n,\theta_0)\left(\widehat{T}^n-\theta_0\right) = \varphi^{-1}(n,\theta_0)\left(\rho^n-\theta_0\right)+\varphi^{-1}(n,\theta_0)\varphi(n,\rho^n)\varkappa^*_{\rho^n}\left(\widehat{\mathcal{Z}}^n(\rho^n)\right)$$
$$= \beta^n(\rho^n,\theta_0)+\left[1+a^n(\rho^n,\theta_0)\right]\varkappa^*_{\rho^n}\left(\widehat{\mathcal{Z}}^n(\rho^n)\right)$$

with

$$a^n(\theta,\theta_0)=\varphi^{-1}(n,\theta_0)\varphi(n,\theta)-1.$$

Lemma 4.55 allows us to reduce the assertion of the theorem to the following one: for each $\varepsilon>0$ and $K\in\mathcal{K}$

$$\sup_{\theta_0\in T_0}\sup_{\theta\in U^n(\theta_0,K)}\mathbf{P}^n_{\theta_0}\left(\left\|\beta+(1+a)\varkappa^*_\theta\left(\widehat{\mathcal{Z}}^n(\theta)\right)-\varkappa^*_{\theta_0}\left(\widehat{\mathcal{Z}}^n(\theta_0)\right)\right\|>\varepsilon\right)\to 0, \tag{4.64}$$

as $n\to\infty$ with $\beta=\beta^n(\theta,\theta_0)$ and $a=a^n(\theta,\theta_0)$.

In order to verify (4.64) we prove the following two assertions:

$$\sup_{\theta_0\in T_0}\sup_{\theta\in U^n(\theta_0,K)}\mathbf{P}^n_{\theta_0}\left(\left\|\varkappa^*_\theta\left(\widehat{\mathcal{Z}}^n(\theta)\right)-\varkappa^*_{\theta_0}\left(T_\beta\widehat{\mathcal{Z}}^n(\theta_0)\right)\right\|>\varepsilon\right)\to 0, \tag{4.65}$$

and

$$\sup_{\theta_0\in T_0}\sup_{\theta\in U^n(\theta_0,K)}\mathbf{P}^n_{\theta_0}\left(\left\||a|\,\varkappa^*_\theta\left(\widehat{\mathcal{Z}}^n(\theta)\right)\right\|>\varepsilon\right)\to 0, \qquad n\to\infty. \tag{4.66}$$

These two statement imply (4.64) in view of the equivariance property of the estimators $\varkappa^*_\theta$. Indeed

$$\varkappa^*_{\theta_0}\left(T_\beta\widehat{\mathcal{Z}}^n(\theta_0)\right)=\varkappa^*_{\theta_0}\left(T_\beta\mathcal{Z}(\theta_0,\widehat{\lambda}^n_{\theta_0})\right)=\varkappa^*_{\theta_0}\left(\mathcal{Z}(\theta_0,\widehat{\lambda}^n_{\theta_0})\right)-\beta=\varkappa^*_{\theta_0}\left(\widehat{\mathcal{Z}}^n(\theta_0)\right)-\beta.$$

Using once again Corollary 4.2 we can change in (4.66) $\mathbf{P}^n_{\theta_0}$ by $\mathbf{P}^n_\theta$. Now, by the condition (Φ') and the equalities $\mathcal{L}\left(\widehat{\lambda}^n_\theta \mid \mathbf{P}^n_\theta\right)=P_{0,\theta}$, the assertion (4.66) is equivalent to the following one:

$$\sup_{\theta\in T}P_{0,\theta}\left(\left\|\varkappa^*_\theta\left(\mathcal{Z}(\theta)\right)\right\|>\delta^{-1}\right)\to 0, \quad \delta\to 0,$$

which is exactly the condition (iii) of Group 3. Thus (4.66) is proved.

In order to prove (4.65) it suffices to check that

$$\sup_{\theta_0\in T_0}\sup_{\theta\in U^n(\theta_0,K)}\mathbf{P}^n_{\theta_0}\left(\left\|\varkappa^*_\theta\left(\widehat{\mathcal{Z}}^n(\theta)\right)-\varkappa^*_{\theta_0}\left(\widehat{\mathcal{Z}}^n(\theta)\right)\right\|>\varepsilon\right)\to 0, \tag{4.67}$$

$$\sup_{\theta_0\in T_0}\sup_{\theta\in U^n(\theta_0,K)}\mathbf{P}^n_{\theta_0}\left(\left\|\varkappa^*_{\theta_0}\left(\widehat{\mathcal{Z}}^n(\theta)\right)-\varkappa^*_{\theta_0}\left(T_\beta\widehat{\mathcal{Z}}^n(\theta_0)\right)\right\|>\varepsilon\right)\to 0 \tag{4.68}$$

as $n\to\infty$. Given $\varepsilon>0$ and $K\in\mathcal{K}$, choose, due to (i) of Group 3, a number $\delta>0$ such that

$$\sup_{\theta_0\in T_0}\left\|\varkappa^*_{\theta_0}(\mathcal{Z})-\varkappa^*_{\theta_0}(\mathcal{Z}')\right\|<\varepsilon$$

if $\|\mathcal{Z} - \mathcal{Z}'\|_K < \delta$. Then for each $\theta_0 \in T_0$, $\theta \in U^n(\theta_0, K)$

$$\mathbf{P}^n_{\theta_0}\left(\left\|\varkappa^*_{\theta_0}\left(\widehat{\mathcal{Z}}^n(\theta)\right) - \varkappa^*_{\theta_0}\left(T_\beta \widehat{\mathcal{Z}}^n(\theta_0)\right)\right\| > \varepsilon\right)$$
$$\leq \mathbf{P}^n_{\theta_0}\left(\left\|\widehat{\mathcal{Z}}^n(\theta) - T_\beta \widehat{\mathcal{Z}}^n(\theta_0)\right\|_K > \delta\right)$$

and (4.68) follows from the connectedness equation, see Theorem 4.2.

Moreover, again by Corollary 4.2 one can substitute in the assertion (4.67) $\mathbf{P}^n_{\theta_0}$ by $\mathbf{P}^n_\theta$. Then the assertion follows from (ii) of Group 3, (Φ') and from the equalities $\mathcal{L}\left(\widehat{\mathcal{Z}}^n(\theta) \mid \mathbf{P}^n_\theta\right) = \mathcal{L}\left(\mathcal{Z}(\theta) \mid P_{0,\theta}\right)$. The theorem is proved.

4.6 Connectedness Equation for Transitive and Weakly Transitive Limit Experiments

In Section 4.4 we showed that under the uniform version of the condition of local λ-convergence of experiments $\mathbb{E}^n$, $n \geq 1$, to a family of experiments $\boldsymbol{\mathcal{E}} = (\mathcal{E}(\theta), \theta \in \Theta)$ the random fields $\widehat{\mathcal{Z}}^n(\theta) = \mathcal{Z}(\theta) \circ \widehat{\lambda}^n_\theta = \left(Z_\alpha(\theta, \widehat{\lambda}^n_\theta), \alpha \in \mathcal{A}\right)$ satisfy the connectedness equation of the form

$$\widehat{\mathcal{Z}}^n(\theta) \approx T_{\beta^n(\theta,\theta_0)} \widehat{\mathcal{Z}}^n(\theta_0)$$

with $\beta^n(\theta, \theta_0) = \varphi^{-1}(n, \theta_0)\,(\theta - \theta_0)$ (see Theorem 4.3).

In the present section we consider the problem of how to state the relation between the random elements $\widehat{\lambda}^n_\theta$ for different θ. In the case when the limit experiments $\mathcal{E}(\theta)$ are transitive or weakly transitive (see below) this problem can be solved explicitly.

We start with some definitions. Let $\mathcal{E} = (X, \mathcal{X}; P_\alpha, \alpha \in \mathcal{A})$ be a canonical experiment satisfying the homogeneity condition, i.e. X is a Polish space, $\mathcal{X}$ is the Borel σ-field and all the measures P_α are mutually absolutely continuous. The space $\mathcal{A}$ is supposed to be a Banach space.

Let $G = (G_\alpha, \alpha \in \mathcal{A})$ be a group of measurable mappings $G_\alpha : X \to X$, $\alpha \in \mathcal{A}$, such that G_0 is the identity mapping and the group operation "$*$" is defined in such a way that

$$G_\alpha * G_\beta = G_{\alpha+\beta}, \quad \alpha, \beta \in \mathcal{A}.$$

Definition 4.7 An experiment $\mathcal{E}$ is called *transitive* if $P_\alpha = P \circ G_\alpha^{-1}$, $\alpha \in \mathcal{A}$, for some measure P on $(X, \mathcal{X})$, i.e.

$$P_\alpha(B) = P(G_\alpha x \in B), \quad B \in \mathcal{X}, \alpha \in \mathcal{A}.$$

Obviously $P_0 = P$. Denote, as usual,

$$Z_\alpha(x) = \frac{dP_\alpha}{dP_0}(x), \quad \alpha \in \mathcal{A}, x \in X.$$

Lemma 4.2 *Let an experiment $\mathcal{E}$ be transitive and homogeneous. Then for each $\alpha, \beta \in \mathcal{A}$*

$$T_\beta Z_\alpha(x) = Z_\alpha(G_\beta x), \quad x \in X, \tag{4.69}$$

where

$$T_\beta Z_\alpha(x) = \frac{Z_{\alpha+\beta}(x)}{Z_\beta(x)}.$$

Proof. By the group property

$$P_{\alpha+\beta} = P \circ G_{\alpha+\beta}^{-1} = \left(P \circ G_\alpha^{-1}\right) \circ G_\beta^{-1} = P_\alpha \circ G_\beta^{-1}, \quad \alpha, \beta \in \mathcal{A}.$$

Hence, for any $x \in X$, $\alpha, \beta \in \mathcal{A}$

$$T_\beta Z_\alpha(x) = \frac{dP_{\alpha+\beta}}{dP_\beta}(x) = \frac{d(P_\alpha \circ G_\beta^{-1})}{d(P_0 \circ G_\beta^{-1})}(x) = \frac{dP_\alpha}{dP_0}(G_\beta x) = Z_\alpha(G_\beta x),$$

and the lemma follows. □

Actually, we do not use below the property of transitivity for the experiments $\mathcal{E}(\theta)$, $\theta \in \Theta$, but we only need the relation (4.69). This motivates the following definition.

Definition 4.8 An experiment $\mathcal{E}$ is *weakly transitive* if it is homogeneous and there exists a group of mappings $\mathcal{G} = (G_\alpha, \alpha \in \mathcal{A})$ such that the property (4.69) holds.

Remark 4.17 The typical example of weakly transitive but non-transitive experiments are the (γ, Γ)-models.

Let experiments $\mathbb{E}^n$, $n \geq 1$, locally λ-converge to $\mathcal{E} = (\mathcal{E}(\theta), \theta \in \Theta)$ and let the experiments $\mathcal{E}(\theta)$, $\theta \in \Theta$, be weakly transitive with the corresponding groups of transformations

$$\mathcal{G}(\theta) = (G_\alpha(\theta), \alpha \in \mathcal{A}).$$

By (4.69)

$$T_\beta Z_\alpha(\theta, x) = Z_\alpha(\theta, G_\beta(\theta)x), \quad \alpha, \beta \in \mathcal{A}, \, x \in X, \, \theta \in \Theta.$$

We shall write this relation in the following symbolical form

$$T_\beta \mathcal{Z}(\theta) = \mathcal{Z}(\theta) \circ G_\beta(\theta), \quad \beta \in \mathcal{A}, \, \theta \in \Theta.$$

With these notations the connectedness equation stated above can be rewritten in the form

$$\mathcal{Z}(\theta) \circ \widehat{\lambda}_\theta^n \approx \mathcal{Z}(\theta_0) \circ G_\beta(\theta_0) \widehat{\lambda}_{\theta_0}^n \tag{4.70}$$

with $\beta = \beta^n(\theta, \theta_0)$.

Now we aim to show that (4.70) implies (under some technical conditions) the following connectedness equation for $\widehat{\lambda}_\theta^n$:

$$\widehat{\lambda}_\theta^n \approx G_\beta(\theta_0)\widehat{\lambda}_{\theta_0}^n, \quad \beta = \beta^n(\theta, \theta_0).$$

First we introduce assumptions (G) and (Z) we use to state this relation.

(G) The experiments $\mathcal{E}(\theta)$, $\theta \in \Theta$, are weakly transitive and the corresponding function $G_\beta(\theta)x$ is continuous in the set of arguments β, θ, x.

The sense of the next condition is as follows: given $x, y \in X$ if $\mathcal{Z}(\theta, x) \approx \mathcal{Z}(\theta, y)$ for different $\theta \in \Theta$, then $\rho(x, y) \approx 0$. For the sake of simplicity we formulate this condition for the case of a finite-dimensional space $\mathcal{A}$.

Let $f = f(x)$ be a measurable vector-function defined on X with values in R^d. Set for each $\varepsilon > 0$

$$I^\varepsilon f(x) = \inf_{\{y:\rho(x,y)\geq\varepsilon\}} \|f(x) - f(y)\|, \quad x \in X. \tag{4.71}$$

Let S be a finite set in $\mathcal{A}$. Denote

$$Z_S(\theta, x) = \{Z_\alpha(\theta, x), \alpha \in S\}$$

i.e. $Z_S(\theta, x)$ is, for given θ and S, the vector-function from X in R^d, d being the cardinality of the set S.

In accordance with (4.71)

$$I^\varepsilon Z_S(\theta, x) = \inf_{\{y:\rho(x,y)\geq\varepsilon\}} \|\mathcal{Z}(\theta, x) - \mathcal{Z}(\theta, y)\|_S, \quad x \in X,$$

where $\|\mathcal{Z}\|_S = \sup_{\alpha\in S} |Z_\alpha|$ for $\mathcal{Z} = (Z_\alpha, \alpha \in \mathcal{A})$.

The required condition reads as follows:

(Z) There is a finite set S in $\mathcal{A}$ such that for each $\varepsilon > 0$

$$\sup_{\theta\in T} P_{0,\theta}\left(I^\varepsilon Z_S(\theta, x) < \delta\right) \to 0, \quad \delta \to 0.$$

This condition means that smallness of $\|\mathcal{Z}(\theta, x) - \mathcal{Z}(\theta, y)\|_S$ implies, with large $P_{0,\theta}$-probability, smallness of $\rho(x, y)$.

Theorem 4.8 *Let conditions (K), (T_0), (Φ') from Section 4.4 be fulfilled, experiments $\mathbb{E}^n$, $n \geq 1$, satisfy the condition of $(\mathcal{K}, T)$-uniform local λ-convergence to experiments $\mathcal{E} = (\mathcal{E}(\theta), \theta \in \Theta)$ which in turn satisfy the conditions (i)-(iii) from Remark 4.6 in Section 4.4. Let also conditions (G),(Z) hold. Then for each $K \in \mathcal{K}$ and any $\varepsilon > 0$*

$$\sup_{\theta_0\in T_0}\ \sup_{\theta\in U^n(\theta_0,K)} \mathbf{P}_{\theta_0}^n\left(\rho\left(\widehat{\lambda}_\theta^n, G_\beta(\theta_0)\widehat{\lambda}_{\theta_0}^n\right) > \varepsilon\right) \to 0, \quad n \to \infty \tag{4.72}$$

with $\beta = \beta^n(\theta, \theta_0)$.

Taking into account the assertion (i) of Theorem 4.6 and the continuity of the mappings $G_\beta(\theta)$ we obtain the following corollary.

Corollary 4.3 *Under the conditions of the theorem above,*

$$\sup_{\theta_0 \in T_0} \sup_{\theta \in U^n(\theta_0, K)} \mathbf{P}^n_{\theta_0}\left(\rho\left(\lambda^n_\theta, G_\beta(\theta_0)\lambda^n_{\theta_0}\right) > \varepsilon\right) \to 0, \quad n \to \infty.$$

Proof of Theorem 4.8. Due to Corollary 4.2 the assertion (4.72) can be proved with $\mathbf{P}^n_\theta$ instead of $\mathbf{P}^n_{\theta_0}$.

Let a set S be such that the property (Z) holds. Define for $\theta_0 \in T_0$, $\theta \in T$, $n \geq 1$,

$$s^n(\theta, \theta_0) = \left\| \mathcal{Z}(\theta, \widehat{\lambda}^n_\theta) - \mathcal{Z}(\theta, G_\beta(\theta_0)\widehat{\lambda}^n_{\theta_0}) \right\|_S.$$

We aim to show that the condition

$$\sup_{\theta_0 \in T_0} \sup_{\theta \in U^n(\theta_0, K)} \mathbf{P}^n_{\theta_0}\left(s^n(\theta, \theta_0) > \delta\right) \to 0, \quad n \to \infty, \qquad \forall \delta > 0 \tag{4.73}$$

implies (4.72). Indeed, if $\rho\left(\widehat{\lambda}^n_\theta, G_\beta(\theta_0)\widehat{\lambda}^n_{\theta_0}\right) > \varepsilon$, then it follows directly from the definition (4.71) that

$$s^n(\theta, \theta_0) \geq I^\varepsilon Z_S(\theta, \widehat{\lambda}^n_\theta).$$

Hence, the event $\left\{\rho\left(\widehat{\lambda}^n_\theta, G_\beta(\theta_0)\widehat{\lambda}^n_{\theta_0}\right) > \varepsilon\right\}$ is impossible if one has for some $\delta > 0$ simultaneously $I^\varepsilon Z_S(\theta, \widehat{\lambda}^n_\theta) > \delta$ and $s^n(\theta, \theta_0) < \delta$.

Using the equality $\mathcal{L}\left(\widehat{\lambda}^n_\theta \mid \mathbf{P}^n_\theta\right) = P_{0,\theta}$ we obtain

$$\begin{aligned} \mathbf{P}^n_\theta\left(\rho\left(\widehat{\lambda}^n_\theta, G_\beta(\theta_0)\widehat{\lambda}^n_{\theta_0}\right) > \varepsilon\right) &\leq \mathbf{P}^n_\theta\left(I^\varepsilon Z_S(\theta, \widehat{\lambda}^n_\theta) < \delta\right) + \mathbf{P}^n_\theta\left(s^n(\theta, \theta_0) > \delta\right) \\ &= P^n_{0,\theta}\left(I^\varepsilon Z_S(\theta, \lambda) < \delta\right) + \mathbf{P}^n_\theta\left(s^n(\theta, \theta_0) > \delta\right). \end{aligned} \tag{4.74}$$

Now (Z) and (4.74) yield the required assertion (4.72).

It remains to prove (4.73). By Theorem 4.3 and Corollary 4.2 one can claim that as $n \to \infty$

$$\sup_{\theta_0 \in T_0} \sup_{\theta \in U^n(\theta_0, K)} \mathbf{P}^n_\theta\left(\left\| \mathcal{Z}(\theta, \widehat{\lambda}^n_\theta) - \mathcal{Z}(\theta, G_\beta(\theta_0)\widehat{\lambda}^n_{\theta_0}) \right\|_S > \gamma\right) \to 0.$$

Therefore, to prove (4.73) we have to check only that for any $\gamma > 0$ as $n \to \infty$

$$\sup_{\theta_0 \in T_0} \sup_{\theta \in U^n(\theta_0, K)} \mathbf{P}^n_{\theta_0}\left(\left\| \mathcal{Z}(\theta_0, G_\beta(\theta_0)\widehat{\lambda}^n_{\theta_0}) - \mathcal{Z}(\theta, G_\beta(\theta_0)\widehat{\lambda}^n_{\theta_0}) \right\|_S > \gamma\right) \to 0. \tag{4.75}$$

(Here we again substituted $\mathbf{P}^n_\theta$ by $\mathbf{P}^n_{\theta_0}$.)

Since $\mathcal{L}\left(\widehat{\lambda}^n_{\theta_0} \mid \mathbf{P}^n_{\theta_0}\right) = P_{0,\theta_0}$, the statement (4.75) is equivalent to the property

$$\sup_{\theta_0 \in T_0} \sup_{\theta \in U^n(\theta_0, K)} P_{0,\theta_0}\left(\|\mathcal{Z}(\theta_0) \circ G_\beta(\theta_0) - \mathcal{Z}(\theta) \circ G_\beta(\theta_0)\|_S > \gamma\right) \to 0.$$

But this fact follows from (Φ') and the continuity of the functions $Z_\alpha(\theta_0, x)$ and $G_\beta(\theta_0)x$ in the set of arguments, due to the concluding lemma below.

Lemma 4.3 *Let T' be a compact set, $\{P_\theta, \theta \in T'\}$ be a tight family of measures on $(X, \mathcal{X})$ and let $f_\theta(x)$ be a continuous function in the set of arguments. Then for each $\varepsilon > 0$*

$$\sup_{\{\theta,\theta' \in T' : \|\theta - \theta'\| < \delta\}} P_\theta\left(|f_\theta(x) - f_{\theta'}(x)| > \varepsilon\right) \to 0, \quad \delta \to 0. \tag{4.76}$$

Proof. Given $\varepsilon > 0$ take a compact C in X such that

$$\sup_{\theta \in T'} P_\theta\left(X \setminus C\right) \leq \varepsilon. \tag{4.77}$$

The functions $f_\theta(x)$ are uniformly continuous on $T' \times C$. Thus, there is a number $\delta > 0$ such that $|f_\theta(x) - f_{\theta'}(x)| < \varepsilon$ for $\|\theta - \theta'\| < \delta$ and (4.76) follows from (4.77). □

Example 4.3 Let $\mathbb{E}^n \overset{(\lambda,\varphi)}{\longrightarrow} \boldsymbol{\mathcal{E}}$ where all the experiments $\mathcal{E}(\theta)$ from the family $\boldsymbol{\mathcal{E}}$ coincide with the standard Gaussian shift experiment

$$\mathcal{E}(\theta) = \mathcal{E}_G = \left(R^d, \mathcal{B}\left(R^d\right); P_\alpha, \alpha \in R^d\right)$$

where P_α is the Gaussian measure with mean α and unit covariance matrix. Obviously

$$P_\alpha = P_0 \circ G_\alpha^{-1}, \quad \alpha \in R^d,$$

where G_α is the shift operator in the space R^d on the vector -α, $G_\alpha x = x - \alpha$, i.e.

$$P_\alpha(A) = P_0(A - \alpha); \quad A \in \mathcal{B}\left(R^d\right).$$

The likelihood for $\mathcal{E}_G$ is of the form

$$Z_\alpha(x) = \exp\left\{(\alpha, x) - |\alpha|^2/2\right\}$$

and, clearly, it is continuous in the set of arguments, hence there is no problem to check the condition (Z).

Let now the LAN condition (i.e. the condition of local λ-convergence to the Gaussian shift) be fulfilled uniformly, with the corresponding random elements $\lambda^n_{\theta_0}$ and localizing sequences $\varphi(n, \theta_0)$. Due to Theorem 4.8 one can claim that the following asymptotic expansion holds uniformly in $\alpha \in R^d$ and $\theta_0 \in \Theta$

$$\lambda^n_{\theta_0 + \varphi(n,\theta_0)\alpha} - G_\alpha \lambda^n_{\theta_0} \to 0 \tag{4.78}$$

under the measure $\mathbf{P}^n_{\theta_0}$, or

$$\lambda^n_{\theta_0+\varphi(n,\theta_0)\alpha} - \lambda^n_{\theta_0} + \alpha \to 0.$$

Example 4.4 Let an experiment $\mathbb{E}^n$ describe the case of n i.i.d. observations with a common density $f(x,\theta)$ with respect to some measure μ where $\theta \in \Theta$ and Θ is an open subset in R^d. Let the function $f(x,\theta)$ be absolutely continuous in θ, for each $\theta \in \Theta$

$$\int |\nabla f(x,\theta)|^2 f^{-1}(x,\theta)\, d\mu < \infty$$

and the matrix function I_θ of the form

$$I_\theta = \int \nabla f(x,\theta) \nabla^\top f(x,\theta)\, f^{-1}(x,\theta)\, d\mu$$

is continuous in θ (here $\nabla f(x,\theta) = \partial f(x,\theta)/\partial\theta$). Then, under the choice $\varphi(n,\theta) = n^{-1/2}$ the condition of local λ-convergence to the family $\mathcal{E} = (\mathcal{E}(\theta_0),\ \theta_0 \in \Theta)$ is satisfied uniformly on compacts for the experiments $\mathbb{E}^n$, where each experiment $\mathcal{E}(\theta_0)$ is a Gaussian shift with the information matrix I_{θ_0}, i.e.

$$\mathcal{E}(\theta_0) = \left(R^d, \mathcal{B}\left(R^d\right);\ P_{\alpha,\theta_0},\alpha \in R^d\right)$$

with the standard Gaussian measure P_{0,θ_0} and

$$Z(\theta_0,x) = \frac{dP_{\alpha,\theta_0}}{dP_{0,\theta_0}}(x) = \exp\left\{(x,\Gamma_{\theta_0}\alpha) - |\Gamma_{\theta_0}\alpha|^2/2\right\},$$

where $\Gamma_{\theta_0} = I^{1/2}_{\theta_0}$.

Evidently, the experiment $\mathcal{E}(\theta_0)$ can be transformed by the reparametrization $\alpha \to \Gamma_{\theta_0}\alpha$ into the standard Gaussian shift and the choice of localizing sequences in the form $\varphi(n,\theta_0) = n^{-1/2}\Gamma^{-1}_{\theta_0}$ reduces the considered situation to the case of Example 4.3. We however, consider this case separately.

The experiment $\mathcal{E}(\theta_0)$ is transitive with respect to the operators G_{α,θ_0} of the form

$$G_{\alpha,\theta_0}x = x - \Gamma_{\theta_0}\alpha$$

and the corresponding connectedness equation is of the form

$$\lambda^n_{\theta_0+\varphi(n,\theta_0)\alpha} - \lambda^n_{\theta_0} + \Gamma_{\theta_0}\alpha \to 0.$$

Example 4.5 Let $\mathbb{E}^n \overset{(\lambda,\varphi)}{\longrightarrow} \mathcal{E} = (\mathcal{E}(\theta_0),\ \theta_0 \in \Theta)$ and the experiments $\mathcal{E}(\theta_0)$ be mixed normal,

$$\mathcal{E}(\theta_0) = \left(R^d \times \mathcal{M}^d, \mathcal{B}\left(R^d \times \mathcal{M}^d\right);\ P_{\alpha,\theta_0},\ \alpha \in R^d\right),$$

where $\mathcal{M}^d$ is the set of positive symmetric $d \times d$-matrix and the measures P_{α,θ_0} are defined by the following conditions:

(i) $P_{0,\theta_0}(d\gamma, d\Gamma) = P_0'(d\gamma)P_0''(d\Gamma)$, $\gamma \in R^d$, $\Gamma \in \mathcal{M}^d$, P_0' being the standard Gaussian measure in R^d, and P_0'' is some probability distribution on $\mathcal{M}^d$;

(ii) $P_{\alpha,\theta_0}(d\gamma, d\Gamma) = \exp\left\{(x, \Gamma\alpha) - |\Gamma\alpha|^2/2\right\} P_{0,\theta_0}(d\gamma, d\Gamma)$.

One can directly check that each experiment $\mathcal{E}(\theta_0)$ is transitive with respect to the operators G_α (not depending on α) of the form

$$G_\alpha(\gamma, \Gamma) = \gamma - \Gamma\alpha,$$

i.e. $P_{\alpha,\theta_0} = P_{0,\theta_0} \circ G_\alpha^{-1}$, $\alpha \in R^d$.

We represent the random elements $\lambda_{\theta_0}^n$ entering in the definition of local λ-convergence $\mathbb{E}^n \xrightarrow{(\lambda,\varphi)} \mathcal{E}$ in the form

$$\lambda_{\theta_0}^n = \left(\gamma_{\theta_0}^n, \Gamma_{\theta_0}^n\right).$$

Now the connectedness equation (4.78) can be rewritten as follows

$$\Gamma_{\theta_0+\varphi(n,\theta_0)\alpha}^n - \Gamma_{\theta_0}^n \rightarrow 0, \tag{4.79}$$

$$\lambda_{\theta_0+\varphi(n,\theta_0)\alpha}^n - \lambda_{\theta_0}^n + \Gamma_{\theta_0}^n\alpha \rightarrow 0, \tag{4.80}$$

the convergence being under the measure $\mathbf{P}_{\theta_0}^n$.

For the case when the distribution of the random matrix Γ in the experiment $\mathcal{E}(\theta_0)$ is degenerated (i.e. the measure P_{0,θ_0}'' is concentrated at some point Γ_{θ_0}) we arrive to the situation of Example 2 and the condition (4.79) means just continuity of Γ_{θ_0} in $\theta_0 \in \Theta$.

The next example includes all the previous ones as particular cases. It is related to the situation when limit experiments are $(\gamma, \Gamma; \varphi)$-models (see Chapter 3).

Example 4.6 Let $\mathbb{E}^n \xrightarrow{(\lambda,\varphi)} \mathcal{E} = (\mathcal{E}(\theta_0), \theta_0 \in \Theta)$ and

$$\mathcal{E}(\theta_0) = \left(R^d \times \mathcal{M}^d, \mathcal{B}\left(R^d \times \mathcal{M}^d\right); P_{\alpha,\theta_0}, \alpha \in R^d\right),$$

$$\frac{dP_{\alpha,\theta_0}}{dP_{0,\theta_0}}(\gamma, \Gamma) = \exp\left\{(x, \Gamma\alpha) - |\Gamma\alpha|^2/2\right\} = \frac{\varphi(\gamma - \Gamma\alpha)}{\varphi(\gamma)} = Z_\alpha(\gamma, \Gamma),$$

$\varphi(\cdot)$ being the standard normal density. In the terminology of Chapter 3 the experiments $\mathcal{E}(\theta_0)$ are $(\gamma, \Gamma; \varphi)$-models.

Let the operators G_α be determined in the same way as in Example 3, i.e.

$$G_\alpha(\gamma, \Gamma) = \gamma - \Gamma\alpha, \quad \alpha \in R^d, (\gamma, \Gamma) \in R^d \times \mathcal{M}^d.$$

Then the experiment $\mathcal{E}(\theta_0)$ is weakly transitive with respect to these operators since for any $\beta \in R^d$

$$\begin{aligned} T_\beta Z_\alpha(\gamma, \Gamma) &= \frac{Z_{\alpha+\beta}}{Z_\beta}(\gamma, \Gamma) = \frac{\varphi(\gamma - \Gamma(\alpha+\beta))}{\varphi(\gamma)} \frac{\varphi(\gamma)}{\varphi(\gamma - \Gamma\beta)} \\ &= \frac{\varphi(\gamma - \Gamma\beta - \Gamma\alpha)}{\varphi(\gamma - \Gamma\beta)} = Z_\alpha\left(G_\beta(\gamma, \Gamma)\right), \quad \alpha \in R^d, (\gamma, \Gamma) \in R^d \times \mathcal{M}^d. \end{aligned}$$

If now $\lambda^n_{\theta_0} = \left(\gamma^n_{\theta_0}, \Gamma^n_{\theta_0}\right)$ are the random elements entering in the definition of local λ-convergence, then Theorem 4.8 and its corollaries provide for these elements the same connectedness equations (4.79), (4.80) as in Example 3.

4.7 Global Estimation under Local λ-Convergence to Transitive and Weakly Transitive Experiments

The main result of this section repeats mainly the statement of Theorem 4.7 concerning the properties of the global estimators $\widehat{T}^n$, $n \geq 1$. However, in the case of convergence to transitive or weakly transitive experiments one can consider the estimators $\varkappa^*_\theta$ entering in the definition of $\widehat{T}^n$ as functions of the observation $x \in X$, using the connectedness equation of type (4.70).

Let $\varkappa^*_\theta$ be an estimator for the experiments $\mathcal{E}(\theta)$, $\theta \in \Theta$. For the sake of simplicity we suppose these estimators to be non-randomized, but all the considerations remain valid for the general randomized case.

The main subject of investigation are the global estimators

$$\widehat{T}^n = \rho^n + \varphi(n, \rho^n)\varkappa^*_{\rho^n}\left(\widehat{\lambda}^n_{\rho^n}\right), \quad n \geq 1$$

where ρ^n are some preliminary estimators obeying the properties of $\varphi(n,\theta)$-consistency. We shall prove the following assertion

$$\lim_n \sup_{\theta_0 \in T_0} \mathbf{P}^n_{\theta_0}\left(\left\|\varphi^{-1}(n,\theta_0)\left(\widehat{T}^n - \theta_0\right) - \varkappa^*_{\theta_0}\left(\widehat{\lambda}^n_{\theta_0}\right)\right\| > \varepsilon\right) = 0, \quad \forall \varepsilon > 0. \quad (4.81)$$

Theorem 4.9 *Let the following conditions hold:*

(i) *Conditions* (K), (T_0), (Φ');
(ii) *The condition of* $(\mathcal{K}, T)$*-uniform local* λ*-convergence to the experiments* $\mathcal{E} = (\mathcal{E}(\theta_0),\ \theta_0 \in \Theta)$ *which are weakly transitive and satisfy the conditions* (G), (Z) *and (i)-(iii) from Remark 4.6 of Section 4.4,;*
(iii) *The preliminary estimators* ρ^n *satisfy the conditions of consistency and discreteness;*
(iv) *The estimators* $\varkappa^*_\theta(x)$ *are continuous in both variables* $\theta \in \Theta$ *and* $x \in X$,

$$\sup_{\theta \in T} P_{0,\theta}\left(\|\varkappa^*_\theta\| > h\right) \to 0, \quad h \to \infty,$$

and for each $\theta \in \Theta$ *and* $\beta \in \mathcal{A}$ *the equivariance property holds*

$$\varkappa^*_\theta\left(G_\beta(\theta)x\right) = \varkappa^*_\theta(x) - \beta, \quad x \in X.$$

Then (4.81) is fulfilled.

Proof. Let $\beta^n = \beta^n(\rho^n, \theta_0) = \varphi^{-1}(n, \theta_0)\,(\rho^n - \theta_0)$. Then

$$\begin{aligned}\varphi^{-1}(n,\theta_0)\left(\widehat{T}^n - \theta_0\right) &= \varphi^{-1}(n,\theta_0)\,(\rho^n - \theta_0) + \varphi^{-1}(n,\theta_0)\varphi(n,\rho^n)\varkappa^*_{\rho^n}\left(\widehat{\lambda}^n_{\rho^n}\right) = \\ &= \beta^n + [1 + a^n]\,\varkappa^*_{\rho^n}\left(\widehat{\lambda}^n_{\rho^n}\right)\end{aligned}$$

with $a^n = a^n(\theta, \theta_0) = \varphi^{-1}(n, \theta_0)\varphi(n, \rho^n) - 1$.

By Lemma 4.2, in order to establish (4.81), it suffices to show that for any $\varepsilon > 0$ and $K \in \mathcal{K}$ as $n \to \infty$

$$\sup_{\theta_0 \in T_0}\ \sup_{\theta \in U^n(\theta_0, K)} \mathbf{P}^n_{\theta_0}\left(\left\|\beta^n + (1 + a^n)\varkappa^*_\theta\left(\widehat{\lambda}^n_\theta\right) - \varkappa^*_{\theta_0}\left(\widehat{\lambda}^n_{\theta_0}\right)\right\| > \varepsilon\right) \to 0. \tag{4.82}$$

For this we show first that as $n \to \infty$

$$\sup_{\theta_0 \in T_0}\ \sup_{\theta \in U^n(\theta_0, K)} \mathbf{P}^n_{\theta_0}\left(\left\|\varkappa^*_\theta\left(\widehat{\lambda}^n_\theta\right) - \varkappa^*_\theta\left(G_{\beta^n}(\theta_0)\widehat{\lambda}^n_{\theta_0}\right)\right\| > \varepsilon\right) \to 0. \tag{4.83}$$

It is easy to see that the following two relations are sufficient for (4.83):

$$\sup_{\theta_0 \in T_0}\ \sup_{\theta \in U^n(\theta_0, K)} \mathbf{P}^n_{\theta_0}\left(\left\|\varkappa^*_\theta\left(\widehat{\lambda}^n_\theta\right) - \varkappa^*_{\theta_0}\left(\widehat{\lambda}^n_\theta\right)\right\| > \varepsilon\right) \to 0, \tag{4.84}$$

$$\sup_{\theta_0 \in T_0}\ \sup_{\theta \in U^n(\theta_0, K)} \mathbf{P}^n_{\theta_0}\left(\left\|\varkappa^*_{\theta_0}\left(\widehat{\lambda}^n_{\theta_0}\right) - \varkappa^*_{\theta_0}\left(G_{\beta^n}(\theta_0)\widehat{\lambda}^n_{\theta_0}\right)\right\| > \varepsilon\right) \to 0. \tag{4.85}$$

By Corollary 4.2 we can replace in (4.84) $\mathbf{P}^n_{\theta_0}$ by $\mathbf{P}^n_\theta$ and and the assertion follows from Lemma 4.3, the conditions (iii) of the theorem, (Φ') and the equality $\mathcal{L}\left(\widehat{\lambda}^n(\theta) \mid \mathbf{P}^n_\theta\right) = P_{0,\theta}$.

Similarly, one can check (4.85) using also the connectedness equation from Theorem 4.8 and the continuity of the estimators $\varkappa^*_\theta(x)$.

Now we prove (4.82). The assertion (4.83) and the equivariance property of the estimators $\varkappa^*_\theta(x)$ yield (as $n \to \infty$)

$$\sup_{\theta_0 \in T_0}\ \sup_{\theta \in U^n(\theta_0, K)} \mathbf{P}^n_{\theta_0}\left(\left\|\varkappa^*_\theta\left(\widehat{\lambda}^n_\theta\right) - \varkappa^*_{\theta_0}\left(\widehat{\lambda}^n_{\theta_0}\right) + \beta^n\right\| > \varepsilon/2\right) \to 0. \tag{4.86}$$

Moreover, (Φ') and condition (iii) of the theorem imply (similarly to the end of the proof of Theorem 4.7):

$$\sup_{\theta_0 \in T_0}\ \sup_{\theta \in U^n(\theta_0, K)} \mathbf{P}^n_{\theta_0}\left(\left\||a^n|\varkappa^*_\theta\left(\widehat{\lambda}^n_\theta\right)\right\| > \varepsilon/2\right) \to 0. \tag{4.87}$$

Now (4.86) and (4.87) imply (4.82) and hence, the required assertion (4.81). The theorem is proved. □

In the construction of the global estimators $\widehat{T}^n$ we use the random elements $\widehat{\lambda}^n_\theta$ which are obtained from λ^n_θ by the Lemma about "Reconstruction". The natural question arising here is of whether one can use in the construction of global estimators $\widehat{T}^n$ the original random elements λ^n_θ instead of $\widehat{\lambda}^n_\theta$. Generally speaking, the answer to this question is negative. But under the situation considered in the present section, when the functions $Z_\alpha(\theta, x)$, $G_\beta(\theta)x$ and the estimators $\varkappa_\theta^*(x)$ are continuous in the set of arguments, this exchange is possible, as follows from the uniform closeness of λ_θ^n and $\widehat{\lambda}_\theta^n$ (the statement (i) of Theorem 4.6) and Lemma 4.3.

Therefore, under the conditions of Theorem 4.9 for the estimators $\widetilde{T}^n$ with

$$\widetilde{T}^n = \rho^n + \varphi(n, \rho^n)\varkappa_{\rho^n}^*\left(\lambda_{\rho^n}^n\right), \quad n \geq 1$$

the following assertion holds:

$$\lim_n \sup_{\theta_0 \in T_0} \mathbf{P}_{\theta_0}^n \left(\left\| \varphi^{-1}(n, \theta_0)\left(\widetilde{T}^n - \theta_0\right) - \varkappa_{\theta_0}^*\left(\lambda_{\theta_0}^n\right) \right\| > \varepsilon \right) = 0, \quad \forall \varepsilon > 0.$$

This implies, in particular, that the asymptotic distribution of $\varphi^{-1}(n, \theta_0)\left(\widetilde{T}^n - \theta_0\right)$ converges, uniformly in $\theta_0 \in T_0$, to the distribution of the estimator $\varkappa_{\theta_0}^*$ in the limit experiment $\mathcal{E}(\theta_0)$.

Example 4.7 Let the condition LAN hold, i.e. experiments $\mathbb{E}^n$ locally λ-convergence (uniformly on compacts) to the Gaussian shift experiment $\mathcal{E}_G$ corresponding to the observation γ of the form

$$\gamma = \alpha + \varepsilon,$$

where $\alpha \in R^d$, $\varepsilon \sim \mathcal{N}(0, \mathbf{1}_d)$, and $\mathbf{1}_d$ is the unit $d \times d$-matrix. The minimax estimator for the experiment $\mathcal{E}_G$ coincides with the observation γ (see Section 3.3 of Chapter 3),

$$\varkappa^*(\gamma) = \gamma,$$

and this estimator is obviously equivariant,

$$\varkappa^*(G_\alpha \gamma) = \varkappa^*(\gamma - \alpha) = \varkappa^*(\gamma) - \alpha.$$

The corresponding global estimator T^n is of the form

$$T^n = \rho^n + \varphi(n, \rho^n)\varkappa^*\left(\lambda_{\rho^n}^n\right) = \rho^n + \varphi(n, \rho^n)\lambda_{\rho^n}^n$$

and by Theorem 4.9

$$T^n - \lambda_{\theta_0}^n \to 0$$

under the measure $\mathbf{P}_{\theta_0}^n$ and this convergence is uniform in $\theta_0 \in \Theta$.

Example 4.8 Let $\mathbb{E}^n \xrightarrow{(\lambda,\varphi)} \mathcal{E} = (\mathcal{E}(\theta_0), \theta_0 \in \Theta)$ and the experiments $\mathcal{E}(\theta_0)$ be the Gaussian shift experiments with information matrices I_{θ_0} (see Example 4.4 in Section 4.6). The minimax estimator for $\mathcal{E}(\theta_0)$ is (see Section 3.3 of Chapter 3)

$$\varkappa_{\theta_0}^*(\gamma) = I_{\theta_0}^{-1/2}\gamma = \Gamma_{\theta_0}^{-1}\gamma$$

and

$$\varkappa^*_{\theta_0}(G_\alpha(\theta_0)\gamma) = \varkappa^*_{\theta_0}(\gamma - \Gamma_{\theta_0}\alpha) = \Gamma^{-1}_{\theta_0}(\gamma - \Gamma_{\theta_0}\alpha) = \varkappa^*_{\theta_0}(\gamma) - \alpha,$$

i.e. this estimator is equivariant.

For the corresponding global estimator T^n we obtain

$$T^n = \rho^n + \varphi(n, \rho^n)\varkappa^*_{\rho^n}(\lambda^n_{\rho^n}) = \rho^n + \varphi(n, \rho^n)\Gamma^{-1}_{\rho^n}\lambda^n_{\rho^n}$$

and by Theorem 4.9

$$T^n - \Gamma^{-1}_{\theta_0}\lambda^n_{\theta_0} \xrightarrow{\mathbf{P}^n_{\theta_0}} 0.$$

Example 4.9 Let $\mathbb{E}^n \xrightarrow{(\lambda,\varphi)} \mathcal{E} = (\mathcal{E}(\theta_0), \theta_0 \in \Theta)$ and the experiments $\mathcal{E}(\theta_0)$ be mixed normal (see Example 4.5 from Section 4.6). Due to Section 3.3 of Chapter 3 the minimax estimator for $\mathcal{E}(\theta)$ is

$$\varkappa^*(\gamma, \Gamma) = \Gamma^{-1}\gamma.$$

This estimator is equivariant since

$$\varkappa^*(G_\alpha(\gamma, \Gamma)) = \varkappa^*(\gamma - \Gamma\alpha, \Gamma) = \Gamma^{-1}(\gamma - \Gamma\alpha) = \varkappa^*(\gamma, \Gamma) - \alpha.$$

If the elements $\lambda^n_{\theta_0}$ determining local λ-convergence are written in the form $\lambda^n_{\theta_0} = (\gamma^n_{\theta_0}, \Gamma^n_{\theta_0})$, then the estimators T^n read as follows:

$$T^n = \rho^n + \varphi(n, \rho^n)\varkappa^*(\gamma^n_{\rho^n}, \Gamma^n_{\rho^n}) = \rho^n + \varphi(n, \rho^n)(\Gamma^n_{\rho^n})^{-1}\lambda^n_{\rho^n}. \tag{4.88}$$

By Theorem 4.9

$$T^n - (\Gamma^n_{\theta_0})^{-1}\lambda^n_{\theta_0} \xrightarrow{\mathbf{P}^n_{\theta_0}} 0. \tag{4.89}$$

Example 4.10 Let $\mathbb{E}^n \xrightarrow{(\lambda,\varphi)} \mathcal{E} = (\mathcal{E}(\theta_0), \theta_0 \in \Theta)$ and the experiments $\mathcal{E}(\theta_0)$ be $(\gamma, \Gamma; \varphi)$-models (see Example 4.6 from Section 4.6). The structure of minimax estimators for experiments $\mathcal{E}(\theta_0)$ is in general unknown, but one can consider a "good" estimators like the generalized Bayes one: $\varkappa^*(\gamma, \Gamma) = \Gamma^{-1}\gamma$. Similarly to the case of Gaussian and mixed Gaussian models this estimator is equivariant and, as in the previous example, the representation (4.88) and the asymptotic expansion (4.89) remain valid.

In the contrary to the preceding examples for which the estimators T^n are asymptotic minimax, in the latest case we are not able to claim the same thing. However, one can show using the result of Theorem 3.1, Section 3.3 of Chapter 3 that the estimators T^n are asymptotically Bayessian for any uniformly continuous prior density $p(\theta)$ (for more details see, for example, Ibragimov and Khasminskii, 1981).

Chapter 5

Statistical Inference for Autoregressive Models of the First Order

5.1 Parameter Estimation

We say that a random sequence $X = (X_0, X_1, \dots)$ obeys the first-order autoregressive equation if

$$X_i = \theta X_{i-1} + \varepsilon_i, \quad i \geq 1, \tag{5.1}$$

where θ is (usually) an unknown parameter, X_0 is a random variable and $\varepsilon = (\varepsilon_i\,,\, i \geq 1)$ is a sequence of random variables with a known distribution which can be treated as "a random noise".

This model is described for each $n \geq 1$ by the statistical experiment

$$\mathbb{E}^n = (\Omega^n, \mathcal{F}^n; \mathbf{P}_\theta^n, \theta \in \Theta)$$

where $\Omega^n = R^{n+1} = \{x : x = (x_0, x_1, \dots, x_n),\, x_i \in R\}$, $\mathcal{F}^n = \mathcal{B}(R^{n+1})$, $\mathbf{P}_\theta^n$ is the probability measure on $(\Omega^n, \mathcal{F}^n)$ describing the distribution of X_0 and $\varepsilon_1, \dots \varepsilon_n$. The unknown autoregressive parameter θ is assumed to belong to some subset Θ of the real line.

To simplify our notation, we shall omit in what follows in this section the super-index n in the notation for $\mathbf{P}_\theta^n$ and for the corresponding expectation $\mathbf{E}_\theta^n$.

Below we study the problem of "effective" (in one or another sense) estimation of the unknown parameter θ. The model (5.1) provides a useful example which illustrates general results and methods of the preceding chapters and shows new effects arising from the *dependence* of observations.

To begin with, we recall some classical results which are based on the direct probabilistic analysis of the model (5.1). First, we study some *general properties* of the random sequence $X = (X_0, X_1, \dots)$ obeying the first order autoregressive equation.

It follows from (5.1) that

$$X_n = \varepsilon_n + \theta\varepsilon_{n-1} + \dots + \theta^{n-1}\varepsilon_1 + \theta^n X_0. \tag{5.2}$$

It is obvious that probabilistic properties of the sequence X depend essentially on the joint distribution of $X_0, \varepsilon_1, \varepsilon_2, \ldots$ In what follows we assume that these variables are independent zero mean Gaussian random variables, i.e. $\varepsilon_i \sim \mathcal{N}(0, \sigma^2)$ with known $\sigma^2 > 0$. However, some results remain valid in a much more general situation with independent random errors ε_i having a finite second moment.

By (5.2) we obtain

$$\mathbf{D}_\theta X_n = \sigma^2 \left(1 + \theta^2 + \ldots + \theta^{2(n-1)}\right) + \theta^{2n} \mathbf{D} X_0 \tag{5.3}$$

and

$$\mathbf{E}_\theta X_n X_{n-k} = \sigma^2 \theta^k \left(1 + \theta^2 + \theta^{2(n-k-1)}\right) + \theta^k \theta^{2(n-k)} \mathbf{D}_\theta X_0.$$

Therefore, if $|\theta| < 1$ and $X_0 \sim \mathcal{N}\left(0, \frac{\sigma^2}{1-\theta^2}\right)$, then

$$\mathbf{E}_\theta X_n = 0, \quad \mathbf{D}_\theta X_n = \frac{\sigma^2}{1-\theta^2}, \quad \mathbf{E}_\theta X_n X_{n-k} = \frac{\sigma^2 \theta^k}{1-\theta^2}$$

and $X = (X_0, X_1, \ldots)$ is a stationary process.

If $|\theta| \geq 1$, then the sequence $X = (X_0, X_1, \ldots)$ is *explosive* in the sense that $\mathbf{D}_\theta X_n \to \infty$ as $n \to \infty$. In the case of $\theta = 1$ one has

$$X_n = X_0 + [\varepsilon_n + \ldots + \varepsilon_1]$$

and in the case of $\theta = -1$

$$X_n = (-1)^n X_0 + [\varepsilon_n - \varepsilon_{n-1} + \ldots \pm \varepsilon_1].$$

Hence, for these two cases ($\theta = \pm 1$) the sequence $X = (X_0, X_1, \ldots)$ is a realization of a *random walk*.

The above simple consideration shows that the features of the random sequence $X = (X_0, X_1, \ldots)$ are rather different for different values of the parameter θ. And one can expect that probabilistic properties of estimators of the parameter θ are different for each of the three cases: $|\theta| < 1$, $|\theta| = 1$ and $|\theta| > 1$.

Now we turn to the estimation problem. For simplicity, we assume that $X_0 = 0$ and $\sigma^2 = 1$.

A natural estimator for θ by observations $X_1, \ldots, X_n$ is the maximum likelihood estimator

$$\widehat{\theta}_n = \arg\max_\theta p_\theta(X_1, \ldots, X_n).$$

Here $p_\theta(X_1, \ldots, X_n)$ is the joint density of the variables $X_1, \ldots, X_n$ obeying (5.1). Obviously

$$p_\theta(X_1, \ldots, X_n) = (2\pi)^{-n/2} \exp\left\{-\frac{1}{2}\sum_{i=1}^n (X_i - \theta X_{i-1})^2\right\}, \tag{5.4}$$

and one obtains by standard calculation

$$\widehat{\theta}_n = \frac{\sum_{i=1}^n X_{i-1}X_i}{\sum_{i=1}^n X_{i-1}^2}. \tag{5.5}$$

Using (5.1) one gets

$$\widehat{\theta}_n = \theta + \frac{\sum_{i=1}^n X_{i-1}\varepsilon_i}{\sum_{i=1}^n X_{i-1}^2} \quad \mathbf{P}_\theta\text{-a.s.} \tag{5.6}$$

Denote $M_n = \sum_{i=1}^n X_{i-1}\varepsilon_i$. Then, for each value of the parameter θ, the sequence $M = (M_n)_{n\geq 1}$ is a martingale (with respect to the filtration $(\mathcal{F}_n)_{n\geq 1}$ with $\mathcal{F}_n = \sigma(X_1, \dots, X_n)$ under the measure $\mathbf{P}_\theta$) with the *quadratic* characteristic $\langle M\rangle = (\langle M\rangle_n)_{n\geq 1}$,

$$\langle M\rangle_n = \sum_{i=1}^n X_{i-1}^2.$$

(The quadratic characteristic $\langle M\rangle$ can be defined, see Shiryaev (1980, Chapter 7), as a *predictable* non-decreasing random process such that $M^2 - \langle M\rangle = \left(M_n^2 - \langle M\rangle_n\right)_{n\geq 1}$ is a martingale.)

Therefore, from (5.6)

$$\widehat{\theta}_n - \theta = \frac{M_n}{\langle M\rangle_n}.$$

Notice that in view of (5.4) the quantity $\mathbf{E}_\theta \sum_{i=1}^n X_{i-1}^2$ is exactly the *Fisher information* $I_n(\theta)$ about the parameter θ contained in $X_1, \dots, X_n$:

$$I_n(\theta) = \mathbf{E}_\theta\left[-\frac{\partial^2 \log p_\theta(x_1, \dots, x_n)}{\partial\theta^2}\right].$$

In other words,

$$I_n(\theta) = \mathbf{E}_\theta\langle M\rangle_n.$$

(This equality explains why $\langle M\rangle$ is often called "the stochastic Fisher information".) Direct calculations based on (5.3) show that

$$I_n(\theta) = \begin{cases} \frac{1}{1-\theta^2}\left(n - \frac{1-\theta^{2n}}{1-\theta^2}\right), & |\theta| \neq 1, \\ \frac{1}{2}n(n-1), & |\theta| = 1. \end{cases} \tag{5.7}$$

Therefore, as $n \to \infty$,

$$I_n(\theta) \sim d_n^2(\theta) = \begin{cases} \frac{n}{1-\theta^2}, & |\theta| < 1, \\ \frac{n^2}{2}, & |\theta| = 1, \\ \frac{\theta^{2n}}{(\theta^2-1)^2} & |\theta| > 1. \end{cases} \tag{5.8}$$

Since $\langle M\rangle_n \to \infty$ ($\mathbf{P}_\theta$-a.s.) for every $-\infty < \theta < \infty$, we obtain by the law of large numbers for square integrable martingales (see, e.g. Shiryaev (1980, Chapter 7))

$$\frac{M_n}{\langle M\rangle_n} \to 0, \qquad (\mathbf{P}_\theta\text{-a.s.}), \quad -\infty < \theta < \infty$$

This implies that the maximum likelihood estimators $\widehat{\theta} = (\widehat{\theta}_n)_{n\geq 1}$ are *strongly consistent*:

$$\mathbf{P}_\theta\left(\lim \widehat{\theta}_n = \theta\right) = 1, \qquad -\infty < \theta < \infty.$$

Similarly to the case of independent observations in the classical estimation theory, one may expect the deviations $\widehat{\theta}_n - \theta$, normalized by their variances $\sqrt{I_n(\theta)}$, to have a non-degenerated limit distribution. By (5.8) the limit distribution for $\sqrt{I_n(\theta)}\left(\widehat{\theta}_n - \theta\right)$ coincides with the limit distribution for $D_n(\theta) = d_n(\theta)(\widehat{\theta}_n - \theta)$,

$$D_n(\theta) = \begin{cases} \sqrt{\frac{n}{1-\theta^2}}\left(\widehat{\theta}_n - \theta\right), & |\theta| < 1, \\ \frac{n}{\sqrt{2}}\left(\widehat{\theta}_n - \theta\right), & |\theta| = 1, \\ \frac{\theta^n}{(\theta^2-1)}\left(\widehat{\theta}_n - \theta\right) & |\theta| > 1. \end{cases}$$

The latter variables can be represented also in the form

$$D_n(\theta) = d_n(\theta)\left(\widehat{\theta}_n - \theta\right) = \frac{M_n/d_n(\theta)}{\langle M\rangle_n/d_n^2(\theta)} \equiv \frac{U_n}{V_n}.$$

In order to investigate the corresponding limit behavior, one may use the classical method based on the direct analysis of the asymptotic properties of the joint characteristic functions $f_n(s,t)$ of the variables U_n, V_n which are quadratic forms of the Gaussian variables $X_1, \dots, X_n$,

$$f_n(s,t) = \mathbf{E}_0 \exp\left\{isU_n + itV_n\right\}.$$

It was shown in Rao (1978) that the functions f_n converge to a function $f(s,t) = \lim_{n\to\infty} f_n(s,t)$ which is of the following form

$$f(s,t) = \begin{cases} e^{it-s^2/2}, & |\theta| < 1, \\ \exp\left\{-2^{-1/2}\theta is\right\}\left[\cos 2\sqrt{it} - \frac{\theta it}{\sqrt{2is}}\sin 2\sqrt{is}\right]^{-1/2}, & |\theta| = 1, \\ \left(1+s^2-2it\right)^{-1/2} & |\theta| > 1. \end{cases} \tag{5.9}$$

Since

$$\mathbf{P}_\theta\left(D_n(\theta) \leq z\right) = \mathbf{P}_\theta\left(U_n - zV_n \leq 0\right), \qquad -\infty < z < \infty,$$

by (5.9)

$$\lim_n \mathbf{E}_\theta\, e^{i\lambda(U_n - zV_n)}$$
$$= \begin{cases} e^{i\lambda z - \lambda^2/2}, & |\theta| < 1, \\ \exp\left\{-2^{-1/2}\theta i\lambda\right\}\left[\cos 2\sqrt{-i\lambda z} - \frac{\theta(-i\lambda z)}{\sqrt{2i\lambda}} \sin 2\sqrt{i\lambda}\right]^{-1/2}, & |\theta| = 1, \\ \left(1 + s^2 - 2it\right)^{-1/2} & |\theta| > 1. \end{cases} \quad (5.10)$$

For the case of $|\theta| < 1$, this means that the random variables $U_n - zV_n$ are asymptotically normal $\mathcal{N}(-z, 1)$. Therefore, for $|\theta| < 1$,

$$\lim_n \mathbf{P}_\theta \left(D_n(\theta) \le z\right) = \lim_n \mathbf{P}_\theta \left(U_n - zV_n \le 0\right)$$
$$= \frac{1}{\sqrt{2\pi}} \int_{-\infty}^{0} e^{-\frac{(x+z)^2}{2}} dx = \frac{1}{\sqrt{2\pi}} \int_{-\infty}^{z} e^{-\frac{x^2}{2}} dx = \Phi(z)$$

or

$$\lim_n \mathbf{P}_\theta \left(\sqrt{\frac{n}{1-\theta^2}} \left(\widehat{\theta}_n - \theta\right) \le z\right) = \Phi(z).$$

(This result was proved for the case of $|\theta| < 1$ in Anderson(1959) and in Rao (1978) by the method explained above.)

Now we turn to the case of $|\theta| > 1$. We will show that

$$\lim_n \mathbf{P}_\theta \left(D_n(\theta) \le z\right) = \lim_n \mathbf{P}_\theta \left(\frac{\theta^n}{(\theta^2 - 1)} \left(\widehat{\theta}_n - \theta\right) \le z\right) = \mathrm{Ch}(z)$$

where $\mathrm{Ch}(z)$ is the Cauchy distribution with the density $\frac{1}{\pi(1+z^2)}$. The simplest way to check this statement is based on (5.10) and the following remark. Let ξ and η be two independent random variables on some probability space $(\Omega, \mathcal{F}, P)$ with the distribution χ^2 with degree of freedom equals one. This means that

$$Ee^{i\lambda\xi} = Ee^{i\lambda\eta} = (1 - 2i\lambda)^{-1/2}.$$

If $\zeta = a\xi - b\eta$, then

$$Ee^{i\lambda\zeta} = \left[(1 - 2i\lambda a)(1 + 2i\lambda b)\right]^{-1/2}.$$

In particular, if a and b are such that

$$2a = \sqrt{1 + z^2} - z, \qquad 2b = \sqrt{1 + z^2} + z, \tag{5.11}$$

then one has for the corresponding variable $\zeta = \zeta(z)$

$$Ee^{i\lambda\zeta(z)} = \left(1 + 2iz\lambda + \lambda^2\right)^{-1/2}.$$

Comparing with (5.10) we conclude that ($|\theta| > 1$)

$$\lim_n \mathbf{P}_\theta \left(D_n(\theta) \le z\right) = P\left(\zeta(z) \le 0\right) = P\left(a\xi - b\eta \le 0\right)$$

with a, b from (5.11). But

$$P(a\xi - b\eta \le 0) = P\left(\xi \le \frac{b}{a}\eta\right) = \frac{1}{2\pi}\int_0^\infty \left[\int_0^{\frac{b}{a}y} \frac{1}{\sqrt{xy}} e^{-\frac{x}{2}-\frac{y}{2}} dx\right] dy.$$

This implies by (5.11) that ($|\theta| > 1$)

$$\lim_n \mathbf{P}_\theta\left(D_n(\theta) \le z\right) = \frac{1}{\pi}\int_{-\infty}^{z} \frac{dx}{1+x^2} = \mathrm{Ch}(z).$$

The latter result can be obtained in another way: Notice that $f(s,t) = (1 + s^2 - 2it)^{-1/2}$ is the characteristic function of the pair $(\xi\eta, \eta^2)$ where ξ and η are independent standard Gaussian random variables. Hence

$$\mathcal{L}\left(U_n, V_n \mid \mathbf{P}_\theta\right) \to \mathcal{L}(\xi\eta, \eta^2), \quad |\theta| > 1,$$

and

$$\lim_n \mathbf{P}_\theta\left(D_n(\theta) \le z\right) = \lim_n \mathbf{P}_\theta\left(\frac{U_n}{V_n} \le z\right)$$
$$= P\left(\frac{\xi\eta}{\eta^2} \le z\right) = P\left(\frac{\xi}{\eta} \le z\right) = \mathrm{Ch}(z).$$

Finally we consider the boundary (or critical) case of $|\theta| = 1$. Let $H(\theta)$ be the limit distribution of $D_n(\theta)$ under $\mathbf{P}_\theta$

$$H_\theta(z) = \lim_n \mathbf{P}_\theta\left(D_n(\theta) \le z\right) = \lim_n \mathbf{P}_\theta\left(\frac{n}{\sqrt{2}}\left(\widehat{\theta}_n - \theta\right) \le z\right)$$

and let $h_\theta(z) = \frac{dH_\theta(z)}{dz}$ be the corresponding density. Starting from (5.9) one can derive the closed form expression for $h_\theta(z)$, see Dickey and Fuller (1979, 1981). Numerical simulations made in Evans and Savin (1981) show that

$$\begin{cases} h_1(-0.2) \approx 0.3158, & H_1(-0.2) \approx 0.617, \\ h_1(0) \approx 0.3413, & H_1(0) \approx 0.683, \\ h_1(0.2) \approx 0.3566, & H_1(0.2) \approx 0.753. \end{cases}$$

A rather unexpected fact is that for $\theta = \pm 1$ the density $h_\theta(z)$ is *non-symmetric* ($h_\theta(z) \ne h_\theta(-z)$, $z > 0$), whereas for $|\theta| \ne 1$ the corresponding densities are symmetric.

Actually the limit distribution $H_\theta(z)$ is exactly the distribution of the random variable

$$\theta \frac{w^2(1) - 1}{2^{3/2}\int_0^1 w^2(s)ds} \tag{5.12}$$

where $w = (w(s),\ s \geq 1)$ is a standard Wiener process. The simplest way to check this fact is as follows. Let, for example, $\theta = 1$. Then

$$M_n = \sum_{i=1}^{n} X_{i-1}\varepsilon_i = \sum_{i=1}^{n}\sum_{j=1}^{i-1} \varepsilon_j \varepsilon_i$$

and

$$\langle M \rangle_n = \sum_{i=1}^{n} \left(\sum_{j=1}^{i-1} \varepsilon_j \right)^2 .$$

Since

$$\mathcal{L}\left(\frac{\varepsilon_i}{\sqrt{n}}\right) = \mathcal{L}\left(w\left(\frac{i}{n}\right) - w\left(\frac{i-1}{n}\right)\right), \qquad i \leq n,$$

it holds

$$\frac{M_n}{n} = \sum_{i=1}^{n} \left(\sum_{j=1}^{i-1} \frac{\varepsilon_j}{\sqrt{n}} \right) \frac{\varepsilon_i}{\sqrt{n}} \stackrel{d}{=} \sum_{i=1}^{n} w\left(\frac{i-1}{n}\right) \Delta w\left(\frac{i}{n}\right)$$

and

$$\frac{\langle M \rangle_n}{n^2} = \sum_{i=1}^{n} \frac{1}{n} \left(\sum_{j=1}^{i-1} \varepsilon_j \right)^2 \stackrel{d}{=} \sum_{i=1}^{n} w^2\left(\frac{i-1}{n}\right) \frac{1}{n}$$

where $\Delta w\left(\frac{i}{n}\right) = w\left(\frac{i}{n}\right) - w\left(\frac{i-1}{n}\right)$ and "$\stackrel{d}{=}$" means the equality in distribution.

General results on weak convergence of random processes (see, e.g., Jacod and Shiryaev (1987), Liptser and Shiryaev (1988)) provide that

$$\mathcal{L}\left(\sum_{i=1}^{n} w\left(\frac{i-1}{n}\right) \Delta w\left(\frac{i}{n}\right), \sum_{i=1}^{n} w^2\left(\frac{i-1}{n}\right) \frac{1}{n} \right)$$
$$\to \mathcal{L}\left(\int_0^1 w(s)dw(s), \int_0^1 w^2(s)ds \right)$$

where $\int_0^1 w(s)dw(s)$ is the stochastic Ito integral equal to $\frac{1}{2}(w^2(1) - 1)$. Therefore,

$$\lim_n \mathbf{P}_1 \left(D_n(\theta) \leq z \right) = \lim_n \mathbf{P}_1 \left(\frac{n}{\sqrt{2}} \left(\widehat{\theta}_n - 1 \right) \leq z \right) = \lim_n \mathbf{P}_1 \left(\frac{M_n/n}{\sqrt{2}\langle M \rangle_n / n^2} \leq z \right)$$
$$= P\left(\frac{\int_0^1 w(s)dw(s)}{\sqrt{2}\int_0^1 w^2(s)ds} \leq z \right) = P\left(\frac{w^2(1) - 1}{2\sqrt{2}\int_0^1 w^2(s)ds} \leq z \right)$$

where P is the Wiener measure. We end up with the following result.

Theorem 5.1 *Let $X_0 = 0$, $\varepsilon = (\varepsilon_n, n \geq 1)$ be a sequence of independent random variables with $\varepsilon_i \sim \mathcal{N}(0,1)$. Let $\widehat{\theta}_n$ be the maximum likelihood estimator from (5.5) and $I_n(\theta)$ be the Fisher information from (5.7). Then*

$$\lim_n \mathbf{P}_\theta \left(\sqrt{I_n(\theta)}\left(\widehat{\theta}_n - \theta\right) \leq z\right) = \begin{cases} \Phi(z), & |\theta| < 1, \\ H_\theta(z), & |\theta| = 1, \\ Ch(z), & |\theta| > 1. \end{cases} \tag{5.13}$$

Here we denote by $\Phi(z)$ the standard normal distribution, by $\mathrm{Ch}(z)$ the Cauchy distribution and by $H_\theta(z)$ the distribution of the random variable from (5.12).

Remark 5.1 Above we supposed that $X_0 = 0$ and $\varepsilon = (\varepsilon_n, n \geq 1)$ is a sequence of *Gaussian* random variables. In the case of $|\theta| \leq 1$ the result (5.13) remains valid also under the assumptions that X_0 is an arbitrary constant or a random variable not depending on $(\varepsilon_n, n \geq 1)$ with a finite second moment, and $(\varepsilon_n, n \geq 1)$ is an arbitrary sequence of i.i.d. random variables with $\mathbf{E}\varepsilon_n = 0$ and $\mathbf{E}\varepsilon_n^2 = 1$.

But in the case of $|\theta| > 1$, the limit distribution $\lim_n \mathbf{P}_\theta (D_n(\theta) \leq z)$ depends, in general, even if random variables $(\varepsilon_n, n \geq 1)$ are independent identically distributed, on the particular distribution of each ε_i (see Koul and Pflug (1990) or Sections 5.2, 5.3 below).

The variety of the limit distributions for the normalized deviation $\sqrt{I_n(\theta)}(\widehat{\theta}_n - \theta)$ for different θ motivates the following natural question: whether one can find a normalizing factor (other than $\sqrt{I_n(\theta)}$) for the deviation $\widehat{\theta}_n - \theta$ which leads to a *unique* limit distribution for each value of the parameter θ. First we show that normalizing $\widehat{\theta}_n - \theta$ by the stochastic Fisher information the *three* limit distributions in (5.13) are reduced to *two*. Indeed, let

$$\widetilde{D}_n(\theta) = \sqrt{\langle M\rangle_n}(\widehat{\theta}_n - \theta).$$

It is clear that

$$\widetilde{D}_n(\theta) = \sqrt{\langle M\rangle_n}(\widehat{\theta}_n - \theta) = \frac{M_n}{\sqrt{\langle M\rangle_n}} = \frac{M_n d_n^{-1}(\theta)}{\sqrt{\langle M\rangle_n d_n^{-2}(\theta)}} = \frac{U_n}{\sqrt{V_n}}.$$

(Recall that $D_n(\theta) = U_n/V_n$.)

It follows from (5.10) that for $|\theta| < 1$

$$\mathcal{L}(U_n, V_n \mid \mathbf{P}_\theta) \to \mathcal{L}(\xi, 1)$$

where $\xi \sim \mathcal{N}(0,1)$. For $|\theta| > 1$ one has

$$\mathcal{L}(U_n, V_n \mid \mathbf{P}_\theta) \to \mathcal{L}(\xi\eta, \eta^2)$$

where ξ and η are independent standard normal. Therefore,

$$\lim_n \mathbf{P}_\theta\left(\widetilde{D}_n(\theta) \le z\right) = \lim_n \mathbf{P}_\theta\left(\frac{U_n}{\sqrt{V_n}} \le z\right)$$

$$= P\left(\frac{\xi\eta}{\sqrt{\eta^2}} \le z\right) = P\left(\xi \operatorname{sign}\eta \le z\right) = P(\xi \le z) = \Phi(z)$$

so that for $|\theta| \ne 1$, the limit distribution of $\widetilde{D}_n(\theta)$ is standard normal. But in the case of $|\theta| = 1$ one gets

$$\lim_n \mathbf{P}_\theta\left(\widetilde{D}_n(\theta) \le z\right) = \lim_n \mathbf{P}_\theta\left(\frac{M_n}{\sqrt{\langle M\rangle_n}} \le z\right) = \lim_n \mathbf{P}_\theta\left(\frac{M_n n^{-1}}{\sqrt{\langle M\rangle_n n^{-2}}} \le z\right)$$

$$= P\left(\frac{\int_0^1 w(s)dw(s)}{\sqrt{2}\int_0^1 w^2(s)ds} \le z\right) = P\left(\frac{w^2(1)-1}{2\sqrt{2}\int_0^1 w^2(s)ds} \le z\right).$$

We arrive at the following result.

Theorem 5.2 *Let $X_0 = 0$, $\varepsilon = (\varepsilon_n, n \ge 1)$ be a sequence of independent random variables with $\varepsilon_i \sim \mathcal{N}(0,1)$. Let $\widehat{\theta}_n$ be the maximum likelihood estimator from (5.5) and $\langle M\rangle_n$ be the Fisher information due to (5.7). Then*

$$\lim_n \mathbf{P}_\theta\left(\sqrt{\langle M\rangle_n}(\widehat{\theta}_n - \theta) \le z\right) = \begin{cases} \Phi(z), & |\theta| \ne 1, \\ H_\theta(z), & |\theta| = 1. \end{cases}$$

The above results show that the maximum likelihood estimators $\widehat{\theta}_n$ are strongly consistent and the normalized deviations $\sqrt{I_n(\theta)}(\widehat{\theta}_n - \theta)$ and $\sqrt{\langle M\rangle_n}(\widehat{\theta}_n - \theta)$ have the limit distributions described by Theorems 5.1 and 5.2.

The following theorem contains some additional information about the properties of the estimators $\widehat{\theta}_n$.

Theorem 5.3 *Mikulski and Monsour (1991). As n $\to \infty$*

$$\sup_\theta \mathbf{E}_\theta|\widehat{\theta}_n - \theta| \to 0.$$

Particularly the estimators $\widehat{\theta}_n$ are asymptotically uniformly consistent, that is, as $n \to \infty$

$$\sup_\theta \mathbf{P}_\theta\left(|\widehat{\theta}_n - \theta| > \varepsilon\right) \to 0, \qquad \forall\varepsilon > 0.$$

Proof. One has

$$\mathbf{E}_\theta|\widehat{\theta}_n - \theta| = \mathbf{E}_\theta\left|\frac{M_n}{\langle M\rangle_n}\right| \le \left[\mathbf{E}_\theta M_n^2\, \mathbf{E}_\theta\left(\frac{1}{\langle M\rangle_n}\right)^2\right]^{1/2}$$

$$= \left[\mathbf{E}_\theta\langle M\rangle_n\, \mathbf{E}_\theta\frac{1}{\langle M\rangle_n^2}\right]^{1/2} = \left[I_n(\theta)\, \mathbf{E}_\theta\frac{1}{\langle M\rangle_n^2}\right]^{1/2}. \tag{5.14}$$

It follows from (5.7) that for $|\theta| < 1$

$$I_n(\theta) = \mathbf{E}_\theta \langle M \rangle_n < \frac{n}{1-\theta^2}. \tag{5.15}$$

Now we estimate the value $\mathbf{E}_\theta \frac{1}{\langle M \rangle_n^2}$. From the definition of the gamma-function,

$$\int_0^\infty s^{k-1} e^{-s\xi} ds = \frac{\Gamma(k)}{\xi^k}, \quad k = 1, 2, \dots, \quad \xi \geq 0.$$

Hence

$$\mathbf{E}_\theta \frac{1}{\langle M \rangle_n^k} = \frac{1}{\Gamma(k)} \mathbf{E}_\theta \int_0^\infty s^{k-1} e^{-s\langle M \rangle_n} ds.$$

Since the variables $(X_1, \dots, X_n)$ are Gaussian ($X_0 = 0$), one easily derives that

$$\mathbf{E}_\theta\, e^{-s\langle M \rangle_n} = \left(A_n(s, \theta^2)\right)^{-1/2}$$

where $A_1 = 1$, $A_2 = 1 + 2s$ and for $n \geq 3$

$$A_n = (1 + 2s)A_{n-1} + \theta^2 (A_{n-1} - A_{n-2}).$$

This implies that

$$A_n = \sum_{k=0}^{n-2} e_k(n, s) \theta^{2k}$$

with

$$e_0(n, s) = (1 + 2s)^{n-1}$$

and

$$e_k(n, s) = \sum_{j=1}^{k} C_{k-1}^{j-1} C_{n-k-1}^{j} (2j)^j (1 + 2s)^{n-k-j-1}, \quad k = 1, \dots, n-2.$$

Using these representations we obtain for $|\theta| < 1$ (see Mikulski and Monsour (1991) for more details)

$$A_n > \frac{1}{2}\left[\frac{1+\theta^2}{2} + \left\{\frac{(1-\theta^2)}{2} + s\right\}^{1/2}\right]^{n-1} \tag{5.16}$$

and

$$\mathbf{E}_\theta \frac{1}{\langle M \rangle_n^2} = \int_0^\infty \frac{s\, ds}{A_n^{1/2}}$$
$$< \sqrt{2}\left\{\frac{4(1-\theta^2)}{(n-3)(n-5)} + \frac{48(1-\theta^2)}{(n-3)(n-5)(n-7)} + \frac{192}{(n-3)(n-5)(n-7)(n-9)}\right\}.$$

By (5.14)–(5.16) we conclude that in the case of $|\theta| < 1$, for each $\delta > 0$, there exists such $n_0(\delta)$ that

$$\mathbf{E}_\theta|\widehat{\theta}_n - \theta| \leq (1+\delta)^{1/2}\frac{6n}{(n-3)(n-5)}, \quad n \geq n_0(\delta). \tag{5.17}$$

Now we turn to the case of $|\theta| \geq 1$. Applying twice the Cauchy-Schwarz inequality we obtain

$$\begin{aligned}
\mathbf{E}_\theta|\widehat{\theta}_n - \theta| &= \mathbf{E}_\theta\left|\frac{M_n}{\langle M\rangle_n}\right| \\
&= \mathbf{E}_\theta\frac{\sum_{i=1}^n X_{i-1}\varepsilon_i}{\sum_{i=1}^n X_{i-1}^2} \\
&\leq \mathbf{E}_\theta\frac{\left(\sum_{i=1}^n \varepsilon_i^2\right)^{1/2}}{\left(\sum_{i=1}^n X_{i-1}^2\right)^{1/2}} \\
&\leq \left[\mathbf{E}_\theta\left(\sum_{i=1}^n \varepsilon_i^2\right)\mathbf{E}_\theta\frac{1}{\sum_{i=1}^n X_{i-1}^2}\right]^{1/2} \\
&\leq \left[n\mathbf{E}_\theta\frac{1}{\langle M\rangle_n}\right]^{1/2}.
\end{aligned}$$

Since $\mathbf{E}_\theta\frac{1}{\langle M\rangle_n}$ decreases as $|\theta|$ increases, it holds for $|\theta| \geq 1$ and $n \geq 6$

$$\mathbf{E}_\theta\frac{1}{\langle M\rangle_n} \leq \mathbf{E}_1\frac{1}{\langle M\rangle_n} < \int_0^\infty \frac{\sqrt{2}ds}{\left[1+\sqrt{2s}\right]^{\frac{n-1}{2}}} = \frac{4\sqrt{2}}{(n-3)(n-5)}. \tag{5.18}$$

Here we have used the following straightforward inequality

$$A_n(s,1) > \frac{1}{2}\left[1+\sqrt{2s}\right]^{n-1};$$

see Mikulski and Monsour (1991). The statements (5.17) and (5.18) imply for $|\theta| \geq 1$ and $n \geq 6$

$$\mathbf{E}_\theta\left|\widehat{\theta}_n - \theta\right| \leq \left[\frac{6n}{(n-3)(n-5)}\right]^{1/2}. \tag{5.19}$$

Combining (5.17) and (5.19) we obtain for n large enough that

$$\sup_\theta \mathbf{E}_\theta\left|\widehat{\theta}_n - \theta\right| \leq \frac{c}{\sqrt{n}}$$

with some fixed constant c. Theorem 5.3 is proved. □

Now we turn to the question of (asymptotic) efficiency of the maximum likelihood estimators $\left(\widehat{\theta}_n,\, n \geq 1\right)$. First we introduce a "natural" *class* of estimators. Later on we show that the MLE is asymptotically efficient among all estimators in this class.

We will say that a sequence of estimators (θ_n) belongs to the class $U(\theta)$ if, given θ, the bias

$$b_\theta(\theta_n) = \mathbf{E}_\theta \left[\widehat{\theta}_n - \theta\right]$$

is differentiable in θ and satisfies the following two conditions:

$$b_\theta(\theta_n) \to 0, \quad \frac{db_\theta(\theta_n)}{d\theta} \to 0, \quad n \to \infty.$$

Notice that the maximum likelihood estimators $(\widehat{\theta}_n)$ belong to the class $U(\theta)$ for each $|\theta| \neq 1$. In fact, first of all, one has by Theorem 5.3 that $b_\theta(\theta_n) \to 0$. Next, differentiability of $b_\theta(\widehat{\theta}_n)$ follows from differentiability of the density

$$p_\theta\left(X_1, \ldots, X_n\right) = (2\pi)^{-n/2} \exp\left\{-\frac{1}{2}\sum_{i=1}^{n}\left(X_i - \theta X_{i-1}\right)^2\right\}$$

$$= (2\pi)^{-n/2} \exp\left\{-\frac{1}{2}\left(1+\theta^2\right)\langle M\rangle_n - 2\theta\sum_{i=1}^{n} X_i X_{i-1} + X_n^2\right\}.$$

Finally,

$$1 + \frac{db_\theta(\widehat{\theta}_n)}{d\theta} = \mathbf{E}_\theta \langle M\rangle_n (\widehat{\theta}_n - \theta)^2. \tag{5.20}$$

Due to Theorem 5.2 the variables $\sqrt{\langle M\rangle_n}(\widehat{\theta}_n - \theta)$ are asymptotically standard normal. By Lemma 5 of Mikulski and Monsour (1991) for each $m \geq 1$ there exist $n(m)$ and a constant C_m such that

$$\mathbf{E}_\theta \left|\sqrt{\langle M\rangle_n}\left(\widehat{\theta}_n - \theta\right)\right|^m \leq C_m < \infty, \quad n \geq n(m).$$

Thus the family $\left\{\left|\sqrt{\langle M\rangle_n}(\widehat{\theta}_n - \theta)\right|^k, n \geq 1\right\}$ is uniformly integrable and the weak convergence $\sqrt{\langle M\rangle_n}(\widehat{\theta}_n - \theta) \to \xi \sim \mathcal{N}(0,1)$ implies also that for $|\theta| \neq 1$ and each $k \geq 1$

$$\lim_n \mathbf{E}_\theta \left|\sqrt{\langle M\rangle_n}\left(\widehat{\theta}_n - \theta\right)\right|^m = E\xi^k.$$

Since $E\xi^2 = 1$, it holds $\mathbf{E}_\theta|\sqrt{\langle M\rangle_n}(\widehat{\theta}_n - \theta)|^2 \to 1$ and this, along with (5.20), gives $\frac{db_\theta(\theta_n)}{d\theta} \to 0$ as $n \to \infty$. Therefore, the maximum likelihood estimators belong to the class $U(\theta)$ for each $|\theta| \neq 1$.

Theorem 5.4 *Mikulski and Monsour (1991). The following statements are fulfilled:*

(1) *For each $|\theta| \neq 1$ the maximum likelihood estimator $(\widehat{\theta}_n)$ is asymptotically efficient in the class $U(\theta)$ in the sense that if $(\theta_n) \in U(\theta)$, then*

$$\limsup_n \frac{\mathbf{E}_\theta \langle M \rangle_n (\widehat{\theta}_n - \theta)^2}{\mathbf{E}_\theta \langle M \rangle_n (\theta_n - \theta)^2} \leq 1. \tag{5.21}$$

(2) *In the case of $|\theta| < 1$ the estimators $(\widehat{\theta}_n)$ are asymptotically efficient also in the ("classical") sense: for every estimator $(\theta_n) \in U(\theta)$ one has*

$$\limsup_n \frac{\mathbf{D}_\theta \widehat{\theta}_n}{\mathbf{D}_\theta \theta_n} \leq 1. \tag{5.22}$$

Proof. The relation (5.21) is based on the generalization of the Rao-Cramer inequality given in Mikulski and Monsour (1991). For the considered autoregressive framework (5.1) this assertion can be rewritten in the form

$$\mathbf{E}_\theta \langle M \rangle_n (\theta_n - \theta)^2 \geq \frac{\left[1 + \frac{db_\theta(\theta_n)}{d\theta}\right]^2}{1 + \frac{db_\theta(\widehat{\theta}_n)}{d\theta}} \tag{5.23}$$

where θ_n is an arbitrary estimator and $\widehat{\theta}_n$ is the maximum likelihood estimator. Moreover, if $\theta_n = \widehat{\theta}_n$ then one meets the equality in (5.23),

$$\mathbf{E}_\theta \langle M \rangle_n (\widehat{\theta}_n - \theta)^2 = 1 + \frac{db_\theta(\widehat{\theta}_n)}{d\theta}.$$

The inequality (5.23) can be proved in the following way. By the Cauchy-Schwarz inequality

$$\begin{aligned}
&\left|\mathbf{E}_\theta \langle M \rangle_n \left(\widehat{\theta}_n - \theta\right)(\theta_n - \theta)\right|^2 \\
&\quad = \left|\mathbf{E}_\theta \left(\langle M \rangle_n^{1/2} \left(\widehat{\theta}_n - \theta\right)\right)\left(\langle M \rangle_n^{1/2} (\theta_n - \theta)\right)\right|^2 \\
&\quad \leq \left(\mathbf{E}_\theta \langle M \rangle_n \left(\widehat{\theta}_n - \theta\right)^2\right)\left(\mathbf{E}_\theta \langle M \rangle_n (\theta_n - \theta)^2\right).
\end{aligned} \tag{5.24}$$

Due to (5.4)

$$\frac{\partial \log p_\theta(X_1, \ldots, X_n)}{\partial \theta} = -\theta \langle M \rangle_n + \sum_{i=1}^n X_i X_{i-1} = M_n$$

and

$$\frac{db_\theta(\theta_n)}{d\theta} = \mathbf{E}_\theta(\theta_n - \theta)\frac{\partial \log p_\theta(X_1, \ldots, X_n)}{\partial \theta} - 1.$$

This yields

$$\begin{aligned}
\mathbf{E}_\theta\langle M\rangle_n(\widehat{\theta}_n-\theta)(\theta_n-\theta) &= \mathbf{E}_\theta M_n(\theta_n-\theta)\\
&= \mathbf{E}_\theta(\theta_n-\theta)\frac{\partial \log p_\theta(X_1,\dots,X_n)}{\partial\theta}\\
&= 1+\frac{db_\theta(\theta_n)}{d\theta}. \qquad (5.25)
\end{aligned}$$

If, in particular, $\widehat{\theta}_n=\theta_n$, then (see (5.20))

$$\mathbf{E}_\theta\langle M\rangle_n(\widehat{\theta}_n-\theta)^2 = 1+\frac{db_\theta(\widehat{\theta}_n)}{d\theta}.$$

This fact along with (5.24) and (5.25) implies that

$$\mathbf{E}_\theta\langle M\rangle_n\left(\widehat{\theta}_n-\theta\right)^2 \geq \frac{\left[1+\frac{db_\theta(\theta_n)}{d\theta}\right]^2}{1+\frac{db_\theta(\widehat{\theta}_n)}{d\theta}}$$

and

$$\frac{\mathbf{E}_\theta\langle M\rangle_n\left(\widehat{\theta}_n-\theta\right)^2}{\mathbf{E}_\theta\langle M\rangle_n\left(\theta_n-\theta\right)^2} \leq \frac{\left[1+\frac{db_\theta(\theta_n)}{d\theta}\right]^2}{\left[1+\frac{db_\theta(\widehat{\theta}_n)}{d\theta}\right]^2}. \qquad (5.26)$$

The definition of the class $U(\theta)$ yields, for each $(\theta_n)\in U(\theta)$ and $|\theta|\neq 1$, that $1+\frac{db_\theta(\theta_n)}{d\theta}\to 1$. Above we have stated also that $1+\frac{db_\theta(\widehat{\theta}_n)}{d\theta}\to 1$. This and (5.26) imply the required inequality (5.21).

A proof of the property (5.22) can be found in Mikulski and Monsour (1991, Theorem 4). □

Remark 5.2 It is worth noting that in the case of $|\theta|=1$ the maximum likelihood estimators $(\widehat{\theta}_n)$ do not belong to the class $U(\theta)$. This follows from the fact that by Theorem 5.2, say for $\theta=1$,

$$\lim_n \mathbf{P}_\theta\left(\sqrt{\langle M\rangle_n}\left(\widehat{\theta}_n-\theta\right)\leq z\right) = P\left(\frac{w^2(1)-1}{2\sqrt{\int_0^1 w^2(s)ds}}\leq z\right).$$

Hence

$$1+\frac{db_\theta(\widehat{\theta}_n)}{d\theta} = \mathbf{E}_\theta\langle M\rangle_n(\widehat{\theta}_n-\theta)^2 \to E\left|\frac{w^2(1)-1}{2\sqrt{\int_0^1 w^2(s)ds}}\right|^2 = 1.142$$

see Anderson(1959). Therefore, $\frac{db_\theta(\widehat{\theta}_n)}{d\theta}\not\to 0$, $n\to\infty$.

5.2 Convergence of Statistical Experiments for First Order Autoregressive Models

Denote by $\mathbb{E}^n = (\Omega^n, \mathcal{F}^n; \mathbf{P}_\theta^n, \theta \in R)$ the statistical experiment corresponding to the first order autoregressive model

$$X_i = \theta X_{i-1} + \varepsilon_i\,, \quad i = 1, \dots, n, \tag{5.27}$$

with $X_0 = 0$, $\Omega^n = R^n$, $\mathcal{F}^n = \mathcal{B}(R^n)$, $n \geq 1$.

In this section we consider the problem of convergence of the experiments $\mathbb{E}^n$ (more precisely, of their reparametrizied versions). We study separately the cases of Gaussian and non-Gaussian errors $(\varepsilon_i, i \geq 1)$ and we distinguish between *stable, unstable* and *explosive* models in accordance with the value $|\theta_0|$ of a given localization point.

In what follows we suppose that the errors $(\varepsilon_i, i \geq 1)$ are independent identically distributed random variables from a distribution $F(x)$ on the real line with a (positive) density $f = f(x)$ with respect to the Lebesgue measure.

Denote

$$Z^n(\theta, \theta_0) = \frac{d\mathbf{P}_\theta^n}{d\mathbf{P}_{\theta_0}^n}(X), \qquad X = (X_1, \dots X_n).$$

Using (5.27) one has

$$Z^n(\theta, \theta_0) = \prod_{i=1}^n \frac{f(X_i - \theta X_{i-1})}{f(X_i - \theta_0 X_{i-1})}. \tag{5.28}$$

For $\theta \neq \theta_0$ the families of measures $(\mathbf{P}_\theta^n)$ and $(\mathbf{P}_{\theta_0}^n)$ are asymptotically singular. Therefore, in order to get a nontrivial limit for the experiments $\mathbb{E}^n$ we use the idea of reparametrization. We start with the case of Gaussian errors when $f(x) = \varphi(x) = \frac{1}{\sqrt{2\pi}} e^{-x^2/2}$.

5.2.1 *Gaussian errors*

Let us fix some $\theta_0 \in R$ as a localization point and represent θ in the form

$$\theta = \theta_0 + \varphi_n(\theta_0)\alpha \tag{5.29}$$

where the localizing sequence $\varphi_n(\theta_0)$ is defined by the equalities $\varphi_n(\theta_0) = d_n^{-1}(\theta_0)$ and $d_n(\theta_0)$ are defined similarly to (5.8):

$$\varphi_n(\theta_0) = \begin{cases} \sqrt{\frac{1-\theta_0^2}{n}}, & |\theta_0| < 1, \\ \sqrt{2}n^{-1}, & |\theta_0| = 1, \\ \frac{\theta_0^2 - 1}{|\theta_0|^n}, & |\theta_0| > 1. \end{cases} \tag{5.30}$$

Given the experiment $\mathbb{E}^n$ and a localization point θ_0 define for each $n \geq 1$ a new "localized" experiment

$$\mathcal{E}^n(\theta_0) = \left(R^n, \mathcal{B}(R^n);\ P^n_{\alpha,\theta_0}, \alpha \in R\right) \tag{5.31}$$

with $P^n_{\alpha,\theta_0} = \mathbf{P}^n_{\theta_0+\varphi_n(\theta_0)\alpha}$.

Let $M_n = \sum_{i=1}^n X_{i-1}\varepsilon_i$, $\langle M\rangle_n = \sum_{i=1}^n X_{i-1}^2$. Then

$$Z^n_\alpha(\theta_0) \equiv Z^n(\theta_0 + \varphi_n(\theta_0)\alpha;\theta_0) = \exp\left\{\alpha\varphi_n(\theta_0)M_n - \frac{\alpha^2}{2}\varphi_n^2(\theta_0)\langle M\rangle_n\right\}.$$

Obviously, each experiment $\mathcal{E}^n(\theta_0)$ is a $(\gamma,\Gamma;\varphi)$-model. Indeed, if

$$\gamma_n = \frac{M_n}{\sqrt{\langle M\rangle_n}}, \quad \Gamma_n^2 = \varphi_n^2(\theta_0)\langle M\rangle_n,$$

then

$$Z^n_\alpha(\theta_0) = \frac{\varphi(\gamma_n - \Gamma_n\alpha)}{\varphi(\gamma_n)} = \exp\left\{\alpha\gamma_n\Gamma_n - \frac{\alpha^2}{2}\Gamma_n^2\right\}. \tag{5.32}$$

In accordance with the results of Section 5.1

$$\mathcal{L}\left(\varphi_n(\theta_0)M_n, \varphi_n^2(\theta_0)\langle M\rangle_n \mid \mathbf{P}^n_{\theta_0}\right) \to \begin{cases} \mathcal{L}(\xi, 1), & |\theta_0| < 1, \\ \mathcal{L}\left(\sqrt{2}\int_0^1 W_t\,dW_t, 2\int_0^1 W_t^2\,dt\right), & |\theta_0| = 1, \\ \mathcal{L}(\xi\eta, \eta^2), & |\theta_0| > 1, \end{cases} \tag{5.33}$$

where ξ, η are independent standard Gaussian variables and $W = (W_t, t \geq 0)$ is a standard Wiener process.

We introduce the following notation:

$$(M, \langle M\rangle) = \begin{cases} (\xi, 1), & |\theta_0| < 1, \\ \left(\sqrt{2}\int_0^1 W_t\,dW_t, 2\int_0^1 W_t^2\,dt\right), & |\theta_0| = 1, \\ (\xi\eta, \eta^2), & |\theta_0| > 1. \end{cases}$$

and

$$\Gamma^2 = \langle M\rangle, \quad \gamma = \frac{M}{\sqrt{\langle M\rangle}}.$$

Now

$$(\gamma, \Gamma) = \begin{cases} (\xi, 1), & |\theta_0| < 1, \\ \left(\frac{\int_0^1 W_t\,dW_t}{\left[\int_0^1 W_t^2\,dt\right]^{1/2}}, \left[2\int_0^1 W_s^2\,ds\right]^{1/2}\right), & |\theta_0| = 1, \\ (\xi \operatorname{sign}\eta, |\eta|), & |\theta_0| > 1, \end{cases} \tag{5.34}$$

and the convergence (5.33) can be rewritten in the form

$$\mathcal{L}\left(\gamma_n\Gamma_n, \Gamma_n^2 \mid \mathbf{P}_{\theta_0}^n\right) \to \mathcal{L}(\gamma\Gamma, \Gamma^2)$$

which in its turn implies

$$\mathcal{L}\left(\gamma_n, \Gamma_n \mid \mathbf{P}_{\theta_0}^n\right) \to \mathcal{L}(\gamma, \Gamma). \tag{5.35}$$

In view of (5.32) the weak convergence (5.35) implies that the reparametrizied experiments $\mathcal{E}^n(\theta_0)$ converge (in the weak and λ-sense as $n \to \infty$) to the statistical experiments $\mathcal{E}(\theta_0)$ which are the (γ, Γ)-models defined in (5.34). Directly from this representation one can observe that the limit experiments $\mathcal{E}(\theta_0)$ have different structures for different values of the localization point θ_0. The case of $|\theta_0| < 1$ is called *stable*, $|\theta_0| = 1$, *unstable* and $|\theta_0| > 1$, *explosive.*
Stable case. If $|\theta_0| < 1$, then the experiment $\mathcal{E}(\theta_0)$ coincides with the standard Gaussian shift experiment

$$\mathcal{E}(\theta_0) = \mathcal{E}_G = (R, \mathcal{B}(R);\ P_\alpha\,, \alpha \in R)$$

where $P_\alpha = P_\alpha(d\gamma)$ is the Gaussian measure on the real line with mean α and the unit covariance. In other words, the experiment $\mathcal{E}_G$ corresponds to the model

$$\gamma = \alpha + \varepsilon, \quad \varepsilon \sim \mathcal{N}(0,1).$$

Explosive case. For the case $|\theta_0| > 1$, it turns out that the experiment $\mathcal{E}(\theta_0)$ is mixed normal,

$$\mathcal{E}(\theta_0) = \mathcal{E}_{MG} = (R \times R_+, \mathcal{B}(R \times R_+);\ P_\alpha(d\gamma, d\Gamma)\,, \alpha \in R)$$

where $P_0(d\gamma, d\Gamma) = P_0'(d\gamma)P_0''(d\Gamma)$, P_0' is the standard Gaussian measure, $P_0'(d\gamma) = \varphi(\gamma)d\gamma$, $\gamma \in R$, and $P_0''(d\Gamma)$ is a probability distribution of the random variable $\Gamma = |\eta|$, $\eta \sim \mathcal{N}(0,1)$. For an arbitrary $\alpha \in R$

$$P_\alpha(d\gamma, d\Gamma) = \exp\left\{\alpha\gamma\Gamma - \frac{\alpha^2}{2}\Gamma^2\right\} P_0(d\gamma, d\Gamma).$$

Unstable (or critical) case. In the situation with $|\theta_0| = 1$ the limit experiment $\mathcal{E}(\theta_0)$ can be described as follows:

$$\mathcal{E}(\theta_0) = (\mathbf{C}[0,1], \mathcal{B}(\mathbf{C}[0,1])\,;\ P_\alpha\,, \alpha \in R)$$

where $P_0 = P_0(dx)$ is the Wiener measure on $(\mathbf{C}[0,1], \mathcal{B}(\mathbf{C}[0,1]))$ and $P_\alpha = P_\alpha(dx)$ is the distribution of the diffusion process $X = (X_t)$ obeying the stochastic equation

$$dX_t = \alpha\sqrt{2}X_t dt + dW_t\,, \quad 0 \le t \le 1.$$

By Girsanov's formula (see e.g. Liptser and Shiryaev(1974)),

$$\frac{dP_\alpha}{dP_0}(x) = \exp\left\{\alpha\gamma\Gamma - \frac{\alpha^2}{2}\Gamma^2\right\}$$

where (γ, Γ) are from (5.34) for $|\theta_0| = 1$.

If one sets $\lambda_n = (\gamma_n, \Gamma_n)$, $\lambda = (\gamma, \Gamma)$, then we have, in view of (5.32), for all θ_0

$$Z_\alpha^n(\theta_0) = Z_\alpha(\theta_0, \lambda_n) \tag{5.36}$$

with

$$Z_\alpha(\theta_0, \lambda) = \exp\left\{\alpha\gamma\Gamma - \frac{\alpha^2}{2}\Gamma^2\right\}.$$

(Note that γ and Γ depend on θ_0.)

The statement (5.35) can be rewritten now as

$$\mathcal{L}\left(\lambda_n \mid \mathbf{P}_{\theta_0}^n\right) \to \mathcal{L}(\lambda).$$

Therefore, λ-convergence holds

$$\mathcal{E}^n(\theta_0) \xrightarrow{\lambda} \mathcal{E}(\theta_0).$$

Moreover, since (5.36) is the exact equality rather than an asymptotic relation (as in the definition of λ-convergence), the uniform (in all $\alpha \in R$) λ-convergence holds

$$\mathcal{E}^n(\theta_0) \xrightarrow{\lambda(R)} \mathcal{E}(\theta_0).$$

Using the terminology of Chapter 4 one can say that the experiments $\mathbb{E}^n$ satisfy the condition of local λ-convergence to the family $\boldsymbol{\mathcal{E}} = (\mathcal{E}(\theta_0), \theta_0 \in R)$:

$$\mathbb{E}^n \xrightarrow{(\lambda,\varphi)} \boldsymbol{\mathcal{E}}, \tag{5.37}$$

where the localizing sequence $\varphi_n(\theta_0)$ is determined by (5.30).

The local λ-convergence (5.37) allows us to state lower bounds for the risks of arbitrary estimators $\widetilde{\theta}_n$ of the parameter θ (Theorem 4.1). However, in order to be able to apply stronger results of Section 5.4 on asymptotic minimax properties of the global estimators one has to check the uniform (in the localization point θ_0) version of the condition of local λ-convergence (5.37). In the case of the autoregressive model (5.27) this uniformity fails near the critical points ± 1. That is why we will consider families $\boldsymbol{\mathcal{E}} = (\mathcal{E}(\theta_0), \theta_0 \in T_0)$ where T_0 is a compact set not containing ± 1.

Set also $\mathcal{K} = \{R\}$, i.e. the class $\mathcal{K}$ consists of one set R (see Definition 4.4 in Chapter 4).

Theorem 5.5 *Let T_0 be a compact on the real line not containing the points ± 1. Then, under the choice of the localizing sequences $\varphi_n(\theta_0)$ of (5.30) the experiments $\mathbb{E}^n$, $n \geq 1$, $(\{R\}, T_0)$-uniformly locally λ-converge to the family $\boldsymbol{\mathcal{E}} = (\mathcal{E}(\theta_0), \theta_0 \in T_0)$ with $\mathcal{E}(\theta_0) = \mathcal{E}_G$ for $|\theta| < 1$ and $\mathcal{E}(\theta_0) = \mathcal{E}_{MG}$ for $|\theta| > 1$.*

Proof. The condition (i) of Definition 4.4 in Chapter 4 is obviously fulfilled by (5.36). To check the condition (ii) of this definition one should verify that the weak convergence (5.35) is uniform in $\theta_0 \in T_0$.

We explain the idea of the proof of this property in the explosive case. It suffices to consider the situation when T_0 is a compact on $(1,\infty)$ of the form $[r,R]$ with $1<r<R<\infty$. We shall construct explicitly $\widehat{\gamma}_n, \widehat{\Gamma}_n$ satisfying

$$\mathcal{L}\left(\widehat{\gamma}_n, \widehat{\Gamma}_n \mid \mathbf{P}^n_{\theta_0}\right) = \mathcal{L}(\gamma,\Gamma), \tag{5.38}$$

$$\Gamma_n - \widehat{\Gamma}_n \xrightarrow{\mathbf{P}^n_{\theta_0}} 0, \tag{5.39}$$

$$\gamma_n - \widehat{\gamma}_n \xrightarrow{\mathbf{P}^n_{\theta_0}} 0, \tag{5.40}$$

such that the conditions (5.39),(5.40) hold uniformly in $\theta_0 \in [r,R]$. (Note that these statistics provide for the case under consideration an explicit construction of the elements $\widehat{\lambda}_n = (\widehat{\gamma}_n, \widehat{\Gamma}_n)$ in Lemma about "Reconstruction", see Theorem 5.1 and Lemma 5.1 in Chapter 2).

Notice first that one has with probability 1 under the measure $\mathbf{P}^n_{\theta_0}$

$$\Gamma_n^2 = \frac{(\theta_0^2-1)^2}{\theta_0^{2n}} \sum_{i=1}^n X_{i-1}^2, \qquad \Gamma_n\gamma_n = \frac{\theta_0^2-1}{\theta_0^n} \sum_{i=1}^n X_{i-1}\varepsilon_i.$$

We shall see that the value Γ_n is determined mostly by the initial observations $X_1, X_2, \ldots$ whereas the value of γ_n depends mostly on the terminal observations $X_n, X_{n-1}, \ldots$. Hence, γ_n and Γ_n are weakly dependent. To make it clearer, set

$$Z_n = \sum_{k=1}^{[n/2]} \theta_0^{-k}\varepsilon_k$$

and

$$\sigma_n^2 = \mathbf{E}^n_{\theta_0} Z_n^2 = \frac{1-\theta_0^{-2[n/2]}}{\theta_0^2-1}.$$

Set also

$$\overline{\Gamma}_n = \sqrt{\theta_0^2-1}\,|Z_n| = \sqrt{\theta_0^2-1}\left|\sum_{k=1}^{[n/2]} \theta_0^{-k}\varepsilon_k\right|$$

and

$$\overline{\gamma}_n = \sqrt{\theta_0^2-1}\,\mathrm{sign} Z_n \sum_{k=[n/2]+1}^{n} \theta_0^{-(n-k+1)}\varepsilon_k.$$

Now, by definition, the variables $\overline{\Gamma}_n$ and $\overline{\gamma}_n \mathrm{sign} Z_n$ are independent and normally distributed with zero mean and the covariances

$$\begin{aligned}
\mathbf{E}^n_{\theta_0}\overline{\Gamma}_n^2 &= (\theta_0^2-1)\,\mathbf{E}^n_{\theta_0} Z_n^2 = 1-\theta_0^{-2[n/2]} \to 1, \qquad n\to\infty,\\
\mathbf{E}^n_{\theta_0}\overline{\gamma}_n^2 &= (\theta_0^2-1)\sum_{k=[n/2]+1}^{n} \theta_0^{-2(n-k+1)} = 1-\theta_0^{-2[n/2]} \to 1, \qquad n\to\infty.
\end{aligned}$$

Finally, set

$$\widehat{\Gamma}_n = \overline{\Gamma}_n \frac{\sigma_n^{-1}}{\sqrt{\theta_0^2 - 1}}, \qquad \widehat{\gamma}_n = \overline{\gamma}_n \frac{\sigma_n^{-1}}{\sqrt{\theta_0^2 - 1}}. \tag{5.41}$$

This definition yields automatically (5.38).

We rewrite the quantities $\widehat{\Gamma}_n$ and $\widehat{\gamma}_n$ in terms of the observed variables $X_1, \ldots, X_n$. Since $X_k - \theta_0 X_{k-1} = \varepsilon_k$ $(\mathbf{P}^n_{\theta_0}\text{-a.s.})$ it holds

$$Z_n = \sum_{k=1}^{[n/2]} \theta_0^{-k}(X_k - \theta_0 X_{k-1})$$

and

$$\begin{aligned} \widehat{\Gamma}_n &= \sigma_n^{-1}|Z_n|, \\ \widehat{\gamma}_n &= \sigma_n^{-1}\operatorname{sign} Z_n \sum_{k=[n/2]+1}^{n} \theta_0^{-(n-k+1)}(X_k - \theta_0 X_{k-1}). \end{aligned}$$

In view of (5.41), it sufficies to prove (5.39) and (5.40) and their uniform versions with $\overline{\Gamma}_n$ and $\overline{\gamma}_n$ instead of $\widehat{\Gamma}_n$ and $\widehat{\gamma}_n$. For any $n, k \geq 1$

$$\theta_0^{-n} X_{k-1} = \theta_0^{-n} \sum_{i=1}^{k-1} \theta_0^{k-i-1} \varepsilon_i = \theta_0^{-(n-k+1)} \sum_{i=1}^{k-1} \theta_0^{-i} \varepsilon_i .$$

This implies that

$$\mathbf{E}^n_{\theta_0} \left| \theta_0^{-n} X_{k-1} - \theta_0^{-(n-k+1)} Z_n \right|^2 \leq C_1 \theta_0^n , \tag{5.42}$$

$$\mathbf{E}^n_{\theta_0} \left| \theta_0^{-n} X_{k-1}^2 - \theta_0^{-2(n-k+1)} Z_n^2 \right|^2 \leq C_2 \theta_0^{n/2} , \tag{5.43}$$

with some constants C_1, C_2 depending only on r and R.

Based on (5.43) we conclude that the following convergence holds uniformly in $\theta_0 \in [r, R]$ as $n \to \infty$:

$$\mathbf{E}^n_{\theta_0} |\Gamma_n^2 - \overline{\Gamma}_n^2| = \mathbf{E}^n_{\theta_0} \left| (\theta_0^2 - 1)^2 \sum_{k=1}^{n} \theta_0^{-2n} X_{k-1}^2 - (\theta_0^2 - 1) Z_n^2 \right| \to 0,$$

and (5.39) follows.

Next, by (5.42), the following holds uniformly in $\theta_0 \in [r, R]$ as $n \to \infty$:

$$\begin{aligned} &\mathbf{E}^n_{\theta_0} |\Gamma_n \gamma_n - \overline{\Gamma}_n \overline{\gamma}_n| \\ &= \mathbf{E}^n_{\theta_0} \left| (\theta_0^2 - 1) \sum_{k=1}^{n} \theta_0^{-n} X_{k-1} \varepsilon_k - (\theta_0^2 - 1) \sum_{k=[n/2]+1}^{n} \theta_0^{-(n-k+1)} Z_n \varepsilon_k \right| \to 0, \end{aligned}$$

and (5.40) follows. □

Now we turn to the case of non-Gaussian errors.

5.2.2 *Non-Gaussian errors: stable autoregression*

Theorem 5.6 *The stable case,* $|\theta_0| < 1$. *Let the random errors* (ε_i) *in (5.27) be independent identically distributed random variables with a density* $f = f(x)$ *satisfying the following condition*

(F) $f(x) > 0, \quad x \in R;$
$\int_R x\, f(x)\, dx = 0, \quad \int_R x^2 f(x)\, dx = 1;$
the function f *is absolutely continuous and for some* $\delta > 0$

$$\int_R \left|\frac{f'(x)}{f(x)}\right|^{2+\delta} f(x)\, dx < \infty.$$

Then for each interval $T_0 = [a,b] \subseteq (-1,1)$ *the experiments* $\mathbb{E}^n$, $n \geq 1$ *satisfy the condition of* $(\mathcal{K}, T_0)$*-uniform local* λ*-convergence to the Gaussian experiment* $\mathcal{E}_{G,I}$ *with*

$$\mathcal{E}_{G,I} = (R, \mathcal{B}(R);\, P_\alpha\,, \alpha \in R)$$

where $P_\alpha \sim \mathcal{N}(I^{1/2}\alpha, 1)$, $\mathcal{K}$ *is the class of compacts on the real line and the localizing sequence is defined by* $\varphi_n(\theta_0) = \sqrt{\frac{1-\theta_0^2}{n}}$, $n \geq 1$.

Proof. It can be carried out by standard methods of asymptotic statistics (see, for instance, Ibragimov and Khasminskii (1981, Chapter II)). We discuss only some important steps of the proof.

The limit experiment $\mathcal{E}_{G,I}$ has the likelihood of the form

$$Z_\alpha(x) = \frac{dP_\alpha}{dP_0}(x) = \exp\left\{xI^{1/2}\alpha - \frac{\alpha^2}{2}I\right\}.$$

Due to (5.28)

$$\begin{aligned} Z^n(\theta_0 + \varphi_n(\theta_0)\alpha; \theta_0) &= \frac{d\mathbf{P}^n_{\theta_0+\varphi_n(\theta_0)\alpha}}{d\mathbf{P}^n_{\theta_0}} \\ &= \prod_{k=1}^n \frac{f(X_k - (\theta_0 + \varphi_n(\theta_0)\alpha)X_{k-1})}{f(X_k - \theta_0 X_{k-1})}. \end{aligned}$$

Set

$$\lambda_n = I^{-1/2}\varphi_n(\theta_0) \sum_{k=1}^n X_{k-1} \frac{f'(X_k - \theta_0 X_{k-1})}{f(X_k - \theta_0 X_{k-1})}.$$

The statement of the theorem means that for every compact K the following asymptotic expansion holds uniformly in $\theta_0 \in [a,b] \subset (-1,1)$ and $\alpha \in K$

$$Z^n(\theta_0 + \varphi_n(\theta_0)\alpha; \theta_0) - \exp\left\{\lambda_n I^{1/2}\alpha - \frac{\alpha^2}{2}I\right\} \xrightarrow{\mathbf{P}^n_{\theta_0}} 0 \tag{5.44}$$

as well as the weak convergence, uniform in $\theta_0 \in [a,b]$,

$$\mathcal{L}\left(\lambda_n \mid \mathbf{P}^n_{\theta_0}\right) \to \mathcal{N}(0,1). \tag{5.45}$$

Let us show (5.45). By the condition (F) one has $\int f'(x)\,dx = 0$. Thus

$$\mathbf{E}^n_{\theta_0}\left(X_{k-1}\frac{f'(X_k)}{f(X_k)} \mid \varepsilon_1,\dots,\varepsilon_{k-1}\right) = X_{k-1}\int_{-\infty}^{\infty}\frac{f'(x)}{f(x)}f(x)\,dx = 0$$

and hence the sequence $M = (M_n\,, n \geq 1)$ is a martingale under the measure $\mathbf{P}^n_{\theta_0}$ (with respect to the filtration $(\mathcal{F}_n\,, n \geq 1)$ with $\mathcal{F}_n = \sigma(\varepsilon_1,\dots,\varepsilon_n)$). Furthermore, M is a square-integrable martingale with the quadratic characteristic

$$\langle M\rangle_n = I\sum_{k=1}^{n} X^2_{k-1}.$$

Since $\lambda_n = I^{-1/2}\varphi_n(\theta_0)M_n$ with $\mathbf{P}^n_{\theta_0}$-probability one, it holds by the law of large numbers

$$I^{-1}\varphi^2_n(\theta_0)\langle M\rangle_n = \frac{1-\theta_0^2}{n}\sum_{k=1}^{n} X^2_{k-1} \xrightarrow{\mathbf{P}^n_{\theta_0}} 1$$

and the assertion (5.42) follows from the central limit theorem for martingales (see Liptser and Shiryaev (1988) and Greenwoood and Shiryaev (1989) for its uniform version).

The statement (5.44) can be proved by standard calculation, see Ibragimov and Khasminskii (1981, Chapter 2). □

5.2.3 *Non-Gaussian errors: explosive autoregression*

Now we consider the explosive case $|\theta_0| > 1$. Here we suppose that the density $f = f(x)$ satisfies the following condition

(F') $f(x) > 0, \quad x \in R$;
$\int_R x\,f(x)\,dx = 0, \quad \int_R x^2 f(x)\,dx = 1$;
there is a constant $\delta > 0$ such that for any number $K > 0$ and some $C = C(K)$

$$\sup_{x,y\in[-K,K]} |\ln f(x) - \ln f(y)| \leq C|x-y|^{\delta}.$$

Note that the condition (F') does not necessarily imply that the density f has finite Fisher information.

Let $\mathcal{E}^n(\theta_0)$ be the localized experiment due to (5.27) with $\varphi_n(\theta_0) = \frac{\theta_0^2-1}{|\theta_0|^n}$ and let the experiment

$$\mathcal{E}(\theta_0) = (X, \mathcal{X};\, P_{\alpha,\theta_0}\,, \alpha \in R)$$

be defined in the following way. Set $X = R \times R^\infty$, $\mathcal{X} = \mathcal{B}(X)$, put

$$\zeta = \sum_{k=1}^{\infty} |\theta_0|^{-k} \varepsilon_k$$

and define the measures P_{α,θ_0} on $(X, \mathcal{X})$ by

$$\begin{aligned} P_{0,\theta_0} &= P_\zeta \times P_\varepsilon^\infty = P_\zeta \times P_{\varepsilon_1} \times P_{\varepsilon_2} \times \dots, \\ P_{\alpha,\theta_0}(d\lambda) &= \prod_{k=1}^{\infty} \frac{f(e_k - \alpha\,\varphi_k(\theta_0)z)}{f(e_k)} P_{0,\theta_0}(d\lambda). \end{aligned} \tag{5.46}$$

Here we denote by $\lambda = (z; e_1, e_2, \dots)$ points from $X = R \times R^\infty$.

Remark 5.3 Convergence of the infinite product in (5.46) is guaranteed by condition (F').

Remark 5.4 The experiment $\mathcal{E}(\theta_0)$ is transitive with respect to the transformation operators G_{α,θ_0} with

$$G_{\alpha,\theta_0}(z; e_1, e_2, \dots) = (z; e_1 - \alpha\,\varphi_1(\theta_0)z, e_2 - \alpha\,\varphi_2(\theta_0)z, \dots).$$

In fact, if μ is the Lebesgue measure on the real line and $\widetilde{\mu} = P_\zeta \times \mu^\infty$, then

$$\frac{dP_{\alpha,\theta_0}}{d\widetilde{\mu}}(\lambda) = \prod_{k=1}^{\infty} f(e_k - \alpha\,\varphi_k(\theta_0)z) = \frac{dP_{0,\theta_0}}{d\widetilde{\mu}}(G_{\alpha,\theta_0}\lambda),$$

i.e. $P_{\alpha,\theta_0} = P_{0,\theta_0} \circ G_{\alpha,\theta_0}^{-1}$.

Theorem 5.7 *Let the random errors (ε_i) in (5.27) be independent identically distributed with a density $f = f(x)$ satisfying the condition (F'). Denote by $\mathcal{K}$ the class of compacts on the real line and define localizing sequences by $\varphi_n(\theta_0) = (\theta_0^2 - 1)|\theta_0|^{-n}$. Then for all constants r, R with $1 < r < R < \infty$ and $T_0 = [-R, -r] \cup [r, R]$ the family $\mathbb{E}^n$, $n \geq 1$ satisfies the condition of $(\mathcal{K}, T_0)$-uniform local λ-convergence to the family of experiments $\mathcal{E}(\theta_0)$, $\theta_0 \in T_0$.*

Proof. For the weak convergence of the localized experiments $\mathcal{E}^n(\theta_0)$ to $\mathcal{E}(\theta_0)$ see Koul and Pflug(1990)). Therefore we explain only the idea of the proof assuming $\theta_0 > 1$.

Let

$$\zeta_n = \sum_{k=1}^{n} |\theta_0|^{-k} \varepsilon_k, \quad \zeta = \sum_{k=1}^{\infty} \theta_0^{-k} \varepsilon_k.$$

For any $n, k \geq 1$

$$\theta_0^{-n} X_n = \zeta_n$$

and ζ_n converges exponentially fast to ζ (a.s. under the measure $\mathbf{P}^n_{\theta_0}$). Furthermore, let

$$Z^n_\alpha(\theta_0) = \frac{d\mathbf{P}^n_{\theta_0+\varphi_n(\theta_0)\alpha}}{d\mathbf{P}^n_{\theta_0}}.$$

Then ($\mathbf{P}^n_{\theta_0}$-a.s.)

$$Z^n_\alpha(\theta_0) = \prod_{k=1}^n \frac{f(X_k - (\theta_0 + \varphi_n(\theta_0)\alpha)\, X_{k-1})}{f(X_k - \theta_0 X_{k-1})} = \prod_{k=1}^n \frac{f(\varepsilon_k - \varphi_n(\theta_0)\alpha X_{k-1})}{f(\varepsilon_k)}.$$

Using the equality $\varphi_n(\theta_0) = \theta_0^{-k}\varphi_{n-k}(\theta_0)$ and replacing $k-1$ by $n-k$ we obtain

$$\begin{aligned} Z^n_\alpha(\theta_0) &= \prod_{k=1}^n \frac{f(\varepsilon_k - \varphi_{n-k+1}(\theta_0)\alpha\zeta_{k-1})}{f(\varepsilon_k)} \\ &= \prod_{k=1}^n \frac{f(\varepsilon_{n-k+1} - \varphi_k(\theta_0)\alpha\zeta_{n-k})}{f(\varepsilon_{n-k+1})} \\ &= \prod_{k=1}^n \frac{f(\varepsilon'_k - \varphi_k(\theta_0)\alpha\zeta_{n-k})}{f(\varepsilon'_k)} \end{aligned}$$

with $\varepsilon'_k = \varepsilon_{n-k+1}$.

Now we notice that the main influence in the latter product is delivered by the terms with small k since $\varphi_k(\theta_0)$ decreases to zero exponentially fast. Moreover, the variables ζ_{n-k} for large n are close to ζ, and ζ weakly depends on $\varepsilon'_k = \varepsilon_{n-k+1}$ for large n and small k. These remarks explain the weak convergence of $Z^n_\alpha(\theta_0)$ to

$$Z_\alpha(\theta_0) = \frac{dP_{\alpha,\theta_0}}{dP_{0,\theta_0}} = \prod_{k=1}^\infty \frac{f(e_k - \varphi_k(\theta_0)\alpha\zeta)}{f(e_k)}. \tag{5.47}$$

The above considerations induce us to take the elements $\lambda_n = \lambda_n(\theta_0)$ for proving λ-convergence in the form

$$\lambda_n = \left(\theta_0^{-n} X_n; \varepsilon'_1, \varepsilon'_2, \dots\right).$$

Here we set $\varepsilon'_k = \varepsilon_{n-k+1} = X_{n-k+1} - \theta_0 X_{n-k}$, for $k \le n$ and define for $k > n$ the variables ε'_k independent of each other and of $\varepsilon_1, \dots, \varepsilon_n$ and identically distributed with the density $f = f(x)$. □

5.3 Asymptotic Efficient Minimax Estimation of Autoregression Parameter

In the present section we study the question of asymptotic efficient estimation for the autoregressive model (5.1). As in the previous section we distinguish between the cases of Gaussian and non-Gaussian errors ε_i.

Let $W = W(x)$, $x \in R$, be a loss function satisfying the standard conditions

$$(W) \qquad \begin{array}{l} W(x) \geq 0, \quad W(0) = 0; \\ W(x) = W(-x), \quad x \in R; \\ W(x) \text{ is an increasing function on } R_+. \end{array}$$

Let Θ be some parametric set on the real line and let $\Sigma = (\mathbb{E}^n, n \geq 1)$ be the sequence of experiments corresponding to the autoregressive model (5.1).

Definition 5.1 The *asymptotic minimax risk* for the experiments Σ and the loss function $W = W(x)$ is defined as:

$$\mathcal{R}(\Sigma, W, \Theta) = \liminf_n \inf_{\widetilde{\theta}_n} \sup_{\theta \in \Theta} \mathbf{E}_\theta^n W\left(\varphi_n^{-1}(\theta)(\widetilde{\theta}_n - \theta)\right).$$

where the normalizing multipliers $\varphi_n(\theta)$ are defined in (5.30) and inf is taken over the class of all (nonrandomized) estimators $\widetilde{\theta}_n$ of the parameter θ.

The local λ-convergence $\mathbb{E}^n \stackrel{(\lambda,\varphi)}{\longrightarrow} \mathcal{E} = (\mathcal{E}(\theta_0), \theta_0 \in \Theta)$ stated above in Section 5.2 allows us to claim, due to Theorem 4.1, that

$$\mathcal{R}(\Sigma, W, \Theta) \geq \sup_{\theta_0 \in \Theta} \mathcal{R}(\mathcal{E}(\theta_0), W) \tag{5.48}$$

where $\mathcal{R}(\mathcal{E}(\theta_0), W)$ is the minimax risk in the experiment $\mathcal{E}(\theta_0)$:

$$\mathcal{R}(\mathcal{E}(\theta_0), W) = \inf_{\varkappa} \sup_{\alpha \in R} E_{\alpha,\theta_0} W(\varkappa - \alpha).$$

Below we calculate the lower bound in the right-hand side of (5.48) and prove its accuracy at least for a compact parametric set Θ not containing the points ± 1. We consider separately the cases of Gaussian errors ε_i and errors with a regular density $f = f(x)$. Note that the structure of asymptotically efficient (in the minimax sense) estimators and the value of the lower bound in (5.48) are different for these two cases.

5.3.1 *Gaussian case. Lower bounds*

We start with the case of Gaussian errors ε_i: $\varepsilon_i \sim \mathcal{N}(0, 1)$.

In the *stable* situation ($|\theta| < 1$) the experiment $\mathcal{E}(\theta_0)$ coincides with the Gaussian shift experiment $\mathcal{E}_G$ corresponding to the observation $\gamma \sim \mathcal{N}(\alpha, 1)$ (see Section 5.2 above). For such an experiment and for the loss function W satisfying the condition (W) the minimax estimator $\varkappa^*(\gamma)$ is simply γ and as a consequence

$$\begin{aligned} \mathcal{R}(\mathcal{E}(\theta_0), W) &= \mathcal{R}(\mathcal{E}_G, W) \\ &= E\,W(\xi) \\ &= \frac{1}{\sqrt{2\pi}} \int_{-\infty}^{\infty} W(x)\, e^{-x^2/2} dx \end{aligned}$$

with $\xi \sim \mathcal{N}(0,1)$ (see Chapter 3).

In the explosive case with $|\theta_0| > 1$, the experiment $\mathcal{E}(\theta_0)$ is mixed normal: $\mathcal{E}(\theta_0) = \mathcal{E}_{MG}$. This experiment $\mathcal{E}_{MG}$ is generated by the pair of observations (γ, Γ) with $\Gamma = |\xi|$, $\xi \sim \mathcal{N}(0,1)$ and the conditional distribution of γ given Γ is the $\mathcal{N}(\Gamma\alpha, 1)$-distribution. In this case for any loss function satisfying condition (W), the minimax estimator is (see Strasser (1985, Section 38) or Section 5.3 below)

$$\varkappa^*(\gamma, \Gamma) = \Gamma^{-1}\gamma$$

and

$$\mathcal{R}(\mathcal{E}(\theta_0), W) = \mathcal{R}(\mathcal{E}_{MG}, W) = E\,W(\eta^{-1}\xi)$$

where ξ and η are independent standard Gaussian random variables. The variable $\eta^{-1}\xi$ has the Cauchy distribution and therefore,

$$\mathcal{R}(\mathcal{E}(\theta_0), W) = \frac{1}{\pi}\frac{1}{\sqrt{2\pi}}\int_{-\infty}^{\infty} W(x)\,\frac{dx}{1+x^2}.$$

The above consideration leads to the following result.

Theorem 5.8 *For the model (5.1) with Gaussian errors ε_i and a loss function W satisfying condition (W), the following inequality holds for any estimators $\widetilde{\theta}_n$ and each $\delta > 0$:*

$$\lim_n \sup_{|\theta|\le 1-\delta} \mathbf{E}_\theta^n W\left(\sqrt{\frac{n}{1-\theta^2}}\left(\widetilde{\theta}_n - \theta\right)\right) \ge E\,W(\xi)$$

where $\xi \sim \mathcal{N}(0,1)$ and

$$\lim_n \sup_{|\theta|\ge 1+\delta} \mathbf{E}_\theta^n W\left(\frac{|\theta|^n}{\theta^2-1}\left(\widetilde{\theta}_n - \theta\right)\right) \ge E\,W(\zeta)$$

where ζ has the Cauchy distribution.

5.3.2 *Gaussian case. Upper bounds*

We now consider the question of attainability of the lower bounds from above and about construction of asymptotically efficient estimators on which these lower bounds can be attained.

In the case of Gaussian errors this question can be solved rather easily. The reason is that in the stable case as well as in the explosive case the minimax estimator for each limit experiment $\mathcal{E}(\theta)$ is the maximum likelihood one. Thus, the natural candidates for efficient estimators for the original experiments $\mathbf{E}^n$ are also maximum likelihood estimators which coincide in this situation with the least square estimators

$$\rho_n = \frac{\sum_{k=1}^n X_{k-1}X_k}{\sum_{k=1}^n X_{k-1}^2}. \tag{5.49}$$

In Section 5.1 we have shown that for $|\theta| < 1$

$$\mathcal{L}\left(\varphi_n^{-1}(\theta)\,(\rho_n - \theta) \mid \mathbf{P}_\theta^n\right) \to \mathcal{N}(0,1) \tag{5.50}$$

and for $|\theta| > 1$

$$\mathcal{L}\left(\varphi_n^{-1}(\theta)\,(\rho_n - \theta) \mid \mathbf{P}_\theta^n\right) \to \mathcal{L}(\zeta) \tag{5.51}$$

where $\mathcal{L}(\zeta)$ is the Cauchy distribution.

Actually, one can easily verify that the convergence in (5.50) and (5.51) is uniform on compacts (in θ), thus obtaining the following result:

Theorem 5.9 *Let (ε_i) be standard Gaussian errors, and let a loss function $W = W(x)$ be bounded and continuous. Then for any $\delta > 0$ and $b > 1+\delta$*

$$\lim_n \sup_{|\theta|\le 1-\delta} \mathbf{E}_\theta^n W\left(\sqrt{\frac{n}{1-\theta^2}}\,(\rho_n - \theta)\right) = E\,W(\xi)$$

where $\xi \sim \mathcal{N}(0,1)$ and

$$\lim_n \sup_{1+\delta\le|\theta|\le b} \mathbf{E}_\theta^n W\left(\frac{|\theta|^n}{\theta^2-1}\,(\rho_n - \theta)\right) = E\,W(\zeta)$$

where ζ has the Cauchy distribution.

5.3.3 *Non-Gaussian case*

Now we consider the situation when the errors ε_i in the model (5.1) are not necessary Gaussian but they have a distribution with a regular density $f = f(x)$.

We start with the stable case. As in Section 5.2 we assume that the density f satisfies the condition (F) and has a finite Fisher information I.

In this situation (see Section 5.2) the limit experiments $\mathcal{E}(\theta_0)$ coincide with the Gaussian shift experiment $\mathcal{E}_{G,I}$ corresponding to one observation $\gamma \sim \mathcal{N}\left(I^{1/2}\alpha, 1\right)$.

The minimax estimator $\varkappa^* = \varkappa^*(\gamma)$ for $\mathcal{E}_{G,I}$ is of the following form

$$\varkappa^*(\gamma) = I^{-1/2}\gamma$$

and the minimax risk equals:

$$\mathcal{R}(\mathcal{E}_{G,I}, W) = E\,W(I^{-1/2}\xi) = \frac{1}{\sqrt{2\pi}}\int_{-\infty}^{\infty} W(I^{-1/2}x)\,e^{-x^2/2}dx.$$

This and (5.48) yield the following result.

Theorem 5.10 *Let the density $f = f(x)$ satisfy the condition $(\underset{\sim}{F})$. Then for any $\delta > 0$, any loss function W satisfying (W) and any estimators $\widetilde{\theta}_n$, $n \ge 1$, the following inequality holds*

$$\lim_n \sup_{|\theta|\le 1-\delta} \mathbf{E}_\theta^n W\left(\sqrt{\frac{n}{1-\theta^2}}\left(\widetilde{\theta}_n - \theta\right)\right) \ge E\,W(I^{-1/2}\xi) \tag{5.52}$$

where $\xi \sim \mathcal{N}(0,1)$.

Now we consider the structure of asymptotically efficient estimators which give the equality in (5.52). First of all we note that the same methods used above allow us to obtain the following result on consistency of the least square estimators.

Theorem 5.11 *Let the density* $f = f(x)$ *satisfy the condition* (F). *Then the least square estimators* ρ_n *defined by (5.49) fulfill for any continuous bounded loss function* $W = W(x)$ *and any* $\delta > 0$ *the following conditions:*

$$\lim_n \sup_{|\theta|\le 1-\delta} \mathbf{P}_\theta^n \left(\sqrt{\frac{n}{1-\theta^2}}|\rho_n - \theta| > b\right) \to 0, \quad b \to \infty, \tag{5.53}$$

and

$$\lim_n \sup_{|\theta|\le 1-\delta} \mathbf{E}_\theta^n W\left(\sqrt{\frac{n}{1-\theta^2}}(\rho_n - \theta)\right) = E\,W(\xi)$$

where $\xi \sim \mathcal{N}(0,1)$.

It is easy to check (see, for instance, Ibragimov and Khasminskii(1981, Section I.7)) that under the condition (F) the Fisher information I satisfies $I \ge 1$ and $I = 1$ only for the Gaussian density. Therefore, as usual, $I > 1$, and for functions W obeying (W) one has

$$E\,W(\xi) > E\,W(I^{-1/2}\xi).$$

This means that the estimators ρ_n, $n \ge 1$, are $\varphi_n(\theta)$-consistent (in the sense of (5.53)) but they are not asymptotically efficient.

For construction of asymptotically efficient estimators we use the results of Chapter 4 where the asymptotic efficiency of the global estimators T_n, $n \ge 1$ was stated.

It is natural to take the estimators ρ_n as an *initial* approximation since these estimators satisfy the consistency condition (5.53). Strictly speaking, in order to apply the results of Section 5.4 we need also the discreteness property for ρ_n. Having ρ_n one can easily construct its discrete version choosing a $\frac{1}{\sqrt{n}}$-grid on $[-1,1]$ and taking the point of this grid closest to ρ_n . Aiming to avoid new notation we assume in what follows that this procedure has been made and ρ_n means just the ***discretization*** of the estimators (5.49).

In accordance with Section 5.4 the global estimators T_n have the following form:

$$T_n = \rho_n + \varphi_n(\rho_n)\varkappa^*_{\rho_n}(\lambda_n(\rho_n))$$

where $\varkappa^*_{\theta_0}$ is the minimax estimator for the limit experiment $\mathcal{E}(\theta_0)$, and $\lambda_n(\theta_0)$ is the statistic entering in the condition of local λ-convergence of $\mathbb{E}^n$ to $\boldsymbol{\mathcal{E}} = (\mathcal{E}(\theta_0), \theta_0 \in \Theta)$.

In the case under consideration $\varkappa^*(\lambda) = I^{-1/2}\lambda$ and for $\lambda_n(\theta_0)$ one can use the representation from Section 5.2

$$\lambda_n(\theta_0) = I^{-1/2}\varphi_n(\theta_0)\sum_{k=1}^{n} X_{k-1}\frac{f'(X_k - \theta_0 X_{k-1})}{f(X_k - \theta_0 X_{k-1})}. \tag{5.54}$$

Thus, we obtain that

$$T_n = \rho_n + I^{-1}\frac{1-\rho_n^2}{n}\sum_{k=1}^{n} X_{k-1}\frac{f'(X_k - \rho_n X_{k-1})}{f(X_k - \rho_n X_{k-1})}. \tag{5.55}$$

The next result is a direct corollary of Theorem 4.9.

Theorem 5.12 *Let the density $f = f(x)$ satisfy the condition (F). Then for any continuous and bounded loss function $W = W(x)$ and any $\delta > 0$ the estimators T_n defined by (5.55) are asymptotically efficient (in the minimax sense) in the domain $\Theta = [1-\delta, 1+\delta]$ and*

$$\lim_n \sup_{|\theta|\le 1-\delta} \mathbf{E}_\theta^n W\left(\sqrt{\frac{n}{1-\theta^2}}(T_n - \theta)\right) = E\,W(\xi) \tag{5.56}$$

where $\xi \sim \mathcal{N}(0,1)$.

Remark 5.5 Let $f(x)$ be the Gaussian density ($f(x) = \varphi(x)$). Then (5.54) takes the following form

$$\lambda_n(\theta_0) = \varphi_n(\theta_0)\sum_{k=1}^{n} X_{k-1}\left(X_k - \theta_0 X_{k-1}\right) = \varphi_n(\theta_0)\left[\sum_{k=1}^{n} X_{k-1}X_k - \theta_0\sum_{k=1}^{n} X_{k-1}^2\right]$$

and

$$\lambda_n(\rho_n) = \varphi_n(\theta_0)\left[\sum_{k=1}^{n} X_{k-1}X_k - \frac{\sum_{k=1}^{n} X_{k-1}X_k}{\sum_{k=1}^{n} X_{k-1}^2}\sum_{k=1}^{n} X_{k-1}^2\right] = 0.$$

Therefore, $T_n = \rho_n$. In other words, in this case the global estimators T_n coincide with the maximum likelihood estimators and the assertion (5.56) coincides with the result of Theorem 5.9.

Finally we consider the *explosive* case and *non-Gaussian errors.* Due to the results of the previous section, under the condition (F') the condition of local λ-convergence holds uniformly on compacts $\mathbb{E}^n \overset{(\lambda,\varphi)}{\longrightarrow} \mathcal{E} = (\mathcal{E}(\theta_0),\ \theta_0 \in \Theta)$ and, contrary to the preceding cases, the structure of each experiment $\mathcal{E}(\theta_0)$ depends on the point θ_0.

The experiments $\mathcal{E}(\theta_0)$ are transitive with respect to the action of operators G_{α,θ_0} (see Section 5.2). This implies (see Strasser(1985, Section 39)) that the minimax estimator $\varkappa_{\theta_0}^*$ for $\mathcal{E}(\theta_0)$ is the Pitman estimator, i.e. the generalized Bayes estimator for the uniform prior distribution on the real line.

It is more convenient now to represent $\varkappa^*_{\theta_0}$ as a function of the likelihood process $\mathcal{Z}(\theta_0) = (Z_\alpha(\theta_0),\ \alpha \in R)$ for the experiment $\mathcal{E}(\theta_0)$:

$$\varkappa^*_{\theta_0}(\mathcal{Z}(\theta_0)) = \arg\inf_{\varkappa \in R} \int_{-\infty}^{\infty} W(\varkappa - \alpha)\, Z_\alpha(\theta_0)\, d\alpha. \tag{5.57}$$

In the case of the quadratic loss function ($W(x) = x^2$) the estimator $\varkappa^*_{\theta_0}$ coincides with the posterior mean

$$\varkappa^*_{\theta_0}(\mathcal{Z}(\theta_0)) = \frac{\int_{-\infty}^{\infty} \alpha\, Z_\alpha(\theta_0)\, d\alpha}{\int_{-\infty}^{\infty} Z_\alpha(\theta_0)\, d\alpha}. \tag{5.58}$$

Note that the estimator $\varkappa^*_{\theta_0}$ from (5.57) is *equivariant*, i.e. $\varkappa^*_{\theta_0}(T_\beta Z) = \varkappa^*_{\theta_0}(Z) - \beta$ where, for $Z = (Z_\alpha, \alpha \in R)$, we set $T_\beta Z = (Z_{\alpha+\beta}/Z_\beta, \alpha \in R)$. Moreover, the distribution $\mathcal{L}\left(\varkappa^*_{\theta_0}(\mathcal{Z}(\theta_0)) - \alpha \mid P_{\alpha,\theta_0}\right)$ does not depend on $\alpha \in R$. Thus

$$\mathcal{R}(\mathcal{E}(\theta_0), W) = E_{0,\theta_0} W\left(\varkappa^*_{\theta_0}(\mathcal{Z}(\theta_0))\right) \tag{5.59}$$

and taking into account (5.48) we obtain the following result.

Theorem 5.13 *Let the density $f = f(x)$ satisfy condition (F'). Then for any $\delta > 0$, $b > 1 + \delta$, any loss function W satisfying condition (W) and any estimators $\widetilde{\theta}_n$, $n \geq 1$, the following inequality holds:*

$$\lim_n \sup_{1+\delta \leq |\theta| \leq b} \mathbf{E}^n_\theta W\left(\frac{|\theta|^n}{\theta^2 - 1}\left(\widetilde{\theta}_n - \theta\right)\right) \geq \sup_{1+\delta \leq |\theta| \leq b} \mathcal{R}(\mathcal{E}(\theta_0), W) \tag{5.60}$$

where $\mathcal{R}(\mathcal{E}(\theta_0), W)$ is defined in (5.58) through (5.59).

To construct estimators T_n attaining the lower bound (5.60) we use an approach similar to the stable case.

First of all we note that again under condition (F') the least square estimators ρ_n are strongly consistent: for any $\delta > 0$

$$\lim_n \sup_{1+\delta \leq |\theta| \leq b} \mathbf{P}^n_\theta \left(\frac{|\theta|^n}{\theta^2 - 1}|\rho_n - \theta| > b\right) \to 0, \quad b \to \infty.$$

Hence the estimators ρ_n ,$n \geq 1$, (more precisely, their discretized versions denoted also ρ_n) can be used as an initial approximation for the construction of the global estimators T_n:

$$T_n = \rho_n + \varphi_n(\rho_n)\varkappa^*_{\rho_n}\left(\widehat{\mathcal{Z}}^n(\rho_n)\right)$$

where $\varkappa^*_{\theta_0}$ is the minimax estimator for $\mathcal{E}(\theta_0)$,

$$\widehat{\mathcal{Z}}^n(\theta_0) = \mathcal{Z}(\theta_0) \circ \widehat{\lambda}_n(\theta_0) = \mathcal{Z}\left(\theta_0, \widehat{\lambda}_n(\theta_0)\right),$$

$\mathcal{Z}(\theta_0) = \mathcal{Z}(\theta_0, \lambda) = (Z_\alpha(\theta_0, \lambda),\ \alpha \in R)$ is the likelihood process for the limit experiment $\mathcal{E}(\theta_0)$ (see (5.47)), and $\widehat{\lambda}_n(\theta_0)$ is the modified version of the statistic $\lambda_n(\theta_0)$

entering in the definition of λ-convergence (see Section 2.5 of Chapter 2 and the preceding section).

In the case under consideration we can propose for $\widehat{\lambda}_n(\theta_0)$ an explicit construction setting, for example,

$$\widehat{\lambda}_n(\theta_0) = \left(\widehat{Z}_n; \widetilde{\varepsilon}_1, \widetilde{\varepsilon}_2, \dots\right), \tag{5.61}$$

where $\widetilde{\varepsilon}_k = \varepsilon_{n-k+1} = X_{n-k+1} - \theta_0 X_{n-k}$ for $k \le \left[\frac{n}{2}\right]$ and for $k > \left[\frac{n}{2}\right]$ the variables $\widetilde{\varepsilon}_k$ are independent of one another and of $\varepsilon_1, \dots, \varepsilon_n$ but have the same distribution with the density $f = f(x)$, and

$$\widehat{Z}_n = \sum_{k=1}^{[n/2]} \theta_0^{-k} \varepsilon_k + \sum_{k=[n/2]+1}^{n} \theta_0^{-k} \varepsilon'_k = \sum_{k=1}^{[n/2]} \theta_0^{-k}(X_k - \theta_0 X_{k-1}) + \sum_{k=[n/2]+1}^{n} \theta_0^{-k} \varepsilon'_k$$

where (ε'_k) is one more set of random variables independent of (ε_k) and $(\widetilde{\varepsilon}_k)$, with the density $f(x)$.

It is easy to verify that for each $|\theta_0| > 1$

$$\mathcal{L}\left(\widehat{\lambda}_n(\theta_0) \mid \mathbf{P}^n_{\theta_0}\right) = P_{0,\theta_0}.$$

Theorem 5.14 *Let the density $f = f(x)$ satisfy condition (F'). Then for any continuous bounded loss function $W = W(x)$ the estimators*

$$\widehat{T}_n = \rho_n + \frac{\rho_n^2 - 1}{|\rho_n|^n} \varkappa^*_{\rho_n}\left(\mathcal{Z}\left(\theta_0, \widehat{\lambda}_n(\theta_0)\right)\right)$$

with $\varkappa^*_{\theta_0}(\mathcal{Z})$ from (5.57), $\widehat{\lambda}_n$ from (5.61) and $\mathcal{Z}(\theta_0, \lambda)$ from (5.47) are asymptotically efficient for each compact parametric set Θ outside the interval $[-1, 1]$, i.e.

$$\lim_n \sup_{\theta \infty \Theta} \mathbf{E}^n_\theta W\left(\frac{|\theta|^n}{\theta^2 - 1}\left(\widehat{T}_n - \theta\right)\right)$$
$$= \liminf_n \inf_{\widetilde{\theta}_n} \sup_{\theta \infty \Theta} \mathbf{E}^n_\theta W\left(\frac{|\theta|^n}{\theta^2 - 1}\left(\widetilde{\theta}_n - \theta\right)\right) = \sup_{\theta \infty \Theta} \mathcal{R}\left(\mathcal{E}(\theta_0), W\right) \tag{5.62}$$

where $\mathcal{R}(\mathcal{E}(\theta_0), W)$ is the minimax risk in the experiment $\mathcal{E}(\theta_0)$ defined by the equality (5.59).

Proof. It consists of a simple verification of the conditions of Theorem 4.7 of Chapter 4 and is left to the reader. □

Remark 5.6 In order to prove (5.62) we need the condition of $(\mathcal{K}, T)$-uniform local λ-convergence of $\mathbb{E}^n$ to $\boldsymbol{\mathcal{E}} = (\mathcal{E}(\theta_0), \theta \in \Theta)$ for some neighborhood Θ' of the set Θ. But since Θ is a compact, there is another compact in the explosive domain $|\theta| > 1$ which contains Θ with some neighborhood and Theorem 5.9 can be applied to the set Θ'.

Remark 5.7 If the function $f = f(x)$ is continuous, then the assertion of Theorem 5.14 remains valid if one uses instead of the estimators $\widehat{T}_n$ the estimators

$$T_n = \rho_n + \frac{\rho_n^2 - 1}{|\rho_n|^n} \varkappa^*_{\rho_n} \left(\mathcal{Z}^n(\rho_n)\right)$$

where $\mathcal{Z}^n(\theta_0) = (Z^n_\alpha(\theta_0), \alpha \in R)$ and

$$Z^n_\alpha(\theta_0) = \prod_{i=1}^n \frac{f(X_i - \frac{\theta_0^2 - 1}{|\theta_0|^n} X_{i-1})}{f(X_i - \theta_0 X_{i-1})}.$$

This follows from Theorem 4.3 since continuity of the function $f(x)$ yields continuity of the function $Z_\alpha(\theta_0)$ in the set of arguments.

5.4 Sequential Estimation

Due to the results of Section 5.1, the use of the random factor $s_n^{-1} = \sqrt{\langle M\rangle_n}$ instead of $\sqrt{I_n(\theta)} = \sqrt{\mathbf{E}_\theta \langle M\rangle_n}$ enables us to reduce *three* limit distributions (as in Theorem 5.1) for $\sqrt{I_n(\theta)}(\widehat{\theta}_n - \theta)$ to *two* limit distributions (as in Theorem 5.2)) for $s_n^{-1}(\widehat{\theta}_n - \theta)$. It is natural to ask whether one can go further and achieve, probably by some modification of the maximum likelihood estimator, the situation with *one* universal limit distribution for *all* admissible values of the unknown parameter. The results of this section show that so-called ***sequential maximum likelihood*** estimators possess the following desirable property: being normalized properly, they are asymptotically $\mathcal{N}(0,1)$ distributed for all values of the parameter θ.

We consider again the first-order autoregressive model

$$X_i = \theta X_{i-1} + \varepsilon_i, \qquad i \geq 1, \tag{5.63}$$

where $X_0 = 0$ and $\varepsilon = (\varepsilon_i)_{i \geq 1}$ is a sequence of i.i.d. standard Gaussian variables, $\mathbf{E}\varepsilon_i = 0$, $\mathbf{E}\varepsilon_i^2 = 1$.

Let

$$M_n = \sum_{i=1}^n X_{i-1}\varepsilon_i, \qquad \langle M\rangle_n = \sum_{i=1}^n X_{i-1}^2,$$

and let for every $h > 0$

$$\tau(h) = \inf\{n \geq 1 : \langle M\rangle_n \geq h\}$$

with $\tau(h) = +\infty$ if $\langle M\rangle_n < h$ for all $n \geq 1$.

The random variables $\tau(h)$ are (finite) Markov (stopping) times and a straightforward calculation shows that $\tau(h) \to \infty$ as $h \to \infty$ for each $\theta \in R$ $\mathbf{P}_\theta$-a.s. Moreover, see Lai and Siegmund (1983), Konev and Pergamentshchicov (1986) for

$|\theta| < 1$

$$\lim_{h\to\infty} \frac{\tau(h)}{h} = 1 - \theta^2 \qquad (\mathbf{P}_\theta\text{-a.s.})$$

and for $|\theta| = 1$

$$\frac{\tau(h)}{\sqrt{h}} \xrightarrow{d} \inf\{t \geq 0 : \int_0^t w^2(s)ds \geq 1\}.$$

Above in Section 5.1 we considered maximum likelihood estimators

$$\widehat{\theta}_n = \left(\sum_{i=1}^{\tau(h)} X_{i-1}^2\right)^{-1} \sum_{i=1}^{n} X_{i-1}X_i. \tag{5.64}$$

Now we study *sequential maximum likelihood estimators* (sequential MLE)

$$\widehat{\theta}_{\tau(h)} = \left(\sum_{i=1}^{\tau(h)} X_{i-1}^2\right)^{-1} \sum_{i=1}^{\tau(h)} X_{i-1}X_i, \qquad h > 0. \tag{5.65}$$

Obviously, for each $\theta \in R$ and $h > 0$

$$\widehat{\theta}_{\tau(h)} = \theta + \frac{M_{\tau(h)}}{\langle M\rangle_{\tau(h)}}, \qquad (\mathbf{P}_\theta\text{-a.s.}) \tag{5.66}$$

and

$$\mathbf{E}_\theta(\widehat{\theta}_{\tau(h)} - \theta) = \mathbf{E}_\theta \frac{M_{\tau(h)}}{\langle M\rangle_{\tau(h)}}. \tag{5.67}$$

The definition of $\tau(h)$ means that the denominator $\langle M\rangle_{\tau(h)}$ is "close" to h. (In fact, we have $\langle M\rangle_{\tau(h)} \geq h$, and if the exact equality $\langle M\rangle_{\tau(h)} = h$ holds here then the sequential MLE would have the desirable unbiased property: $\mathbf{E}_\theta \widehat{\theta}_{\tau(h)} = \theta$. Unfortunately, in general, $\langle M\rangle_{\tau(h)} \neq h$ and $\mathbf{E}_\theta \widehat{\theta}_{\tau(h)} \neq \theta$).

A slight modification of the above definition allows to obtain unbiased sequential estimators, see Anscombe (1953), Borisov and Konev (1977), where the following estimators were considered:

$$\theta^*_{\tau(h)} = \frac{1}{h}\left(\sum_{i=1}^{\tau(h)} X_{i-1}X_i + \alpha_{\tau(h)}X_{\tau(h)}X_{\tau(h)+1}\right). \tag{5.68}$$

The correcting multiplier $\alpha_{\tau(h)}$ is to be chosen by the relation

$$\sum_{i=1}^{\tau(h)} X_{i-1}^2 X_i + \alpha_{\tau(h)} X_{\tau(h)}^2 = h. \tag{5.69}$$

This definition implies $0 \le \alpha_{\tau(h)} \le 1$ ($\mathbf{P}_\theta$-a.s.). It was shown in Anscombe (1953), Borisov and Konev (1977) that $\tau(h) < \infty$ ($\mathbf{P}_\theta$-a.s.), $\theta \in R$,

$$\mathbf{E}_\theta \theta^*_{\tau(h)} = \theta, \qquad \theta \in R, \tag{5.70}$$

i.e. $\theta^*_{\tau(h)}$ is the unbiased estimator of θ and, moreover,

$$\sup_\theta \mathbf{E}_\theta (\theta^*_{\tau(h)} - \theta)^2 \le \frac{1}{h}, \qquad \theta \in R \tag{5.71}$$

Note, however, that from the asymptotic point of view (as $h \to \infty$), the estimators $\theta^*_{\tau(h)}$ and $\widehat{\theta}_{\tau(h)}$ possess the same properties.

Now we state the main result describing the asymptotic properties of the sequential estimators $\widehat{\theta}_{\tau(h)}$ as $h \to \infty$. Everywhere we write s_n for $\langle M\rangle_n^{-1/2}$.

Theorem 5.15 *a) For each* $-\infty < \theta < \infty$

$$\sup_x \left| \mathbf{P}_\theta \left(s^{-1}_{\tau(h)} (\widehat{\theta}_{\tau(h)} - \theta) \le x \right) - \Phi(x) \right| \to 0, \qquad h \to \infty. \tag{5.72}$$

b) The convergence in (5.72) is uniform on the parameter θ *in the sense that*

$$\sup_{|\theta|\le 1} \sup_x \left| \mathbf{P}_\theta \left(s^{-1}_{\tau(h)} (\widehat{\theta}_{\tau(h)} - \theta) \le x \right) - \Phi(x) \right| \to 0, \qquad h \to \infty, \tag{5.73}$$

and for all constants r *and* R *with* $1 < r \le R < \infty$

$$\sup_{r\le|\theta|\le R} \sup_x \left| \mathbf{P}_\theta \left(s^{-1}_{\tau(h)} (\widehat{\theta}_{\tau(h)} - \theta) \le x \right) - \Phi(x) \right| \to 0, \qquad h \to \infty. \tag{5.74}$$

Theorem 5.16 *For all constants* r *and* R *with* $1 < r \le R < \infty$ *and each* $h > 0$, *there exist a constant* $C(r, R)$ *and a standard normal random variable* $\zeta^*_h \sim \mathcal{N}(0,1)$ *such that*

$$\sup_{r\le|\theta|\le R} \mathbf{P}_\theta \left(\left| s^{-1}_{\tau(h)} (\widehat{\theta}_{\tau(h)} - \theta) - \zeta^*_h \right| > 3h^{-\gamma} \right) \le C(r, R)\, h^{-\gamma} \tag{5.75}$$

with $\gamma = 1/13$.

The statement (5.73) of Theorem 5.15 is proved in Mikulski and Monsour (1991), see also Grambsch (1983) and Greenwood and Shiryaev (1992). It remains then to prove only the statements related to the case of $|\theta| > 1$.

Remark 5.8 The result of Theorem 5.16 obviously implies the following strengthening of (5.74):

$$\sup_{\theta\le 1} \sup_x \left| \mathbf{P}_\theta \left(s^{-1}_{\tau(h)} (\widehat{\theta}_{\tau(h)} - \theta) \le x \right) - \Phi(x) \right| \le C(r, R)\, h^{-\gamma}.$$

So (5.75) implies the assertion (5.74) of Theorem 5.15.

The proof of the assertion (5.75) is rather difficult and it makes essentially use of the assumption of *independence* and *normality* of the variables $\varepsilon_1, \varepsilon_2, \ldots$.

Before turning to the proof, we present some discussion explaining the idea behind the result. Represent (for given θ with $|\theta| > 1$) the expression $s_n^{-1}(\widehat{\theta}_n - \theta)$ under consideration in the following form:

$$s_n^{-1}(\widehat{\theta}_n - \theta) = \frac{M_n}{\langle M \rangle_n^{1/2}} = \frac{\xi_n}{\eta_n},$$

where $\xi_n = \xi_n(\theta)$ and $\eta_n = \eta_n(\theta)$ are defined by the equalities

$$\xi_n = \frac{M_n}{|\theta|^n}, \qquad \eta_n = \frac{\langle M \rangle_n^{1/2}}{|\theta|^n}. \tag{5.76}$$

Introduce also random variables $z = z(\theta)$, $\zeta_n = \zeta_n(\theta)$ with

$$\begin{aligned} z(\theta) &= \sum_{k=1}^{\infty} \theta^{-k} \varepsilon_k, \\ \zeta_n &= \operatorname{sign} z \sqrt{1 - \theta^2} \sum_{k=1}^{n} \theta^{-(n-k)} \varepsilon_k. \end{aligned} \tag{5.77}$$

Note that z is normal with parameters $(0, \theta^2 - 1)$.

Let the sign "$\sim$" mean asymptotic equivalence in distribution. Then

$$\theta^{-n} X_n = \sum_{k=1}^{n} \theta^{-k} \varepsilon_k \sim z. \tag{5.78}$$

We will make more precise the meaning of this statement later on. By (5.64), this yields, for sufficiently large n,

$$\frac{\xi_n}{\eta_n} \sim \zeta_n.$$

This allows to reduce the study of the asymptotic (as $h \to \infty$) behavior of $\xi_{\tau(h)}/\eta_{\tau(h)}$ to that of $\zeta_{\tau(h)}$.

The variables $\zeta_{\tau(h)}$ coincide up to the sign with the variables

$$\sqrt{1 - \theta^2} \sum_{k=1}^{\tau(h)} \theta^{-(\tau(h)-k)} \varepsilon_k. \tag{5.79}$$

An analysis of the asymptotic behavior of this sum is based on the following remark. If $\tau(h)$ and $\varepsilon = (\varepsilon_n)_{n \geq 1}$ were independent, then the conditional distribution of $\zeta_{\tau(h)}$ upon $\tau(h)$ would be normal $\mathcal{N}(0, 1 - \theta^{-2\tau(h)})$ which tends rapidly to $\mathcal{N}(0,1)$ as $h \to \infty$. In the case under consideration, $\tau(h)$ and $\varepsilon = (\varepsilon_n)_{n \geq 1}$ are, of course, dependent. However, one can notice that the main contribution in the sum $S_{\tau(h)}$ is delivered by the terms ε_k which *immediately precede* $\tau(h)$ but the value of the stopping time $\tau(h)$ is determined mostly by the errors ε_k with *small* k

corresponding to the initial moment of observation. This means that actually, for h large, the variables $\tau(h)$ and the leading summands in $\zeta_{\tau(h)}$ are *weakly dependent.* This fact and the above discussion explain why the limit distribution of $\zeta_{\tau(h)}$ is asymptotically standard normal.

Now we turn to the accurate proof and notation. Denote by $\mathbf{C}^-$ the class of functions $C = C(\theta)$ defined on the set $\{\theta : |\theta| > 1\}$ and monotonously decreasing as $|\theta|$ increases.

Lemma 5.1 *Let τ be an arbitrary stopping time (with respect to the filtration $(\mathcal{F}_n)_{n\geq 1}$ with $\mathcal{F} = \sigma(\varepsilon_1, \dots, \varepsilon_n)$ such that $\tau \geq m$ ($\mathbf{P}_\theta$-a.s.), $|\theta| > 1$. Then*

$$\mathbf{P}_\theta\left(\left|\frac{\xi_\tau}{\eta_\tau} - \zeta_\tau\right| > |\theta|^{-m/6}\right) \leq C(\theta)|\theta|^{-\frac{m}{6}+1} \tag{5.80}$$

with $C(\theta) \in \mathbf{C}^-$.

Proof. Let, say, $\theta > 1$. Denote $\sigma_z^2 = \mathbf{E}\, z^2$ where $\mathbf{E}$ means integration with respect to the Gaussian measure $\mathbf{P}$ on the space of sequences $\varepsilon = (\varepsilon_k)_{k\geq 1}$. Then

$$\sigma_z^2 = \mathbf{E}\, z^2 = \mathbf{E}\left(\sum_{k=1}^{\infty} \theta^{-k}\varepsilon_k\right)^2 = \sum_{k=1}^{\infty} \theta^{-2k} = \frac{\theta^{-2}}{1-\theta^{-2}}.$$

Let $m \geq 1$ and let τ be a stopping time such that $\tau \geq m$ ($\mathbf{P}_\theta$-a.s.). Then for all $a > 0$, $b > 0$

$$\mathbf{P}_\theta\left(\left|\frac{\xi_\tau}{\eta_\tau} - \zeta_\tau\right| > a\right) \leq \mathbf{P}_\theta\left(\inf_{n\geq m} \eta_n < b\right) + \mathbf{P}_\theta\left(\sup_{n\geq m} |\xi_n - \eta_n\zeta_n| > ab\right). \tag{5.81}$$

Denote also

$$\rho^2 = \sum_{k=0}^{\infty} \theta^{-2k} = \frac{1}{1-\theta^{-2}},$$

$$\eta = \frac{|z|}{\sqrt{1-\theta^{-2}}} = \rho|z|.$$

We have

$$\mathbf{P}_\theta\left(\inf_{n\geq m} \eta_n < b\right) \leq \mathbf{P}_\theta\left(\sup_{n\geq m} |\eta_n - \eta| > b\right) + \mathbf{P}_\theta(\eta < 2b). \tag{5.82}$$

To estimate $\mathbf{P}_\theta(\eta < 2b)$, notice that z is a random variable with the normal distribution $\mathcal{N}(0, \frac{\theta^{-2}}{1-\theta^{-2}})$. Using this fact one can easily derive that

$$\mathbf{P}_\theta(\eta < 2b) = \mathbf{P}_\theta(|z|/\sigma_z < 2b/(\varrho\sigma_z)) = \Phi(2b\theta) - \Phi(-2b\theta) \leq 2\theta b. \tag{5.83}$$

Here $\Phi(\cdot)$ is the standard normal distribution function and we have used the inequality $|\Phi'(x)| \leq (2\pi)^{-1/2} \leq 1/2$ for all $x \in R$.

Now we present some estimate for $\mathbf{P}_\theta\{\sup_{n\ge m}|\eta_n-\eta|>b\}$. Let $z_n=z_n(\theta)$ with

$$z_n(\theta)=\theta^{-(n-1)}X_{n-1}=\sum_{k=1}^{n-1}\theta^{-k}\varepsilon_k.$$

Then

$$\mathbf{E}\,z_n^2=\sum_{k=1}^{n-1}\theta^{-2k}\le\mathbf{E}\,z^2=\frac{\theta^{-2}}{1-\theta^{-2}}\qquad(=\sigma_z^2)$$

and

$$\mathbf{E}\,(z-z_n)^2=\sum_{k=n}^{\infty}\theta^{-2k}=\theta^{-2n}\frac{1}{1-\theta^{-2}}=\theta^{-2n}\rho^2.$$

This implies for each $k\ge 1$

$$\begin{aligned}\mathbf{E}_\theta\left|\theta^{-2(k-1)}X_{k-1}^2-z^2\right| &= \mathbf{E}\,(z-z_k)^2\\ &\le \mathbf{E}\,|z_k(z-z_k)|+\mathbf{E}\,|z(z-z_k)|\\ &\le \left(\mathbf{E}\,(z-z_k)^2\right)^{1/2}\left[\left(\mathbf{E}\,z_k^2\right)^{1/2}+\left(\mathbf{E}\,z^2\right)^{1/2}\right]\\ &\le 2\rho\sigma_z\theta^{-k}. \end{aligned}\tag{5.84}$$

Notice now that

$$\eta^2=\frac{z^2}{1-\theta^{-2}}=\frac{z^2(1-\theta^{-2n})}{1-\theta^2}+\frac{z^2\theta^{-2n}}{1-\theta^2}=z^2\sum_{k=1}^{n}\theta^{-2(n-k)}+\frac{z^2\theta^{-2n}}{1-\theta^2}.$$

Therefore, taking into account (5.84), we get

$$\begin{aligned}\mathbf{E}_\theta|\eta_n^2-\eta^2| &\le \mathbf{E}_\theta\left|\theta^{-2(n-1)}\sum_{k=1}^{n}X_{k-1}^2-\sum_{k=1}^{n}\theta^{-2(n-k)}z^2\right|+\mathbf{E}\,z^2\frac{\theta^{-2n}}{1-\theta^{-2}}\\ &\le \sum_{k=1}^{n}\theta^{-2(n-k)}\,\mathbf{E}_\theta\,\theta^{-2(k-1)}\left|X_{k-1}^2-z^2\right|+\sigma_z^2\rho^2\,\theta^{-2n}\\ &\le 2\rho\sigma_z\theta^{-n}\sum_{k=1}^{n}\theta^{-k}+\sigma_z^2\rho^2\,\theta^{-2n}\\ &\le C_1^2(\theta)\theta^{-n}\end{aligned}\tag{5.85}$$

with

$$C_1^2(\theta)=\frac{2\rho\sigma_z}{1-\theta^{-1}}+\rho^2\sigma_z^2\in\mathbf{C}^-.$$

Since the values η and η_n are non-negative, it follows from (5.85),

$$\mathbf{E}_\theta|\eta_n-\eta|^2\le\mathbf{E}_\theta|\eta_n-\eta||\eta_n+\eta|=\mathbf{E}_\theta|\eta_n^2-\eta^2|\le C_1^2(\theta)\,\theta^{-n}\tag{5.86}$$

and

$$\mathbf{E}_\theta|\eta_n - \eta| \le \left(\mathbf{E}_\theta|\eta_n - \eta|^2\right)^{1/2} \le C_1(\theta)\,\theta^{-n/2}. \tag{5.87}$$

Next, the use of (5.86) yields

$$\begin{aligned}\mathbf{P}_\theta\left(\sup_{n\ge m}|\eta_n - \eta| > b\right) &\le \sum_{n\ge m}\mathbf{P}_\theta(|\eta_n - \eta| > b) \le b^{-2}\sum_{n\ge m}\mathbf{E}_\theta|\eta_n - \eta|^2 \\ &\le C_1^2(\theta)\,b^{-2}\sum_{n\ge m}\theta^{-n} \le C_2(\theta)\,b^{-2}\theta^{-m}\end{aligned} \tag{5.88}$$

with

$$C_2(\theta) = C_1^2(\theta)\,\frac{1}{1-\theta^{-1}} \in \mathbf{C}^-.$$

The inequalities (5.88) and (5.83) imply (see (5.82)) that

$$\mathbf{P}_\theta(\inf_{n\ge m}\eta < b) \le C_2(\theta)\,b^{-2}\theta^{-m} + 2\,\theta\, b. \tag{5.89}$$

Thus, we succeeded in proving the lower estimate for the probability $\mathbf{P}_\theta(\inf_{n\ge m}\eta < b)$ in the inequality (5.81). Now we estimate the value $\mathbf{P}_\theta(\sup_{n\ge m}|\xi_n - \eta_n\zeta_n| > a\,b)$ entering in the right-hand side of (5.81). We begin by estimating the mathematical expectation $\mathbf{E}_\theta|\xi_n - \eta_n\zeta_n|$. One has

$$\mathbf{E}_\theta|\xi_n - \eta_n\zeta_n| \le \mathbf{E}_\theta|(\eta - \eta_n)\zeta_n| + \mathbf{E}_\theta|\xi_n - \eta\zeta_n|. \tag{5.90}$$

Since

$$\mathbf{E}_\theta\zeta_n^2 = (1-\theta^{-2})\mathbf{E}_\theta\left(\sum_{k=1}^n \theta^{-(n-k)}\varepsilon_k\right)^2 = (1-\theta^{-2})\sum_{k=1}^n \theta^{-2(n-k)} \le 1,$$

it follows from (5.86)

$$\mathbf{E}_\theta|(\eta - \eta_n)\zeta_n| \le \left[\mathbf{E}_\theta\,\zeta_n^2\;\mathbf{E}_\theta(\eta - \eta_n)^2\right]^{1/2} \le C_1(\theta)\,\theta^{-n/2}. \tag{5.91}$$

Using the above proved equality $\mathbf{E}_\theta(z - z_k)^2 = \theta^{-2k}\rho^2$ we get

$$\begin{aligned}\mathbf{E}_\theta|\xi_n - \eta\,\zeta_n| &= \mathbf{E}_\theta\left|\theta^{-(n-1)}\sum_{k=1}^n X_{k-1}\varepsilon_k - z\sum_{k=1}^n \theta^{-(n-k)}\varepsilon_k\right| \\ &\le \mathbf{E}_\theta\left|\sum_{k=1}^n \theta^{-(n-k)}\varepsilon_k[\theta^{-(k-1)}X_{k-1} - z]\right| \\ &\le \mathbf{E}_\theta\sum_{k=1}^n \theta^{-(n-k)}\left(\mathbf{E}_\theta\varepsilon_k^2\right)^{1/2}\left[\mathbf{E}_\theta(z_k - z)\right]^{1/2} \\ &= \rho\, n\,\theta^{-n}\end{aligned}$$

and, since $n \le \theta^{n/2}/(1-\theta^{-1/2})$,

$$\mathbf{E}_\theta|\xi_n - \eta\,\zeta_n| \le C_3(\theta)\,\theta^{-n/2} \tag{5.92}$$

with

$$C_3(\theta) = \frac{\rho}{1-\theta^{-1/2}} \in \mathbf{C}^-.$$

The inequalities (5.91), (5.92) and (5.90) provide that for all $\delta > 0$

$$\begin{aligned}
\mathbf{P}_\theta\left(\sup_{n\ge m} |\xi_n - \eta_n\zeta_n| > \delta\right) &\le \sum_{n\ge m} \mathbf{P}_\theta\left(|\xi_n - \eta_n\zeta_n| > \delta\right) \\
&\le \delta^{-1}\sum_{n\ge m} \mathbf{E}_\theta|\xi_n - \eta_n\zeta_n| \\
&\le \delta^{-1}[C_1(\theta) + C_3(\theta)]\sum_{n\ge m}\theta^{-n/2} \\
&\le \delta^{-1}C_4(\theta)\,\theta^{-m/2},
\end{aligned} \tag{5.93}$$

with $C_4(\theta) = [C_1(\theta) + C_3(\theta)](1-\theta^{-1/2}) \in \mathbf{C}^-$.

Putting together the estimates (5.93) and (5.89), we conclude from (5.81) that

$$\mathbf{P}_\theta\left(\left|\frac{\xi_\tau}{\eta_\tau} - \zeta_\tau\right| > a\right) \le C_2(\theta)\,b^{-2}\theta^{-m} + 2\,\theta\,b + C_3(\theta)\,(ab)^{-1}\theta^{-m/2}.$$

By inserting here $a = \theta^{-m/6}$ and $b = \theta^{-m/6}$ we get for $\theta > 1$

$$\mathbf{P}_\theta\left(\left|\frac{\xi_\tau}{\eta_\tau} - \zeta_\tau\right| > \theta^{-m/6}\right) \le C(\theta)\,\theta^{-m/6+1}$$

with $C(\theta) \in \mathbf{C}^-$. A similar estimate holds for $\theta < -1$ and Lemma 5.1 is proved. □

In order to formulate the next auxiliary assertion concerning the asymptotic behavior of $\zeta_{\tau(h)}$ as $h \to \infty$, we introduce some additional notation.

Let $\alpha = 2/3$ and $[n\alpha]$ denote the integer part of $n\,\alpha$,

$$\begin{aligned}
\widetilde{X}_n &= \sum_{k=1}^{[n\alpha]} \theta^{n-k}\varepsilon_k, \\
\widetilde{\tau}(h) &= \inf\{n \ge 1 : \sum_{k=1}^{n} \widetilde{X}_{k-1}^2 \ge h\}, \\
\widetilde{\zeta}_n &= \operatorname{sign}\widetilde{X}_n\sqrt{1-\theta^{-2}}\sum_{k=[n\alpha]+1}^{n} \theta^{-(n-k)}\varepsilon_k, \\
\widetilde{\mathcal{F}}_n &= \sigma(\widetilde{X}_k\,,\, 1 \le k \le n) = \sigma(\varepsilon_k\,,\, 1 \le k \le [n\alpha]).
\end{aligned}$$

Lemma 5.2 *For all $m > 1$, $h > 0$ and $|\theta| > 1$*

$$\mathbf{P}_\theta(\tau(h) \ne \widetilde{\tau}(h),\, \tau(h) > m) \le C'(\theta)|\theta|^{-m/6+1} \tag{5.94}$$

and

$$\mathbf{P}_\theta\big(|\zeta_{\tau(h)} - \widetilde{\zeta}_{\widetilde{\tau}(h)}| > |\theta|^{-m/6},\ \tau(h) > m\big) \le C''(\theta)|\theta|^{-m/6+1} \tag{5.95}$$

where $C'(\theta) \in \mathbf{C}^-$, $C''(\theta) \in \mathbf{C}^-$.

Proof. We will assume $\theta > 1$. Let

$$Y_n = \sum_{k=1}^{n} X_{k-1}^2, \qquad \widetilde{Y}_n = \sum_{k=1}^{n} \widetilde{X}_{k-1}^2.$$

Then

$$\tau(h) = \inf\{n \ge 1 : Y_n \ge h\}, \qquad \widetilde{\tau}(h) = \inf\{n \ge 1 : \widetilde{Y}_n \ge h\}.$$

Note that

$$\begin{aligned}
&\mathbf{P}_\theta\big(|\zeta_{\tau(h)} - \widetilde{\zeta}_{\widetilde{\tau}(h)}| > \delta,\ \tau(h) > m\big) \\
&\qquad \le \mathbf{P}_\theta\big(\tau(h) \ne \widetilde{\tau(h)},\ \tau(h) > m\big) + \mathbf{P}_\theta\left(\sup_{n>m} |\zeta_n - \widetilde{\zeta}_n| > \delta\right).
\end{aligned}$$

Now we estimate from above each term in this inequality. Clearly

$$\begin{aligned}
&\mathbf{P}_\theta\left(\sup_{n>m} |\zeta_n - \widetilde{\zeta}_n| > \delta\right) \\
&\quad = \mathbf{P}_\theta\left(\sup_{n>m} \rho^{-1}\left|\operatorname{sign} z \sum_{k=1}^{n} \theta^{-(n-k)}\varepsilon_k - \operatorname{sign} \widetilde{X}_n \sum_{k=[n\,\alpha]+1}^{n} \theta^{-(n-k)}\varepsilon_k\right| > \delta\right) \\
&\quad \le \mathbf{P}_\theta\left(\sup_{n>m} |\operatorname{sign} z - \operatorname{sign} \widetilde{X}_n| > 0\right) + \mathbf{P}_\theta\left(\sup_{n>m} \left|\sum_{k=1}^{[n\alpha]} \theta^{-(n-k)}\varepsilon_k\right| > \rho\,\delta\right) \\
&\quad \le \mathbf{P}_\theta(|z| < \sigma) + \mathbf{P}_\theta\left(\sup_{n>m} |z - \theta^{-n}\widetilde{X}_n| > \delta\right) \\
&\qquad + \sum_{n \ge m} \rho^{-2}\delta^{-2}\,\mathbf{E}_\theta\left|\sum_{k=1}^{[n\alpha]} \theta^{-(n-k)}\varepsilon_k\right|^2 .
\end{aligned} \tag{5.96}$$

Since $\eta = \rho\,|z|$ it holds by (5.83)

$$\mathbf{P}_\theta(|z| < \delta) = \mathbf{P}_\theta(\eta < \rho\delta) \le \theta\,\rho\,\delta. \tag{5.97}$$

The equality $\theta^{-n}\widetilde{X}_n = z_{[n\alpha]}$ yields

$$\begin{aligned}\mathbf{P}_\theta\left(\sup_{n>m}|z-\theta^{-n}\widetilde{X}_n|>\delta\right) &\le \delta^{-2}\sum_{n\ge m}\mathbf{E}_\theta|z-\theta^{-n}\widetilde{X}_n|^2\\ &\le \delta^{-2}\sum_{n\ge m}\theta^{-2n\alpha}\\ &= \frac{1}{1-\theta^{-2\alpha}}\delta^{-2}\theta^{-2m\alpha}.\end{aligned}\tag{5.98}$$

Then

$$\rho^{-2}\mathbf{E}_\theta\left|\sum_{k=1}^{[n\alpha]}\theta^{-(n-k)}\varepsilon_k\right|^2 = (1-\theta^{-2})\sum_{k=1}^{[n\alpha]}\theta^{-2(n-k)} \le \theta^{-2(1-\alpha)n}.$$

This inequality and the estimates (5.97), (5.98) and (5.96) with $\alpha = 2/3$, $\delta = \theta^{-m/6}$ imply

$$\begin{aligned}&\mathbf{P}_\theta\big(\sup_{n>m}|\zeta_n-\widetilde{\zeta}_n|>\theta^{-m/6}\big)\\ &\le \theta\rho\theta^{-m/6}+\frac{\theta^{-m}}{1-\theta^{-4/3}}+\theta^{-m/3}\sum_{n\ge m}\theta^{-2(1-\alpha)n}\\ &\le C''(\theta)\theta^{-m/6+1}\end{aligned}\tag{5.99}$$

with

$$C''(\theta)=\frac{1}{(1-\theta^{-2})^{1/2}}+\frac{1}{1-\theta^{-4/3}}+\frac{1}{1-\theta^{-2/3}}\in\mathbf{C}^-$$

and (5.95) follows.

Now we estimate the probability $\mathbf{P}_\theta\big(\tau(h)\ne\widetilde{\tau}(h),\ \tau(h)>m\big)$. In the same way as for (5.88) we obtain for every $\delta>0$ and $\theta>1$

$$\mathbf{P}_\theta\big(\sup_{n>m}|\widetilde{\eta}_n-\eta|>\delta\big)\le C_2'(\theta)\delta^{-2}\theta^{-m\alpha}\tag{5.100}$$

with some $C_2'\in\mathbf{C}^-$. Next, for each $\delta\in(0,1)$ the use of (5.83) implies

$$\begin{aligned}\mathbf{P}_\theta\big(\eta\in(2\delta,\delta^{-1/2})\big) &\le \mathbf{P}_\theta(\eta<2\delta)+\mathbf{P}_\theta(\eta>\delta^{-1}/2)\\ &\le 2\theta\delta+4\delta^2\mathbf{E}_\theta\eta^2\\ &= 2\theta\delta+4\delta^2\sigma_z^2\rho^2.\end{aligned}\tag{5.101}$$

Denote for fixed values of m, h and δ, by D the random event

$$\left\{\omega:\ \tau(h)>m,\ \sup_{n\ge m}(|\eta_n-\eta|+|\widetilde{\eta}_n-\eta|)\le\delta,\ \eta\in(2\delta,\delta^{-1/2})\right\}$$

and let D^c be the complement of D.

With this notation we have

$$\begin{aligned}
&\mathbf{P}_\theta\big(\tau(h) \neq \widetilde{\tau}(h),\, \tau(h) > m\big)\\
&\quad = \mathbf{P}_\theta\big(\tau(h) \neq \widetilde{\tau}(h),\, (\tau(h) > m) \cap D\big) + \mathbf{P}_\theta\big(\tau(h) \neq \widetilde{\tau}(h),\, (\tau(h) > m) \cap D^c\big)\\
&\quad = \mathbf{P}_\theta\big(\tau(h) \neq \widetilde{\tau}(h),\, D\big) + \mathbf{P}_\theta\left(\sup_{n\geq m} (|\eta_n - \eta| + |\widetilde{\eta}_n - \eta|) > \delta\right)\\
&\qquad + \mathbf{P}_\theta\left(\eta \in (2\delta, \delta^{-1/2})\right). \qquad (5.102)
\end{aligned}$$

The last two terms here can be estimated by (5.99), (5.101) and (5.88) in the following way ($\delta = \theta^{-2m/9}$):

$$\begin{aligned}
&\mathbf{P}_\theta\left(\sup_{n\geq m} (|\eta_n - \eta| + |\widetilde{\eta}_n - \eta|) > \delta\right) + \mathbf{P}_\theta\left(\eta \in (2\delta, \delta^{-1/2})\right)\\
&\qquad \leq C_5(\theta)\, \theta^{-2m/9+1} \qquad (5.103)
\end{aligned}$$

with $C_5(\theta) \in \mathbf{C}^-$.

Next we consider $\mathbf{P}_\theta\big(\tau(h) \neq \widetilde{\tau}(h),\, D\big)$. Notice that

$$\mathbf{P}_\theta\big(\tau(h) < \widetilde{\tau}(h),\, D\big) \leq \mathbf{P}_\theta\left(\bigcup_{n>m} \{\tau(h) = n,\, \widetilde{\tau}(h) > n,\, D\}\right).$$

Here

$$\begin{aligned}
\{\tau(h) = n,\, \widetilde{\tau}(h) > n,\, D\} &\subseteq \{Y_n \geq h,\, \widetilde{Y}_n < h,\, D\}\\
&= \{\eta_n \geq \theta^{-2(n-1)}h,\, \widetilde{\eta}_n < \theta^{-2(n-1)}h,\, D\}.
\end{aligned}$$

It is easy to see that for all $H > 0$

$$\begin{aligned}
&\mathbf{P}_\theta(\eta_n \geq H,\, \widetilde{\eta}_n < H,\, D)\\
&\quad \leq \mathbf{P}_\theta\big(\eta_n \geq H,\, \widetilde{\eta}_n < H,\, |\eta_n - \eta| + |\widetilde{\eta}_n - \eta|) < \delta,\, \eta \in (2\delta, \delta^{-1/2})\big)\\
&\quad \leq \mathbf{P}_\theta\big(\eta \in (H - \delta, H + \delta)\big) I_{\{H \in (\delta, \delta^{-1})\}}\\
&\quad \leq C_6(\theta)\, \theta\, \delta\, I_{\{H \in (\delta, \delta^{-1})\}}
\end{aligned}$$

with $C_6(\theta) \in \mathbf{C}^-$. Hence

$$\mathbf{P}_\theta\big(\tau(h) < \widetilde{\tau}(h),\, D\big) \leq \sum_{n \in I(h)} C_6(\theta)\, \theta\, \delta$$

where $I(h)$ is the set of all integer n such that $h\theta^{-2(n-1)} \in (\delta, \delta^{-1})$, i.e.

$$I(h) = \{n :\, 1 + \log_c(h\delta) \leq n \leq 1 + \log_c(h\delta^{-1})\}$$

with $c = \theta^2$.

Note that the number of elements in the set $I(h)$ is not greater than $1+\log_c \delta^{-2}$. By inserting $\delta = \theta^{-2m/9} = c^{-m/9}$ we obtain

$$\mathbf{P}_\theta\big(\tau(h) < \widetilde{\tau}(h),\, D\big) \le C_6(\theta)\,\theta\,\delta\,(1+\log_b \delta^{-2}) \le C_6(\theta)\,\theta^{-2m/9+1}\left(1+\frac{2m}{9}\right).$$

We have used here that $m\,\theta^{-am} \le (1-\theta^{-a})^{-1}$ for all $a > 0$.

In a similar way one can estimate the value $\mathbf{P}_\theta\big(\tau(h) > \widetilde{\tau}(h),\, D\big)$ using additionally the following inclusion

$$\{\tau(h) = n,\ \widetilde{\tau}(h) < n,\ D\} \subseteq \{Y_{n-1} < h,\ \widetilde{Y}_{n-1} \ge h,\ D\}.$$

This and the inequality (5.103) show that

$$\mathbf{P}_\theta\big(\tau(h) \ne \widetilde{\tau}(h),\, D\big) \le C_7(\theta)\,\theta^{-m/6+1}$$

with $C_7(\theta) \in \mathbf{C}^-$, and the assertion (5.94) follows in view of (5.99) and (5.95). Lemma 5.2 is proved. □

Let $\widetilde{\tau}$ be a Markov time with respect to the filtration $(\widetilde{\mathcal{F}}_n)_{n\ge 1}$. Let us consider the random variable

$$\widetilde{\zeta}_{\widetilde{\tau}} = \operatorname{sign} \widetilde{X}_{\widetilde{\tau}}\sqrt{1-\theta^2}\sum_{k=[\widetilde{\tau}\alpha]+1}^{\widetilde{\tau}} \theta^{-(\widetilde{\tau}-k)}\varepsilon_k. \tag{5.104}$$

Note that $\widetilde{X}_n$ is measurable with respect to the σ-field $\widetilde{\mathcal{F}}_n = \mathcal{F}_{[n\alpha]}$ and the event $\{\widetilde{\tau} = n\}$ belongs to $\widetilde{\mathcal{F}}_n$, but the random variables ε_k with $k > [n\alpha]$ are independent of this σ-field $\widetilde{\mathcal{F}}_n$. This implies that the conditional distribution of $\widetilde{\zeta}_{\widetilde{\tau}}$ under the condition $\widetilde{\tau} = n$ is normal with parameters $(0, \sigma_n^2(\theta))$ where

$$\begin{aligned}
\sigma_n^2(\theta) &= \mathbf{E}_\theta\left(\rho^{-1}\sum_{k=[n\alpha]+1}^{n} \theta^{-(n-k)}\varepsilon_k\right)^2 \\
&= (1-\theta^{-2})\sum_{k=[n\alpha]+1}^{n} \theta^{-2(n-k)} \\
&= (1-\theta^{-2})\sum_{i=0}^{n-[n\alpha]-1} \theta^{-2i} \\
&= 1-\theta^{-2(n-[n\alpha])}.
\end{aligned}$$

Hence, the variable $\widetilde{\zeta}_{\widetilde{\tau}}$ is mixed normal and the variable

$$\widehat{\zeta}_{\widetilde{\tau}} = \frac{\widetilde{\zeta}_{\widetilde{\tau}}}{\sqrt{1-\theta^{-2(\widetilde{\tau}-[\widetilde{\tau}\alpha])}}} = \sigma_{\widetilde{\tau}}^{-1}(\theta)\widetilde{\zeta}_{\widetilde{\tau}}$$

is standard normal, $\mathcal{L}(\widehat{\zeta}_{\widetilde{\tau}} \mid \mathbf{P}_\theta) \sim \mathcal{N}(0,1)$.

Now we derive a bound for the probability of the event $|\widehat{\zeta}_{\widetilde{\tau}} - \widetilde{\zeta}_{\widetilde{\tau}}| > \delta$ on the set $\widetilde{\tau} \geq m$. Obviously

$$
\begin{aligned}
\mathbf{P}_\theta\big(|\widehat{\zeta}_{\widetilde{\tau}} - \widetilde{\zeta}_{\widetilde{\tau}}| > \delta, \widetilde{\tau} \geq m\big) &\leq \mathbf{P}_\theta\big(\sup_{n\geq m} |\widehat{\zeta}_n - \widetilde{\zeta}_n| > \delta\big) \\
&\leq \delta^{-2} \sum_{n\geq m} \mathbf{E}_\theta |\widehat{\zeta}_n - \widetilde{\zeta}_n|^2 \\
&\leq \delta^{-2} \sum_{n\geq m} (1 - \sigma_n(\theta))^2 \\
&\leq \delta^{-2} \sum_{n\geq m} (1 - \sigma_n^2(\theta))^2 \\
&\leq \delta^{-2} \sum_{n\geq m} \theta^{-4(1-\alpha)n} \\
&= \frac{\delta^{-2}}{1 - \theta^{-4/3}} \theta^{-4m/3}.
\end{aligned}
$$

Therefore, the following assertion holds (with $\delta = \theta^{-m/6}$):

Lemma 5.3 *For every Markov time $\widetilde{\tau}$ (with respect to $(\widetilde{\mathcal{F}}_n)_{n\geq 1}$), the random variable $\widehat{\zeta}_{\widetilde{\tau}}$ is normally distributed with mean 0 and variance 1, and*

$$\mathbf{P}_\theta\big(|\widehat{\zeta}_{\widetilde{\tau}} - \widetilde{\zeta}_{\widetilde{\tau}}| > \delta, \widetilde{\tau} \geq m\big) \leq C'''(\theta)\, |\theta|^{-m/6} \tag{5.105}$$

with $C'''(\theta) \in \mathbf{C}^-$.

The next assertion shows that, for each $|\theta| > 1$, the times $\tau(h)$ have "large" values if h is large.

Lemma 5.4 *For all $m > 1$, $h > 0$ and $|\theta| > 1$*

$$\mathbf{P}_\theta\big(\tau(h) \leq m\big) \leq C''''(\theta)\, h^{-1} |\theta|^{2m} \tag{5.106}$$

with $C''''(\theta) \in \mathbf{C}^-$.

Proof. We have

$$
\begin{aligned}
\mathbf{P}_\theta\big(\tau(h) \leq m\big) &= \mathbf{P}_\theta\left(\sum_{k=1}^m X_{k-1}^2 \geq h\right) \\
&\leq h^{-1} \mathbf{E}_\theta \sum_{k=1}^m X_{k-1}^2 \\
&= h^{-1} \theta^{2(m-1)} \sum_{k=1}^m \theta^{-2(m-k)} \mathbf{E}_\theta (\theta^{-(k-1)} X_{k-1})^2
\end{aligned}
$$

and

$$\mathbf{E}_\theta(\theta^{-(k-1)} X_{k-1})^2 = \mathbf{E}_\theta z_k^2 \leq \mathbf{E}_\theta z^2 = \sigma_z^2.$$

Thus

$$\mathbf{P}_\theta\big(\tau(h) \le m\big) \le h^{-1}\theta^{2(m-1)}\sigma_z^2 \sum_{k=1}^{m} \theta^{-2(m-k)} \le \frac{\rho^2}{\theta^2}\,\sigma_z^2\,\frac{\theta^{2m}}{h}$$

and the assertion (5.106) follows. □

The previous intermediate results enable us to complete the proof of the main assertion (5.75). From Lemma 5.1

$$\mathbf{P}_\theta\left(\left|s_{\tau(h)}^{-1}(\widehat{\theta}_{\tau(h)} - \theta) - \zeta_{\tau(h)}\right| > |\theta|^{-m/6},\ \tau(h) > m\right) \le C(\theta)\,\theta^{-m/6+1}. \quad (5.107)$$

By Lemma 5.2

$$\mathbf{P}_\theta\big(|\zeta_{\tau(h)} - \widetilde{\zeta}_{\widetilde{\tau}(h)}| > |\theta|^{-m/6},\ \tau(h) > m\big) \le C''(\theta)\,|\theta|^{-m/6+1}. \quad (5.108)$$

By Lemma 5.3

$$\mathbf{P}_\theta\big(|\widetilde{\zeta}_{\widetilde{\tau}(h)} - \widehat{\zeta}_{\widetilde{\tau}(h)}| > |\theta|^{-m/6},\ \widetilde{\tau}(h) > m\big) \le C'''(\theta)\,|\theta|^{-m/6} \quad (5.109)$$

and

$$\widehat{\zeta}_{\widetilde{\tau}(h)} \sim \mathcal{N}(0,1). \quad (5.110)$$

It was also shown in Lemma 5.2 that

$$\mathbf{P}_\theta\big(\tau(h) \ne \widetilde{\tau}(h),\ \widetilde{\tau}(h) > m\big) \le C'(\theta)\,|\theta|^{-m/6+1}. \quad (5.111)$$

From Lemma 5.4

$$\mathbf{P}_\theta\big(\tau(h) \le m\big) \le C''''(\theta)\,h^{-1}|\theta|^{2m}. \quad (5.112)$$

Putting together all the inequalities (5.107)–(5.112) we obtain (for $\zeta_h^* = \widehat{\zeta}_{\widetilde{\tau}(h)}$)

$$\mathbf{P}_\theta\left(\left|s_{\tau(h)}^{-1}(\widehat{\theta}_{\tau(h)} - \theta) - \zeta_h^*\right| > 3\,|\theta|^{-m/6}\right)$$
$$\le [C(\theta) + C'(\theta) + C''(\theta) + C'''(\theta)]\,|\theta|^{-m/6+1} + C''''(\theta)\,h^{-1}\theta^{2m}. \quad (5.113)$$

Denote $c = \theta^2$, $\gamma = 1/13$ and let

$$m = \lceil \log_c h^{1-\gamma} \rceil$$

where $\lceil a \rceil$ means the closest integer not less than a. Then

$$h^{-1}\theta^{2m} \le h^{-1}\,c^{\log_c h^{1-\gamma}+1} = c\,h^{-\gamma}$$

and for $\gamma = 1/13$

$$|\theta|^{-m/6} \le c^{-1/12\,\log_c h^{1-\gamma}} \le h^{-(1-\gamma)/2} = h^{-1/13}.$$

Therefore, it follows from (5.113) that

$$\mathbf{P}_\theta\left(\left|s_{\tau(h)}^{-1}(\widehat{\theta}_{\tau(h)} - \theta) - \zeta_h^*\right| > 3\,h^{-1/13}\right) \le C_{10}(\theta)\left(|\theta| h^{-1/13} + \theta^2 h^{-1/13}\right) \quad (5.114)$$

with $C_{10}(\theta) \in \mathbf{C}^-$.

If $1 < r < |\theta| < R < \infty$ then, in view of (5.114), there exists a constant $C(r, R)$ such that

$$\mathbf{P}_\theta \left(\left| s_{\tau(h)}^{-1} (\widehat{\theta}_{\tau(h)} - \theta) - \zeta_h^* \right| > 3 h^{-1/13} \right) \leq C(r, R) \, h^{-1/13}.$$

This completes the proof of the theorem.

Bibliography

AKAHIRA, M. AND TAKEUCHI, K. (1986). *Non-Regular Statistical Estimation.* Tokyo.

AKRITAS, M. AND JONSON, R. (1982). Efficiencies of tests and estimators of p-order autoregressive processes when the error distribution is nonnormal. *Ann. Inst. Statist. Math.*, **34**, 579-589.

ANDERSON, T. (1955). The integral of a symmetric unimodal function over a symmetric convex set and some probability inequalities. *Proc. Amer. Math. Soc.*, **6**, 170-176.

ANDERSON, T. (1959). On asymptotic distribution of estimates of parameters of stochastic differential equations. *Ann. Math. Stat.*, **30**, 676-687.

ANSCOMBE, F. (1953). Sequential estimation. *Trans. Roy. Statist. Soc., Ser.B*, **15**, 1-21.

BAHADUR, R. (1954). Sufficiency and statistical decision function. *Ann. Math. Stat.*, **25**, 423-462.

BAHADUR, R. (1960). On the asymptotic efficiency of tests and estimators. *Sankhya*, **22**, 229-252.

BASAWA, I. AND BROCKWELL, P. (1984). Asymptotic conditional inference for regular nonergodic models with an application to autoregressive processes. *Ann. Statist*, **12**, 161-171.

BASAWA, I. AND KOUL, H. (1979). Asymptotic tests of composite hypotheses for nonergodic type stochastic processes. *Stoch. Proc. Appl.*, **9**, 291-305.

BASAWA, I., MALLIK, A., MCCORMICK, W., AND TAYLOR, R. (1989). Bootstraping explosive autoregressive process. *Ann. Statist.*, **17**, 1479-1488.

BASAWA, I. AND RAO, B. P. (1980). *Statistical Inference for Stochastic Processes.* Academic Press.

BASAWA, I. AND SCOTT, D. (1983). *Asymptotic Optimal Inference for Non-ergodic Models.* Springer Verlag, New York.

BERAN, R. (1976). Adaptive estimates for autoregressive processes. *Ann. Inst. Statist. Math.*, **28**, 77-89.

BERK, R. AND BICKEL, P. (1968). On invariance and almost invariance. *Ann. Math. Stat.*, **39**, 1573-1576.

BERNSTEIN, S. (1917). *Probability Theory.* Moscow.

BICKEL, P. (1982). On adaptive estimation. *Ann. Statist.*, **10**, 647-671.

BILLINGSLEY, P. (1968). *Convergence of Probability Measures.* J. Wiley and Sons, New York.

BLACKWELL, D. (1947). Conditional expectation and unbiased sequential estimation. *Ann. Math. Stat.*, **18**, 105-110.

BLACKWELL, D. (1951). Comparison of experiments. *Proc. 2nd Berkeley Symp. Math.*

Stat. Probab., **1**, 93–102.

BLACKWELL, D. (1953). Equivalent comparisons of experiments. *Ann. Math. Stat.*, **24**, 265–272.

BLACKWELL, D. AND BICKEL, P. (1967). A note on Bayes estimates. *Ann. Math. Stat.*, **38**, no. 1, 190–1911.

BLYTH, C. (1951). On minimax statistical decision procedures and their admissibility. *Ann. Math. Stat.*, **22**, 22–42.

BOHNENBLUST, A., SHAPLEY, L., AND SHERMAN, S. (1949). Reconnaissance in game theory. Rand Research Memorandum 1949/208.

BOLL, C. (1955). *Comparison of experiments in the infinite case.* PhD thesis, Stanford Univ., Stanford.

BORISOV, V. AND KONEV, V. (1977). On sequential estimation of parameters of discrete processes. *Avtomatika i Telemechanika*, **10**, 58–65.

BOROVKOV, A. (1984). *Mathematical Statistics.* Nauka, Moscow.

BOUDAR, I. AND MILNES, P. (1981). Amenability. A survey for statistical applications of Hunt-Stein and related conditions on groups. *Z. Wahrsch. verw. Geb.*, **57**, 103–128.

CHAN, N. AND WEI, C. (1987). Asymptotic inference for weakly nonstationary AR(1) processes. *Ann. Statist*, **15**, 1050–1063.

CHAN, N. AND WEI, C. (1988). Limiting distributions of least squares estimates of unstable autoregressive processes. *Ann. Statist*, **16**, 367–401.

CHENTSOV, D. (1972). *Statistical Decisions and Optimal Inference.* Nauka, Moscow.

CHERNOFF, H. (1952). A measure of asymptotic efficiency for tests of an hypothesis based on the sum of observations. *Ann. Math. Stat.*, **23**, 493–502.

CHIBISOV, D. (1968). One theorem on admissible tests and its application to the asymptotic hypothesis testing. *Theory Probab. Appl.*, **12**, 90–103.

COX, D. (1991). Gaussian likelihood estimation for nearly nonstationary AR(1) process. *Ann. Statist.*, **19**, 1129–1154.

COX, D. AND LLATAS, I. (1991). Maximum likelihood type estimation for nearly nonstationarity autoregression time series. *Ann. Statist*, **19**, 1109–1128.

DAVIES, R. (1985). Asymptotic inference when the amount of information is random. In *Proc. Berkeley Conf. in Honor of J.Neiman and J.Kiefer*, no. 2, pages 841–864.

DIACONIS, P. AND FREEDMAN, D. (1986). On the consistency of Bayes estimates. *Ann. Statist.*, **14**, no. 1, 1–25.

DICKEY, D. AND FULLER, W. (1979). Distribution of the estimators for autoregressive time series with a unit root. *J. Amer. Stat. Assoc.*, **74**, 427–431.

DICKEY, D. AND FULLER, W. (1981). Likelihood ratio statistics for autoregressive time series with a unit root. *Econometrica*, **49**, 1057–1072.

DUDLEY, R. (1968). Distances of probability measures and random variables. *Ann. Math. Stat.*, **39**, no. 5, 1563–1572.

EFROIMOVICH, S. (1980). On sequential estimation under the condition of local asymptotic normality. *Theory Probab. Appl.*, **25**, 30–43.

EHM, W. AND MULLER, D. (1983). Factorizing the information contained in an experiment, conditionally on the observed value of a statistic. *Z. Wahsch. verw. Geb.*, **65**, 121–134.

EVANS, G. AND SAVIN, N. (1981). The calculation of the limiting distribution of the least square estimator of the parameter in a random walk model. *Ann. Statist.*, **9**, 1114–1118.

FABIAN, V. AND HANNAN, J. (1982). On estimation and adaptive estimation for locally asymptotically normal families. *Z. Wahsch. verw. Geb.*, **59**, 459–478.

FABIAN, V. AND HANNAN, J. (1987). Local asymptotical behaviour of densities. *Statistics and Decisions*, **5**, 105–138.

FABIUS, J. (1964). Asymptotic behaviour of Bayes estimates. *Ann. Math. Stat.*, **35**, 39–384.

FARELL, R. (1967). Weak limits of sequences of Bayes procedures in estimation theory. In *Proc. 5th Berkeley Symp. Math. Stat. Prob.*, no. 1, 83–111.

FEIGIN, P. (1978). The efficiency criteria problem for stochastic processes. *Stoch. processes and their appl.*, **6**, 115–127.

FISHER, R. (1922). On the mathematical foundations of the theoretical statistics. *Phil. Trans. Roy. Soc, Ser. A*, **222**, 309–368.

FOUNTIS, G. AND DICKEY, D. (1989). Testing for a unit root nonstationarity in multivariate autoregressive time series. *Ann. Statist.*, **17**, no. 1, 419–428.

GIRSHICK, M. AND SAVAGE, L. (1951). Bayes and minimax estimates for quadratic loss function. In *Proc 2th Berkeley Symp. Math. Stat. Prob.*, no. 1, 53–73.

GRAMBSCH, P. (1983). Sequential sampling based on the observed Fisher information to guarantee the accuracy of the maximum likelihood estimator. *Ann. Statist.*, **11**, 68–77.

GREENWOOD, P. AND SHIRYAEV, A. (1985). *Contiguity and the Statistical Invariance Principle*. Gordon and Breach, New York.

GREENWOOD, P. AND SHIRYAEV, A. (1989). On uniform weak convergence of semimartingales with application to estimation problem in first-order autoregressive model. In *Statistics and Control of Stochastic Processes*, Moscow, Nauka, 40–48. (in Russian).

GREENWOOD, P. AND SHIRYAEV, A. (1992). Asymptotic minimaxity of a sequential estimator for a first order autoregressive model. *Stochastics*, **38**, 49–65.

GREENWOOD, P. AND WEFELMEYER, W. (1991). *Efficient Estimating Equations for Nonparametric Filtered Models*, Statistical inference in stochastic processes, *Probab. Pure Appl.* **6**, 107–141.

HAJEK, J. (1970). A characterization of limiting distributions of regular estimators. *Z. Wahsch. verw. Geb.*, **14**, 323–330.

HAJEK, J. (1972). Local asymptotic minimax and admissibility of estimation. In *Proc. 6th Berkeley Symp. Math. Stat. Prob.*, no. 7, 175–194.

HAJEK, J. AND SIDAK, Z. (1967). *Theory of Rank Tests*. Academia, Prague.

HALL, P. AND HEYDE, C. (1980). *Martingale Limit Theory and its Application*. Academic Press, Sidney.

HALMOS, P. AND SAVAGE, L. (1949). Application of the Radon-Nikodym theorem to the theory of sufficient statistics. *Ann. Math. Stat.*, **20**, 225–241.

HANSEN, O. AND TORGENSEN, E. (1974). Comparison of linear normal experiments. *Ann. Statist.*, **2**, 367–373.

HEWITT, E. AND ROSS, A. (1963). *Abstract Harmonic Analysis*, Volume 1. Berlin.

HEYDE, C. (1978). On an optimal asymptotic property of the maximum likelihood estimator of parameters from a stochastic process. *Stoch. processes and their appl.*, **8**, 1–9.

HEYER, H. (1982). *Theory of Statistical Experiments*. Springer Verlag, Berlin.

HOPFNER, R. (1990). Null reccurent birth–and–death processes, limits of certain martingales and local asymptotic mixed normality. *Scand. J. Statist.*, **17**, 201–215.

HOPFNER, R., JACOD, J., AND LADELLI, L. (1990). Local asymptotic normality and mixed normality for Markov statistical models. *Probab. Theory and Rel. Fields*, **86**, 105–129.

IBRAGIMOV, I. AND KHASMINSKII, R. (1974). On sequential estimation. *Theory Probab. Appl.*, **19**, 245–256.

IBRAGIMOV, I. AND KHASMINSKII, R. (1981). *Statistical Estimation. Asymptotic Theory.*

Springer-Verlag, Berlin, Heidelberg, New York.

IBRAGIMOV, I. AND KHASMINSKII, R. (1991). Asymptotically normal families of distribution and efficient estimation. *Ann. Statist.*, **19**, no. 4, 1681–1700.

IGANAKI, N. (1970). On the limiting property of a sequence of estimators with uniformity property. *Ann. Inst. Stat. Math.*, **22**, 1–13.

JACOD, J. (1989a). Convergence of filtered statistical models and Hellinger process. *Stoch. Processes and their appl.*, **32**, 47–68.

JACOD, J. (1989b). Filtered statistical models and Hellinger process. *Stoch. Processes and their appl.*, **32**, 3–45.

JACOD, J. AND SHIRYAEV, A. (1987). *Limit Theorems for Stochastic Processes.* Springer Verlag, New York etc.

JANSSEN, A. (1986). Limits of translation invariant experiments. *J. Multiv. Anal.*, **20**, 129–142.

JANSSEN, A. (1991). Asymptotical linear and mixed normal sequences of statistical experiments. *Sankhya, Ser.A*, **53**, 1–26.

JEGANATHAN, P. (1980). An extention of a result of L. Le Cam concerning asymptotical normality. *Sankhya, Ser.A*, **42**, 146–160.

JEGANATHAN, P. (1981). On a decomposition of the limit distribution of a sequence of estimators. *Sankhya, Ser.A*, **43**, 26–36.

JEGANATHAN, P. (1982). On the asymptotic theory of estimation when the limit of log likelihood is mixed normal. *Sankhya, Ser.A*, **44**, 173–212.

JEGANATHAN, P. (1983). Some asymptotic properties of risk function when the limit of experiments is mixed normal. *Sankhya, Ser.A*, **45**, 66–87.

JEGANATHAN, P. (1988). On the strong approximation of the distribution of estimators in linear stochastic models, I,II. Stationary and explosive AR models. *Ann. Statist.*, **16**, no. 3, 1283–1314.

JEGANATHAN, P. (1990). On the asymptotic behavour of least squares estimators in AR series with roots near the unit circle, I and II. Technical Report 150, Dept. Statist., Univ. Michigan.

KABAILA, P. (1983). On the asymptotic efficiency of estimators or the parameters of ARMA process. *J. Time Ser. Anal.*, **4**, 37–47.

KOLMOGOROV, A. (1974). *Fundamentials of Probability Theory.* Nauka, Moscow.

KONEV, V. AND PERGAMEMTSCHIKOV, S. (1988). On given accuracy estimation of parameters of explosive dynamical systems. *Avtomatika i Telemechanika*, **11**, 130–141.

KONEV, V. AND PERGAMEMTSCHIKOV, S. (1997). The prescribed precision estimators of the auto-regression parameter using the generalized least squares methods. *Theory Probab. Appl.*, **41**, no. 4, 678–694.

KONEV, V. AND PERGAMENTSHIKOV, S. (1986). On the duration of sequential estimation of the parameters of stochastic processes in discrete time. *Stochastics*, **2**, 133–154.

KOUL, H. AND LEVENTAL, S. (1989). Weak convergence of the residual empirical process in explosive autoregression. *Ann. Statist.*, **17**, no. 4, 1784–1794.

KOUL, H. AND PFLUG, G. (1990). Weakly adaptive estimators in explosive autoregression. *Ann. Statist.*, **18**, 939–960.

KRAMKOV, D. (1994). On comparison of some statistical models type of "trend with noise". *Theory Probab. Appl.*, **38**, no. 3, 537–540.

KREISS, J. (1987). On adaptive estimation in stationary ARMA processes. *Ann. Statist.*, **15**, 112–133.

KREISS, J. (1990). Testing linear hypotheses in autoregressions. *Ann. Statist.*, **18**, no. 3, 1470–1482.

KUCHLER, U. AND SORENSEN, M. (1989). Exponential families of stochastic processes. a unifying semimartingale approach. *Int. Stat. Review.*, **57**, no. 2, 123–144.

LAI, T. AND SIEGMUND, D. (1983). Fixed accuracy estimation of an autoregressive parameter. *Ann. Statist.*, **11**, 478–485.

LAPLACE, P. (1810a). Memoire sur les formules qui sont fonction de tres grands nombers et sur leurs application aux probabilites. *Oeuvres de Laplace*, **12**, 301–345.

LAPLACE, P. (1810b). Memoire sur les integrales definies et leurs application aux probabilites. *Oeuvres de Laplace*, **12**, 357–412.

LECAM, L. (1955). An extension of Wald's theory of statistical decision function. *Ann. Math. Stat.*, **26**, 69–81.

LECAM, L. (1960). Locally asymptotically normal families of distributions. *Univ. Calif. Publ. Statist.*, **3**, 27–98.

LECAM, L. (1964). Sufficiency and approximate sufficiency. *Ann. Math. Stat.*, **35**, 1419–1455.

LECAM, L. (1969). *Theorie asymptotique de la decision statistique.* Univ. of Montreal Press, Montreal.

LECAM, L. (1972). Limit of experiments. In *Proc. 6th Berkeley Symp. Math. Stat. Prob.*, no. 1, 245–261.

LECAM, L. (1973). Convergence of estimates under dimensionality restriction. *Ann. Statist.*, **1**, 38–53.

LECAM, L. (1974). *Notes on Asymptotic Methods in Statistical Decision Theory.* Centre de Recherches Mathematiques, Univ. of Montreal, Montreal.

LECAM, L. (1979). *On a theorem of J.Hajek, Contributions to statistics, Jaroslav Hajek Mem.*, Academia, Prague, 119–135.

LECAM, L. (1985). Sur l'approximation de familles de mesures par des familles gaussiennes. *Ann. Inst. H. Poincare*, **27**, no. 3, 225–287.

LECAM, L. (1986). *Asymptotic Methods in Statistical Decision Theory.* Springer–Verlag, New York.

LECAM, L. AND YANG, G. (1988). On the preservation of local asymptotic normality under information loss. *Ann. Statist.*, **16**, 483–520.

LECAM, L. AND YANG, G. (1990). *Asymptotics in Statistics. Some basic concepts.* Springer-Verlag, New York etc.

LEHNMANN, E. (1988). Comparing location experiments. *Ann. Statist.*, **16**, no. 2, 521–533.

LINDSEY, D. (1956). On a measure of the information provided by an experiment. *Ann. Math. Stat.*, **27**, 986–1005.

LINDSEY, D. (1972). *Bayessian Statistics. A Review.* SIAM, Philadelphia.

LIPTSER, R. AND SHIRYAEV, A. (1976, 1978). *Statistics of Random Processes, I,II.* Springer-Verlag, Berlin, Heidelberg, New York.

LIPTSER, R. AND SHIRYAEV, A. (1988). *Theory of Martingales.* Springer-Verlag, Berlin, Heidelberg, New York.

LUSCHGY, H. (1992a). Local asymptotic mixed normality for semimartingale experiments *Prob. Theory and Rel. Fields*, **92**, 151–176.

LUSCHGY, H. (1992b). Comparison of location models for stochastic processes. *Prob. Theory and Rel. Fields*, **93**, 39–66.

MAMMEN, E. (1987). Optimal local Gaussian approximation of an exponential family. *Probab. Theory Related Fields*, **76**, no. 1, 103–119.

MANN, H. AND WALD, A. (1943). On the statistical treatment of linear stochastic difference equations. *Econometrica*, **11**, 173–200.

MIKULSKI, P.W. AND MONSOUR, M.J. (1991). Optimality of the maximum likelihood estimator in first-order autoregressive processes. *J. Time Ser. Anal.*, **12**, No.3,

237-253.

MILBRODT, H. (1983). Global asymptotical normality. *Statistics and Decisions*, **1**, 401–425.

MILBRODT, H. AND STRASSER, H. (1983). *Limits of Triangular Array of Experiments*, Springer, Berlin–Heidelberg–New York.

MILLAR, P. (1983). The minimax principle in asymptotic theory. *Lecture Notes in Mathematics*, **976**, 76–267.

MOUSSATAT, W. (1975). *On the asymptotic theory of statistical experiments and some of its applications*. PhD thesis, Univ. of California, Berkeley.

OOSTERHOFF, G. AND VAN ZWET, W. (1979). *A Note on Contiguity and Hellinger Distance*, in *Contributions to statistics, Jaroslav Hajek Mem. Vol.*, Academia, Prague, 157–166.

PARTHASARATY, T. AND RAGHAVAN, T.E.S. (1971). *Some topics in two-person games*. Amer. Elsevier Publ., New York.

PERGAMENTSCHIKOV, S. (1991). Asymptotic properties of sequential design for estimation of first order autoregressive parameter. *Theory Probab. and Appl.*, **36**, 42–53.

PERGAMENTSCHIKOV, S. AND SHIRYAEV, A. (1992). On sequential estimation of parameter of stochastic differential equation with random coefficients. *Teoria Veroiatn. i ee Prim.*, **37**, 482–501.

PFANZAGL, J. AND WEFELMEYER, W. (1982). *Contributions to a General Asymptotic Statistical Theory*, *Lecture Notes in Statistics*, **13**, Springer, New York.

PFANZAGL, J. AND WEFELMEYER, W. (1985). *Asymptotic Expansions for General Statistical Models*, *Lecture Notes in Statistics*, **31**, Springer, New York.

PHILIPPS, P. (1987a). Time series regression with a unit root. *Econometrica*, **55**, 277–301.

PHILIPPS, P. (1987b). Towards a unified asymptotic theory for autoregression. *Biometrica*, **74**, 535–547.

PITMAN, E. (1979). *Some Basic Theory for Statistical Inference*. Chapman and Hall, Sydney.

POLLARD, D. (1984). *Convergence of Stochastic Processes*. Springer, New York.

RACHEV, S., RÜSCHENDORF L. AND SCHIEF, A. (1992). Uniformities for the convergence in law and in probability. *J. of Theoretical Probab.*, **5**, no. 1, 33–44.

RAO, M. (1978). Asymptotic distribution of an estimator of the boundary parameter of an unstable process. *Ann. Statist.*, **6**, 185–190.

ROUSSAS, G. (1972). *Contiguous Probability Measures. Some application in Statistics.* Cambridge Univ. Press.

SACKS, J. (1963). Generalized Bayes solution in estimation problem. *Ann. Math. Statist.*, **34**, 751–768.

SCHICK, A. (1986). On asymptotically efficient estimation in semiparametric models. *Ann. Statist.*, **14**, 1139–1151.

SCHWARTH, L. (1965). On Bayes procedures. *Z. Warsch. verw. Geb.*, **4**, 10–26.

SHERMAN, S. (1957). On a theorem of Hardy, Littlewood, Polia and Blackwell. *Proc Nat. Acad. Sci. USA*, **37**, 826–831.

SHIRYAEV, A. (1984). *Probability*. Springer-Verlag, Berlin, Heidelberg, New York. (Second Edition 1995.)

SPOKOINY, V. (1986). *One limit theorem for martingales and its application to sequential experimental design*, 99-117. VNIISI, Moscow.

SPOKOINY, V. (1992a). Lower bounds of minimax risk in estimation problems without lan condition. *Stochastics and Stoc. Reports*, **38**, 67–93.

SPOKOINY, V. (1992b). On asymptotically optimal sequential experimental design. *Advances in Sov. Mathematics*, **12**, 135–150.

SPOKOINY, V. (1994). Efficient estimation and experimental design for controlled systems. Asymptotic approach. *Stochastics and Stoc. Reports*, **50**, 129–160.

STEIN, CH. (1951). *Notes on the comparison of experiments*. Univ. of Chicago.

STEIN, CH. (1959). The admissability of Pitman's estimator for a single location parameter. *Ann. Math. Stat.*, **30**, 970–979.

STONE, M. (1961). Non-equivalent comparison of experiments and their use for experiments involving location parameter. *Ann. Statist.*, **32**, 326–332.

STRASSER, H. (1981). Convergence of estimates I,II. *J. Multiv. Analysis*, **11,12**, 127–151 and 152–172.

STRASSER, H. (1982). Local asymptotic minimax properties of Pitman estimate. *Z. Wahsch. verw. Geb.*, **60**, 223–247.

STRASSER, H. (1985a). *Mathematical Theory of Statistics*. de Gruyter.

STRASSER, H. (1985b). Scale invariance of statistical experiments. *Probab. Math. Statist.*, **5**, 1–20.

SWENSEN, A. (1980). *Asymptotic inference for a class of stochastic processes*. PhD thesis, Univ. of California, Berkeley.

SWENSEN, A. (1983). A note on asymptotic inference in a class of non-stationary processes. *Stoch. Proc. Appl.*, **15**, 181–191.

SWENSEN, A. (1985a). The asymptotic distribution of the likelihood ratio for autoregressive time series with a regression trend. *J. Multiv. Anal.*, **16**, 54–70.

SWENSEN, A. (1985b). A note on statistical inference for a class of diffusions and approximate diffisions. *Stoch. Proc. Appl*, **19**, 111–123.

TORGENSEN, E. (1972). Comparison of translation experiments. *Ann. Math. Statist.*, **43**, 1383–1399.

TORGENSEN, E. (1976). Comparison of statistical experiments. *Scand. J. Statist.*, **3**, 186–208.

TORGENSEN, E. (1991). *Comparison of Statistical Experiments*. Cambridge Univ. Press.

VAN DER VAART, A. (1991). An asymptotic representation theorem. *Int. Statist. Review*, **59**, 97–122.

VAN DER VAART, A. AND WELLNER, A. (1996). *Weak convergence and empirical processes. With applications to statistics*. Springer Series in Statistics. New York, Springer.

VON MISES, R. (1931). *Wahscheinlichkeitsrechnung*. Springer–Verlag.

WALD, A. (1950). *Statistical decision functions*. J. Wyley and Sons, New York.

WALD, A. AND WOLFOVITZ, J. (1957). Two methods of randomization in statistics and the theory of games. *Ann. Math. Stat.*, **53**, 581–586.

WHITE, J. (1958). The limiting distribution of the serial correlation coefficient in explosive case. *Ann. Math. Stat.*, **29**, 1188–1197.

WITTENBERG, H. (1964). Limiting distribution of random sums of independent random variables. *Z. Warsch. verw. Geb.*, **3**, 7–18.

Index

List of Symbols

List of Conditions

www.ingramcontent.com/pod-product-compliance
Lightning Source LLC
Chambersburg PA
CBHW081531190326
41458CB00015B/5517
* 9 7 8 9 8 1 0 2 4 1 0 1 8 *